Beihang Postgraduate Series
北京航空航天大学“研究生英文教材”系列丛书

Resonant Sensors
谐振式传感器

Fan Shangchun　Guo Zhanshe
樊尚春　　郭占社

北京航空航天大学出版社

Abstract

This book systematically discussed the pertinent fundamental theory of the resonant sensors. Then some kinds of resonant sensors including resonant press sensor, resonant acceleration sensor, resonant angular rate sensor, resonant mass flow sensor and surface acoustic wave resonant senor are introduced.

It can be selected as reference to the masters, undergraduates and pertinent engineering staffs whose research direction are instrumentation science and technology, control science and engineering, mechanical engineering etc.

图书在版编目(CIP)数据

谐振式传感器 = Resonant Sensors : 英文 / 樊尚春，郭占社著. -- 北京 : 北京航空航天大学出版社，2014.4

ISBN 978-7-5124-1504-1

Ⅰ. ①谐… Ⅱ. ①樊… ②郭… Ⅲ. ①谐振传感器—英文 Ⅳ. ①TP212

中国版本图书馆 CIP 数据核字(2014)第 040900 号

Resonant Sensors

谐振式传感器

Fan Shangchun　Guo Zhanshe

樊尚春　郭占社

责任编辑　张冀青

*

北京航空航天大学出版社出版发行

北京市海淀区学院路 37 号(邮编 100191)　http://www.buaapress.com.cn

发行部电话:(010)82317024　传真:(010)82328026

读者信箱:bhpress@263.net　邮购电话:(010)82316524

北京时代华都印刷有限公司印装　各地书店经销

*

开本:787×960　1/16　印张:20　字数:448 千字

2014 年 4 月第 1 版　2014 年 4 月第 1 次印刷　印数:1 500 册

ISBN 978-7-5124-1504-1　定价:69.00 元

Preface

Sensor is the front end of information technology and the core of the information. It plays an important role in the progress of science and technology. According to resonant phenomenon, resonant sensor is based on the principle that its inherent characteristic of sensitive element is changing regularly with the measurand. With the incomparable advantages of direct digital output, small hysteresis, good reproducibility and high stability, it is recognized as high-performance sensor. Resonant sensor is the focus and hotspot in the research of current sensor technology field, especially the best choice in the system of automatic and intelligent measurement. It has an important application value in aviation, aerospace, petroleum, chemical industry, metallurgy, food, weather, measurement and other fields.

Resonant sensor is developing fast in these years. The author has been focusing on the development and design of high-performance resonant sensor according to the need of defense modernization and the economic construction. The author has devoted himself to the theoretical study, experimental research and engineering practice of resonant sensor for almost 30 years. This book is the summary of achievement and experience made by the author.

Chapter one introduces some fundamental theory of resonator sensor. Such as resonance phenomenon, sensitive mechanism, mechanical quality factor of resonant sensitive elements. In Chapter 2～5, authors focus on several typical resonant sensors including resonant pressure sensor, resonant inertial sensor, resonant mass flow sensor and surface acoustic wave resonant senor. The working principle, sensitive mechanism and typical applications have been described clearly.

The book attaches great importance to both basic theory analysis and engineering application when studying every kind of resonant sensor. Not only scientific problem, but also key technologies are emphasized on each resonant sensor. The book has made some profound and systematic summary according to the theory analysis and practical application. There are also some questions attached to every chapter of the book for readers to master the relevant knowledge and carry out further study.

The book is written by professor Fan Shangchun and associate professor Guo Zhanshe from the Department of Instrument Science and Opto-electronic Engineering of Beihang University.

In the process of the research on resonant sensor, the author has undertaken more than twenty research projects relating to resonant sensor, including the national natural science fund, the national "863 project", the base of national defense scientific research, aviation science funds, aerospace innovation fund, etc. There are some relevant content reference from some experts and scholars in the book. I would like to express my acknowledgements to professor Ding Tianhuai from Tsinghua University and Xia Shanhong researcher from the Institude of Electronics of Chinese Academy of Science, who have reviewed the manuscript of the book and gave us valuable comments and suggestions.

The content of resonant sensor is extensive and profound and it's developing rapidly. Some mistakes and inappropriate content may exist in the book due to the limitation of authors' knowledge. Comments and suggestions from each reader are appreciated.

Authors

28 November, 2013

Contents

Chapter 1 Fundamental Theory of Resonant Sensors ······ 1

1.1 Introduction ······ 1
1.2 Resonance Phenomenon ······ 2
1.3 Basic Structure of the Resonator Sensors ······ 4
1.4 Sensitive Mechanism of the Resonator Sensors ······ 6
1.5 The Closed-loop System of the Resonator Sensors ······ 6
1.5.1 Time-domain Analysis ······ 6
1.5.2 Complex Frequency Domain Analysis ······ 7
1.6 Mechanical Quality Factor of the Resonant Sensitive Elements ······ 8
1.6.1 Definition and Calculation ······ 8
1.6.2 The Analysis of the Affecting Factors of Sensors ······ 10
1.6.3 Methods on Improving the Mechanical Quality Factor ······ 12
1.7 Methods of Measurement of the Frequency Output Resonator Sensors ······ 13
1.7.1 Frequency Method ······ 13
1.7.2 Periodic Method ······ 14
1.8 The Design Features of the Resonator Sensors ······ 14
1.9 The Characteristics and Advantages of the Resonator Sensors ······ 15
Questions and Exercises ······ 16

Chapter 2 Resonant Pressure Sensor ······ 17

2.1 Vibrating String Type Pressure Sensor ······ 17
2.1.1 Structure and Principle ······ 17
2.1.2 Characteristic Equation ······ 19
2.1.3 Excitation Mode ······ 19
2.2 Resonant Diaphragm Pressure Sensor ······ 20
2.2.1 Structure and Principle ······ 20

2.2.2 Characteristic Equation ······ 21

2.3 Electromagnetic Excitation Vibrating Cylinder Pressure Sensor ······ 22

2.3.1 Structure and Principle ······ 22

2.3.2 Characteristic Equation ······ 24

2.3.3 Excitation Mode ······ 25

2.3.4 Characteristics Linearization and Error Compensation ······ 26

2.4 Piezoelectric Exciting Resonant Cylinder Pressure Sensor ······ 29

2.4.1 Structure and Principle ······ 29

2.4.2 Characteristics of Piezoelectric Excitation ······ 30

2.4.3 Transformation of Picking-up Signals ······ 32

2.4.4 The Realization of a Stable Single Modal Self-exciting System ······ 34

2.4.5 Related Problems about Double Modal ······ 36

2.5 Quartz Vibration Beam Pressure Sensor ······ 40

2.5.1 Structure and Principle ······ 41

2.5.2 Characteristic Function ······ 43

2.6 Micromechanical Silicon Resonant Pressure Sensor ······ 44

2.6.1 Sensing Structure and Mathematical Model ······ 45

2.6.2 The Process of Signal Transformation ······ 47

2.6.3 The Temperature Field Model and Thermal Characteristic Analysis of the Beam Resonator ······ 49

2.7 Micro Mechanical Silicon Resonant Pressure Sensor with Differential Output ··· 51

2.7.1 Sensing Structure and Working Mechanism ······ 51

2.7.2 Characteristic Equation ······ 52

Questions and Exercises ······ 53

Chapter 3 Resonant Inertial Sensor ······ 55

3.1 Introduction ······ 55

3.2 Quartz Vibrating Beam Acceleration Sensor ······ 56

3.3 Resonant Silicon Micromachined Acceleration Sensor ······ 57

3.4 Piezoelectric Excitation Resonant Cylindrical Shell Angular Rate Sensor ······ 59

3.5 Static Exciting Hemispherical Resonant Angular Velocity Sensor ······ 65

3.6 Micro Silicon Resonant Angular Velocity Sensor ······ 74

3.6.1 Silicon Capacitive Surface Micromechanical Gyroscope ······ 74

3.6.2 Silicon Micromechanical Gyroscope with Direct Frequency Output ······ 76

Questions and Exercises ······ 77

Chapter 4 Resonant Mass Flow Sensor ········ 79

4.1 Introduction ········ 79
4.2 The Basic Working Principle ········ 81
4.3 Mathematical Model of the Measurement ········ 83
4.3.1 The Measurement of Mass Flow ········ 83
4.3.2 Density Measurement ········ 87
4.3.3 The Measurement of the Two-component Fluid ········ 89
4.4 Signal Detection Scheme and Its Realization ········ 90
4.4.1 Phase Difference Detection of the Mass Flow ········ 90
4.4.2 The Amplitude Ratio Detection of the Mass Flow ········ 94
4.4.3 Density Detection ········ 96
4.5 The Implementations of Coriolis Mass Flow Sensor ········ 96
4.6 The Classification of Coriolis Mass Flow Sensor ········ 100
4.6.1 The Classification According to the Shape of the Measuring Tube ········ 100
4.6.2 Classified According to the Number of Segments of the Measuring Tube ········ 102
4.6.3 Classified According to the Connection of the Measuring Tube with the Double Tube Mode ········ 102
4.6.4 Classified According to the Layout of the Flowing Direction of the Measuring Tube ········ 103
4.7 Main Application Fields and its Characteristics ········ 103
4.7.1 Application Overview ········ 103
4.7.2 Advantages ········ 105
4.7.3 Disadvantages ········ 106
4.8 Interference Factors and Their Restraining ········ 106
4.8.1 Pressure Loss ········ 106
4.8.2 Measuring the Flow of Gas ········ 108
4.8.3 Liquid Containing Gas ········ 109
4.8.4 Liquid Containing Solid ········ 110
4.8.5 The Influence of the Parameters of Fluid Conditions or Physical Properties to the Fluid Measurement ········ 110
4.8.6 The Interference of Nonlinear Vibration and Working Frequency ········ 113
Questions and Exercises ········ 113

Chapter 5 Surface Acoustic Wave Resonant Sensor ······ 115

5.1 Introduction ······ 115
5.2 SAW Interdigital Transducer ······ 116
5.2.1 Basic Characteristics of IDT ······ 117
5.2.2 Basic Analyzing Model of IDT ······ 120
5.3 SAW Resonator ······ 123
5.3.1 Structure and Working Principle ······ 123
5.3.2 Stability of Frequency and Temperature ······ 126
5.4 Typical SAW Sensor ······ 127
5.4.1 SAW Strain Sensor ······ 127
5.4.2 SAW Pressure Sensor ······ 130
5.4.3 SAW Acceleration Sensor ······ 133
5.4.4 SAW Gas Sensor ······ 135
5.4.5 SAW Flow Sensor ······ 139
Questions and Exercises ······ 142

Appendix A Dynamic Characteristics Analysis of Vibrating Cylinder Pressure Transducers (VCPT) ······ 144

Appendix B Mathieu Equation with Application to Analysis of Dynamic Characteristics of Resonant Inertial Sensors ······ 165

Appendix C Vibration Membrane Liquid Density Sensor ······ 179

Appendix D Finite-element Modelling and Simulation on Frequency Characteristics of the Silicon Beam Resonator Attached to an E-type Round Diaphragm for Measuring the Concentrated Force ······ 185

Appendix E Nonlinear Vibration Characteristics of Coriolis Mass Flowmeter ······ 200

Appendix F Modeling of Coriolis Mass Flow Meter of a General Plane-shape Pipe ······ 216

Appendix G Frequency Measurement Study of Resonant Vibratory Gyroscopes ······ 234

Appendix H An Electrothermally Excited Dual Beams Silicon Resonant Pressure Sensor with Temperature Compensation ······ 246

Appendix I Non-linear Dynamics of an Electrothermally Excited Resonant Pressure Sensor ······ 263

Appendix J Design and FEM Simulation Study of the Electro-thermal Excitation Resonant Beam with Slit-structure ······ 288

References ······ 306

Chapter 1
Fundamental Theory of Resonant Sensors

Basic contents:

- Resonance, the resonance phenomenon and the harmonic oscillator
- The fundamental frequency and the resonant frequency
- The resonant state and its evaluation
- *Q* factor (the mechanical quality factor)
- Open-loop characteristic of resonant sensor and its measurement
- The self-exciting closed-loop system
- The amplitude and phase conditions of the closed-loop system

1.1 Introduction

Based on mechanical resonance technology, sensors which are measured by the natural resonant characteristics of the sensing elements with the variation are known as the resonant sensors. As the resonator sensors themselves have the periodic signal output (quasi-digital signal), only a simple digital circuit (neither the A/D nor V/F) can be converted to digital signal easily connected with microprocessor. Meanwhile, the repeatability, resolution and stability of the sensing elements are excellent, so the resonant measurement principle has naturally become the key research contents nowadays.

In the period of the development of analog control system, a number of sensors have been developed. By measuring the changing of the resistors, capacitors or inductors, parameters such as pressure, temperature and displacement could indirectly be measured corresponding output signals could be the voltage or current. These sensors are not suitable to make the digital interface to the digital for digital measurement and control system, and thus need to increase the A/D converter between the sensor and control circuit, which not only reduces the system reliability and response speed, but also increases the cost.

Developing the digital output sensor to adapt to the needs of digital age is a common requirement of many fields. Strictly speaking, in reality, despite the encoder used for detecting linear displacement and angular displacement, there is almost no direct digital sensor. In addition, a family of sensors based on the cyclical trigger mechanism is appeared in succession, such as the trigger sensors, CCDs and optical sensors.

Resonant sensor with the resonant sensitive structure can make use of the resonance vibration frequency, phase and amplitude parameters as sensitive information. As resonant sensor has many advantages can be applicable to measurements in a variety of parameters, such as pressure, force, angle, flow, temperature, humidity, level, density and composition of the gas measurement, this type of sensor has developed into a new family of sensors.

Now the measurement of the practical resonant sensor is mainly based on the natural vibration characteristics of mechanical resonant sensitive structure. According to the characteristics of the resonant sensitive structures, resonant sensors can be divided into two categories: one is based on the traditional process of structural parameters. They have the character of are relatively large dimensions and their commonly used resonant sensitive structures are the resonant tube, resonant beam, resonant film and resonant elbow. They are made of precision-machined precision alloy and the performance is excellent all walks of life satisfaction. The other one is a new type of silicon or quartz resonator sensors which uses the micromachining process to achieve. Micro harmonic oscillator has many species and its size is usually micron or even nanoscale. The research results have been proved that, the micro-institution resonator sensor also has very good repeatability stability and reliability despite the characteristics of the small structure, low power consumption and fast response. Therefore, it has aroused special interest and has become an important branch of the resonator sensor.

This chapter first describes the common basic theory of the resonator sensor based on the mechanical resonant sensitive structure.

1.2 Resonance Phenomenon

The resonant measurement principle is based on the resonant sensing elements, which means it can be achieved by the vibration characteristics of the harmonic oscillator. The harmonic oscillator in the course of its work can be equivalent to a single degree of freedom system (Fig. 1.1(a)). The kinetic equation can be expressed as

$$m\ddot{x} + c\dot{x} + kx - F(t) = 0 \tag{1.1}$$

where

m is the equivalent mass of the vibration system;

c is the equivalent damping coefficient of the vibration system;

k is the equivalent stiffness of the vibration system;

$F(t)$ is the external force.

$m\ddot{x}$, $c\dot{x}$ and kx reflect the inertia, damping and elastic force of the vibration system. Their direction is shown in Fig. 1.1(b).

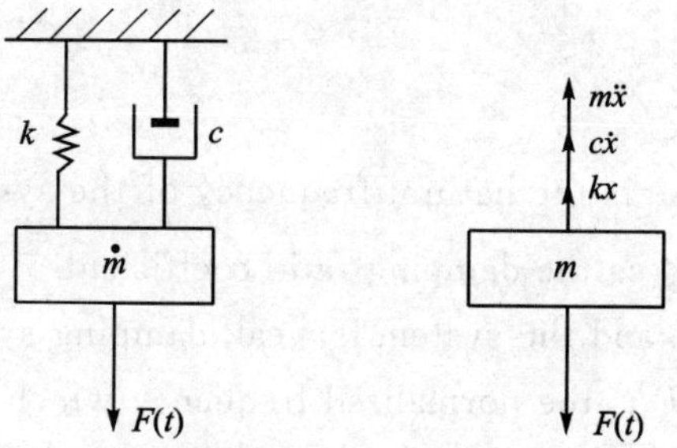

(a) Model of one DOF system (b) Direction of the force

Fig. 1.1 Two order single degree of freedom vibration system

According to the characteristics of the resonant state, when the above vibration system is in the resonant state, the role of external forces should be equal to the damping force of the system, the inertia force should be equal to the elastic force, and the system is under the vibration of its natural frequency, as given by the expression below:

$$\left.\begin{aligned} c\dot{x} - F(t) &= 0 \\ m\ddot{x} + kx &= 0 \end{aligned}\right\} \tag{1.2}$$

At the same time, the external force of the vibration system is ahead of the displacement vector 90°, which is at the same phase of the velocity vector. The sum of the elastic force and inertia force is zero. The natural frequency of the system ω_n can be described as

$$\omega_n = \sqrt{\frac{k}{m}} \tag{1.3}$$

This is just an ideal situation and is difficult to achieve in practical applications, because of the uncertainty of the actual vibration damping force. Therefore, the resonance phenomenon can be recognized from the spectral properties of the system.

When the external force $F(t)$ in Eq. (1.1) is a periodic signal, it can be expressed as

$$F(t) = F_m \sin \omega t \tag{1.4}$$

The system of normalized amplitude and phase response can be respectively shown in Eqs. (1.5) and (1.6).

$$A(\omega) = \frac{1}{\sqrt{(1-P^2)^2 + (2\zeta_n P)^2}} \tag{1.5}$$

$$\varphi(\omega) = \begin{cases} -\arctan \dfrac{2\zeta_n P}{1-P^2} & P \leqslant 1 \\ -\pi + \arctan \dfrac{2\zeta_n P}{P^2-1} & P > 1 \end{cases} \tag{1.6}$$

$$P=\frac{\omega}{\omega_n}$$

where

ω_n is the natural frequency of the system;

ζ_n is the damping ratio coefficient of the system, $\zeta_n = C/2\sqrt{km}$. As for the resonator, $\zeta_n \ll 1$ and the system is weak damping system.

P is the normalized frequency which is relative to the natural frequency of the system.

Fig. 1.2 is for the system's amplitude-frequency characteristic curve and phase frequency response curve.

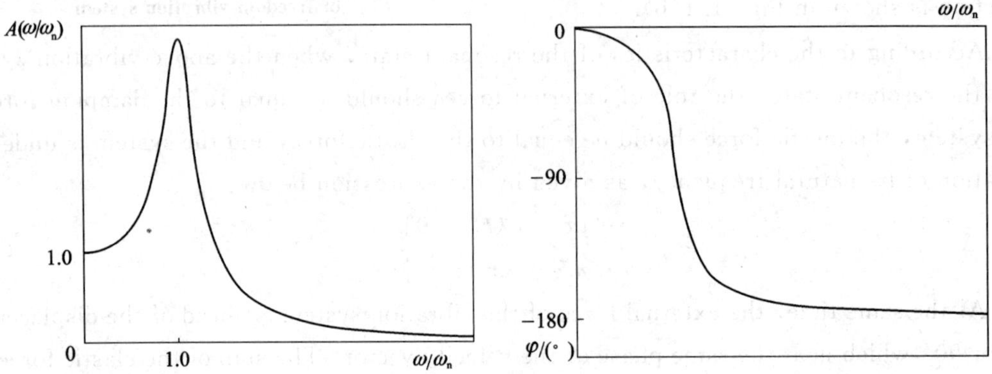

Fig. 1.2 Amplitude-frequency character and phase-frequency character curves for the system

While $P=\sqrt{1-2\zeta_n^2}$, $A(\omega)$ reaches its maximum, there comes

$$A_{\max}=\frac{1}{2\zeta_n\sqrt{1-\zeta_n^2}}\approx\frac{1}{2\zeta_n} \tag{1.7}$$

The phase of the system φ is

$$\varphi=-\arctan\frac{2\zeta_n P}{2\zeta_n^2}\approx-\arctan\frac{1}{\zeta_n}\approx-\frac{\pi}{2} \tag{1.8}$$

Typically, the maximum amplitude gain of the system is defined as the resonant state in engineering, and the corresponding excitation frequency ($\omega_r=\omega_n\sqrt{1-2\zeta_n^2}$) is defined as the resonant frequency of the system.

1.3 Basic Structure of the Resonator Sensors

The vast majority of the resonator sensor is working at the self-exciting closed-loop state. The basic structure is shown in Fig. 1.3.

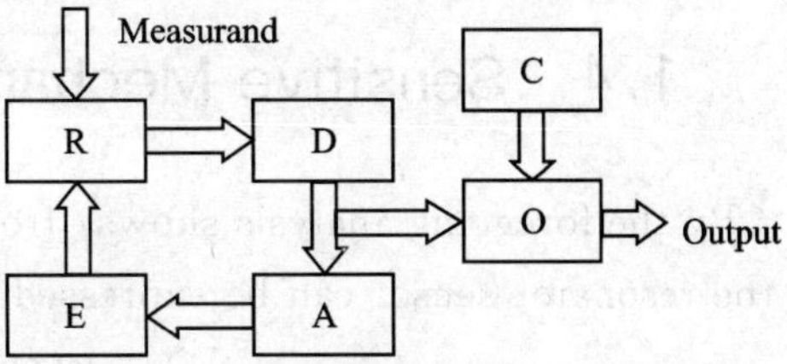

Fig. 1.3 Realization mode of basic measuring principle for resonant type

R is the resonance-sensitive components, also known as the harmonic oscillator. It is the core component of the sensor working in its own inherent vibration modes for continuous vibration. The vibration characteristics have a direct impact on the performance of the resonator sensor.

Harmonic oscillator currently used in many forms, such as the resonant beam, composite tuning fork, resonance tube, resonant film, resonant hemispherical shells and elastic elbow.

D and E respectively represents the signal detector and actuator, which are the necessary components to realize the coversion between the electrical and mechanical signals and provide the conditions for the composition of the self-exciting closed-loop system.

A is the amplifier and inextricably linked with the incentives, which is used to regulate the signal amplitude and phase to make the system working in the self-exciting closed-loop state reliably and stably. The early amplifiers mostly were made of discrete components, but the recent ones mainly use IC process and are dedicated to the design of multi-functional integrated amplifier direction.

O is for the system to sense the output device and is the part for detecting the periodic signal, which can be used for the detection of the frequency, amplitude ratio and phase difference of the periodic signal.

C is the compensation device, which mainly compensates for the temperature error, and sometimes also act as the compensating elements to the zero position and the disturbance caused by the measuring environment.

The six main components above constitute the three important aspects of the resonator sensor.

① The electricity-machine-electricity harmonic oscillator part composed by E, R and D. It is the core of the resonator sensor. Appropriate incentives and the pick-up means for constituting an ideal ERD design for resonator sensor are essential.

② The self-exciting closed-loop part composed by E, R and D. It is the composite condition of the resonator sensor.

③ The signal detection and output part composed by R, D. It is the means to test the measurand.

1.4 Sensitive Mechanism of the Resonator Sensors

By the foregoing analysis shows: from the point of the signal detection view, the output of the resonator sensor can be expressed as

$$x(t) = Af(\omega t + \phi) \tag{1.9}$$

where

A is the amplitude of the detected signal;

ω is the angular frequency of the detection signal;

ϕ is the phase of the detection signal;

$f(\cdot)$ is the normalized periodic function.

When $(n+1)T \geqslant t \geqslant nT$, $|f(\cdot)|_{\max} = 1$; $T = 2\pi/\omega$, it is periodic; A, ω and ϕ are the characteristics parameters of $x(t)$; ϕ possesses the coresidual of 360°(2π).

Obviously, as long as one of the characteristic parameters $x(t)$ of the detection is observably changed by the measurand, it can be detected by testing the above characteristic parameter.

Recently, most resonant sensors are used for the detection of the angular frequency ω, such as the resonance cylinder pressure transducer and resonance membrane pressure sensor.

According to the fact that resonant sensors are sensitive to amplitude or phase, relative (parameter) measurement can be adopted to improve the measurement accuracy by detecting the amplitude ratio or phase difference, such as the resonant direct mass flow sensor.

1.5 The Closed-loop System of the Resonator Sensors

1.5.1 Time-domain Analysis

Fig. 1.4 shows, from the signal actuator into consideration, a transient role in the actuator input signal can be described as

$$u_1(t) = A_1 \sin \omega_V t \tag{1.10}$$

where, A_1 is the amplitude of the excitation voltage signal, which is greater than zero; ω_V is the frequency of the driving voltage signal, which is in close proximity to the natural frequency of the harmonic oscillator; Through the harmonic oscillator, detector and amplifier, $u_1(t)$ turns out to be $u_1^+(t)$ and can be written as

$$u_1^+ = A_2 \sin(\omega_V t + \phi_T) \tag{1.11}$$

where A_2 is the amplitude of the voltage signal $u_1^+(t)$, which is also greater than zero.

When the following conditions are met, the self-exciting closed-loop can be produced with the frequency ω_V.

$$A_2 \geqslant A_1 \tag{1.12}$$

$$\phi_T = 2n\pi \quad (n = 0, \pm 1, \pm 2, \cdots) \tag{1.13}$$

Eqs. (1.12) and (1.13) are called the time-domain amplitude and phase conditions of the system in the self-exciting closed-loop.

The above considerations are the conditions of the self-exciting closed-loop at a certain point closed-loop conditions. As for the resonator sensor, the condition of the self-exciting closed-loop should be met in the working frequency range $[f_L, f_H]$, which brings the special requirements.

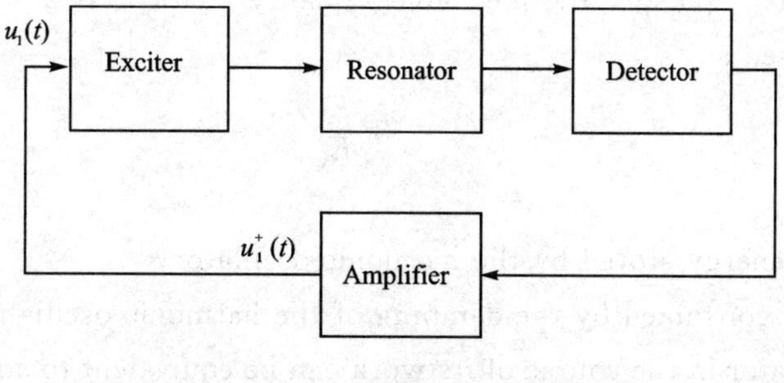

Fig. 1.4 Time-domain analysis for the self-exciting closed-loop condition

1.5.2 Complex Frequency Domain Analysis

In Fig. 1.5, $R(s)$, $E(s)$, $A(s)$ and $D(s)$ are the transfer function of the harmonic oscillator, actuators, amplifiers and vibration pick-up respectively, and s is the Laplace operator. The transfer function $G(s)$ can be described as

$$G(s) = R(s)E(s)A(s)D(s) \tag{1.14}$$

Obviously, when the following conditions are met, the self-exciting closed-loop can be produced with the frequency ω_V.

$$|G(j\omega_V)| \geqslant 1 \tag{1.15}$$

$$\angle G(j\omega_V) = 2n\pi \quad (n = 0, \pm 1, \pm 2, \cdots) \tag{1.16}$$

Eqs. (1.15) and (1.16) are called the complex frequency domain amplitude and phase conditions of the system in the self-exciting closed-loop.

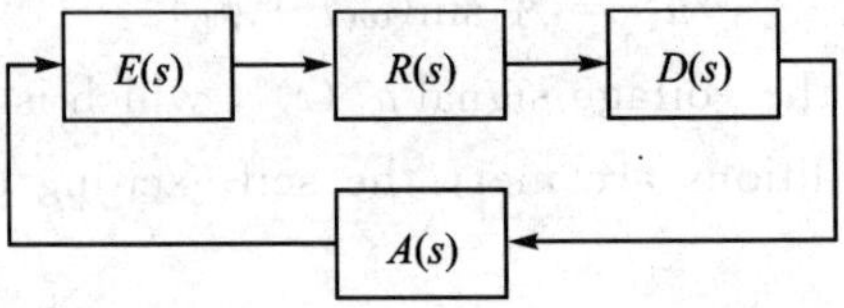

Fig. 1.5 Complex frequency domain analysis for the self-exciting closed-loop condition

1.6 Mechanical Quality Factor of the Resonant Sensitive Elements

1.6.1 Definition and Calculation

As for the resonant sensor, the mechanical quality factor Q is a very important index, which can be expressed as

$$Q = 2\pi \frac{E_S}{E_C} \tag{1.17}$$

where

E_S is the entire energy stored by the harmonic oscillator;

E_C is the energy consumed by the damping of the harmonic oscillator.

Harmonic oscillator in the course of its work can be equivalent to a single degree of freedom system (Fig. 1.1 (a)). The kinetic equation can be described as

$$m\ddot{x} + c\dot{x} + kx - F(t) = 0 \tag{1.18}$$

where, m is the equivalent mass of the vibration system; c is the equivalent damping coefficient of the vibration system; k is the equivalent stiffness of the vibration system; $F(t)$ is the external force. $m\ddot{x}$, $c\dot{x}$ and kx reflect the inertia force, damping force and elastic force of the vibration system. Their directions are shown in Fig. 1.1 (b).

According to the characteristics of the resonant state, when the above vibration system is in the resonant state, the role of external forces should be equal to the damping force of the system, the inertia force should be equal to the elastic force and the system is under the vibration of its natural frequency, as given by the expression below:

$$\left.\begin{aligned} c\dot{x} - F(t) &= 0 \\ m\ddot{x} + kx &= 0 \end{aligned}\right\} \tag{1.19}$$

At the same time, the external force of the vibration system is ahead of the displacement vector 90°, which is at the same phase of the velocity vector. The sum of the elastic force and inertia force is zero. The natural frequency of the system ω_n can be described as

$$\omega_n = \sqrt{\frac{k}{m}} \tag{1.20}$$

This is just an ideal situation and is difficult to achieve in practical applications, because of the uncertainty of the actual vibration damping force. Therefore, the resonance phenomenon can be recognized from the spectral properties of the system.

When the external force $F(t)$ in Eq. (1.21) is a periodic signal, it can be expressed as

$$F(t) = F_m \sin \omega t \tag{1.21}$$

The system of normalized amplitude and phase response can be respectively shown in the Eqs. (1.22) and (1.23).

$$A(\omega) = \frac{1}{\sqrt{(1-P^2)^2 + (2\zeta_n P)^2}} \tag{1.22}$$

$$\phi(\omega) = \begin{cases} -\arctan \dfrac{2\zeta_n P}{1-P^2} & P \leqslant 1 \\ -\pi + \arctan \dfrac{2\zeta_n P}{P^2 - 1} & P > 1 \end{cases} \tag{1.23}$$

$$P = \frac{\omega}{\omega_n}$$

where

ω_n is the natural frequency of the system;

ζ_n is the damping ratio coefficient of the system;

P is the normalized frequency which is relative to the natural frequency of the system.

While $P = \sqrt{1-2\zeta_n^2}$, $A(\omega)$ reaches its maximum, there comes

$$A_{max} = \frac{1}{2\zeta_n \sqrt{1-\zeta_n^2}} \approx \frac{1}{2\zeta_n} \tag{1.24}$$

The phase of the system ϕ is

$$\phi = -\arctan \frac{2\zeta_n P}{2\zeta_n^2} \approx -\arctan \frac{1}{\zeta_n} \approx -\frac{\pi}{2} \tag{1.25}$$

Typically, the maximum amplitude gain of the system is defined as the resonant state in engineering, and the corresponding excitation frequency ($\omega_r = \omega_n \sqrt{1-2\zeta_n^2}$) is defined as the resonant frequency of the system.

As for the resonant sensor, the resonant sensitive structure is an underdamping system, $1 \gg \zeta_n > 0$. The harmonic oscillator amplitude-frequency characteristic that Fig. 1.7 (or Fig. 1.8) shows can be given by:

$$Q \approx \frac{1}{2\zeta_n} \approx A_m \tag{1.26}$$

$$Q \approx \frac{\omega_r}{\omega_2 - \omega_1} \tag{1.27}$$

The corresponding amplitude gain of ω_1 and ω_2 is $\frac{A_m}{\sqrt{2}}$, which is called the half-power point, shown in Fig. 1.6.

The above analysis shows that the Q factor not only reflects the size of damping ratio coefficient and the extent of energy consumption speed in the harmonic oscillator vibration, but also reflects the steepness of resonance peak in the amplitude-frequency characteristic curve which means the capability of frequency selection of the resonant sensitive component.

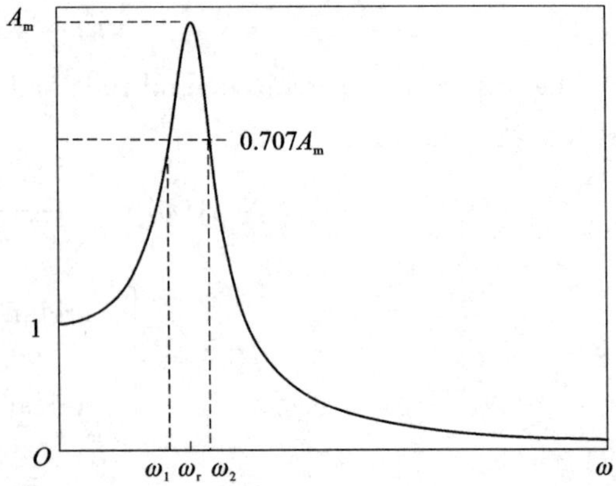

Fig. 1.6 Q factor obtained from the amplitude-frequency character of the resonator

From the perspective of vibration energy, the higher Q factor shows that, in relative to the given harmonic oscillator energy stored weekly, the less energy the damping dissipates, the higher energy storage efficiency the system has, also the stronger the ability resisting to the outside world interfere is; From the perspective of the amplitude-frequency characteristic curve, the higher Q factor means that, the closer the natural frequency and the frequency of the harmonic oscillator get, the better the selected frequency characteristic is, the easier the detection of the resonant frequency gets, the more stable the vibration frequency is, the better the repeatability is. In conclusion, as for the resonant measuring principle, improving the quality factor of the harmonic oscillator is crucial. The essential question for the designer is to increase the Q factor by all kinds of means.

1.6.2 The Analysis of the Affecting Factors of Sensors

The Eq. (1.15) and Fig. 1.8 show that the amplitude condition is easy to meet when Q factor increases. This can be given by

$$\phi(\omega) = -\arctan \frac{P}{Q(1-P^2)} \tag{1.28}$$

$$\frac{\partial\phi(\omega)}{\partial P}=-\frac{Q(1+P^2)}{P^2+Q^2(1-P^2)^2} \tag{1.29}$$

When $P=1$ 时, $\phi=-\pi/2$, $(\partial\phi/\partial P)=-2Q$. Considering in the phase range of $-\pi/2$: $\phi\in\left[-\frac{\pi}{2}-\phi_m, -\frac{\pi}{2}+\phi_m\right]$, when $\phi_m\leqslant\frac{\pi}{4}$, $|\partial\phi/\partial P|$, Q will monotonously increase with it. This suggests that the change in the value of the phase angle caused by the same frequency change increases with the increasing Q factor. When the same phase angle change is required, the large Q factor means that the relative deviation between ω and ω_n is small. In other words, the larger value provides a wide range of phase angle, so that the self-exciting closed-loop system can be composed.

Then we'll discuss the Q factor influence to sensor accuracy. Suppose that the system is working at the frequency range $[f_L, f_H]$ and the phase shift provided by the harmonic oscillator is $[\phi_L, \phi_H]$. So the Eq. (1.23) shows that the corresponding vibration frequency of any phase angle can be given as

$$P\approx 1+\frac{1}{2Q\tan\phi} \tag{1.30}$$

Obviously, as for the large given ϕ and Q, $|P-1|$ decreases, which means that ω is closer to the natural frequency ω_n, random drift of the self-resonance frequency is smaller, the vibration is more stable and the accuracy is higher. In addition, the above formula shows that resonant sensor has the best incentive point. When $P_B=1$ is founded, $\phi_B=-\pi/2$, $\omega=\omega_n$, and the vibration frequency is equal to the natural frequency without the influence of the Q factor. This indicates that: the system with a natural frequency of vibration can set to the best incentive point. When the system is working in the frequency range, considering from the perspective of reducing the interference, the best incentive point can be set in the accordance with the formulas below:

$$|1-P_L|=|1-P_H| \tag{1.31}$$

That is

$$P_L+P_H=2$$

$$\omega_B=\frac{2\omega_L\omega_H}{\omega_L+\omega_H} \tag{1.32}$$

This result is instructive for the design of amplifiers. For the purpose to improve anti-jamming capability of the resonator sensor, the amplifier design should meet the following conditions:

$$\angle E(j\omega_B)+\angle A(j\omega_B)+\angle D(j\omega_B)=\frac{\pi}{2}+2n\pi \qquad (n \text{ is integer}) \tag{1.33}$$

At the same time as far as possible to take a large value of | tan α_L | and | tan α_H |, α_L and α_H are

$$\left.\begin{aligned}\alpha_L &= \angle E(j\omega_L) + \angle A(j\omega_L) + \angle D(j\omega_L)\\ \alpha_H &= \angle E(j\omega_H) + \angle A(j\omega_H) + \angle D(j\omega_H)\end{aligned}\right\} \tag{1.34}$$

1.6.3 Methods on Improving the Mechanical Quality Factor

The above analysis shows that the harmonic oscillator with high Q factor is important for the constitution of a self-exciting closed loop system and the improvement of the system performance, so measures should be taken to increase Q factor of the harmonic oscillator. This is the core issue of the design of the resonant sensors.

Usually the way to improve of Q factor can be considered from the following four areas:

① The materials of high Q factor. Characteristics of the material are determined by the lattice structure and internal molecular motion state, such as Q factor of the quartz material up to $10^6 \sim 10^7$ orders of magnitude and general metal only in $10^4 \sim 10^5$ orders of magnitude.

② The better means of processing. Minimize the residual stress within the harmonic oscillator caused by the processing. For example, as the sensitive components of the resonant cylinder pressure sensor have a wall thickness of 0.08 mm, rotate and tensile technique are adopted. A large residual stress would be emerged and Q factor would reach 3 000～4 000. But by the precision grinding process, the Q factor would reach more than 8 000, which is much higher than the former.

③ Optimize the design of the harmonic oscillator boundary structure and packaging. Prevent the coupling of the harmonic oscillator with external vibration, and effectively isolate the vibration of the harmonic oscillator with the external environment. Thus the way of the tuned decoupling is adopted and the harmonic oscillator should be connected with the outside world through its "nodes".

④ Optimize the working environment of the harmonic oscillator, and make it as far as possible not be impacted by the measured medium.

In general, as for the actual harmonic oscillator, its Q factor decreases 1 to 2 orders of magnitude. This indicates that there is still much work to do in the processing and assembly of the harmonic oscillator.

1.7 Methods of Measurement of the Frequency Output Resonator Sensors

The output frequency of the frequency output resonator sensor is the frequency of the output square wave signal of the closed-loop system. Signal frequency measurement method is typically in two ways: frequency method and periodic method.

1.7.1 Frequency Method

Frequency measurement method is to measure the number of pulses within 1 s, which means that the number of pulses is equal to the input signal frequency. Rectangular wave pulses signal of the sensor are sent to the gate circuit and the "door" switch is controlled by the standard clock frequency, which means that the standard clock frequency signal CP is used as a gating signal. Rectangular wave pulses through the "door" number within 1 s is the input signal frequency, which can be described as $f_{\text{in}} = n_{\text{in}}/T_{\text{CP}}$.

The counter can not be calculated the fractional value of the cycle, so the gating time of 1 s means that the sensor error is 1 Hz. If the frequency of the sensor changes from 4 kHz to 5 kHz (full scale pressure changes), the sensor resolution measured with this method is 0.1% and the measurement time is 1 s. Obviously, this resolution for high-precision resonator sensor is not enough. In order to improve the resolution, it is necessary to extend the measurement time, but this in turn affects the dynamic performance of the sensor. Consequently, as for the conventional resonator sensor, if the output frequency ranges from 100 Hz to 15 kHz, the frequency method should not be adopted. However, for high frequency signals above 100 kHz, the frequency method can be used.

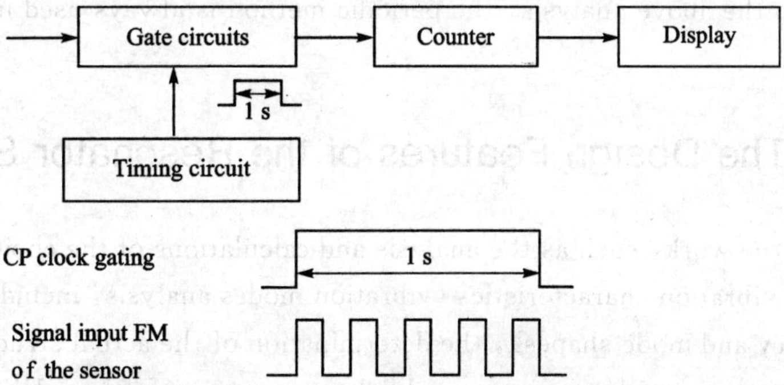

Fig. 1.7 Measuring circuit of the frequency method

1.7.2 Periodic Method

Periodic method is to complete a cycle time required for measuring repetitive signals. The measurement circuit is shown in Fig. 1.8.

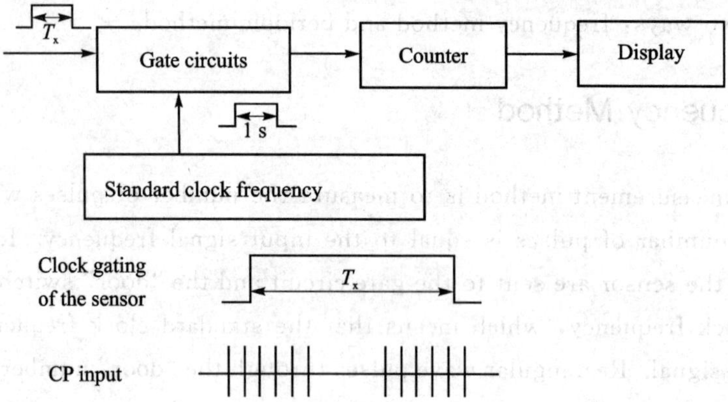

Fig. 1.8 Measuring circuit of the periodic method

The circuit takes the sensor output as the gating signal. Assuming 12 MHz standard frequency signal as input and the output of the sensor of 4 kHz, then the number of pulses of the counter clock in each input pulse cycle is 3 000 ($12\times10^6/(4\times10^3)=3\ 000$) and the measurement period T_{in} is 0.25 ms($T_{in}=n_{in}/f_{CP}=3\ 000/(12\times10^6)$ ms), which means that sensor resolution would be up to 0.1% in 0.25 ms. This suggests that for the above measurement needs, 1/4 000 of the frequency measurement time is required for periodic method. When the gating time is extended to 2.5 ms or 25 ms, the resolution would be 0.01% or 0.001%.

According to the above analysis, the periodic method is always used in the conventional resonator sensor.

1.8 The Design Features of the Resonator Sensors

The core of the work, such as the analysis and calculations of the choice of the harmonic oscillator and its vibration characteristics (vibration modes analysis, including the analysis of resonant frequency and mode shapes), the determination of the actual structure and the parameters of the harmonic oscillator, is to establish a resonant sensor-model to optimize a high

Q factor and high sensitivity of the harmonic oscillator, please see the related content of Chapter 2～5.

The problems that the choice of detection and excitation source along with the matching problem with the harmonic oscillator include the choice of the relative position of the harmonic oscillator and the size of incentive energy. This part of the content will be discussed in details below.

The receiving, processing and conversion of the detection signal and the design of the amplifier in amplitude and phase conditions that should be comprehensively considered in full-scale integrated for the resonant sensors which are sensitive to the frequency. But the resonant sensors which are sensitive to the amplitude ratio and phase difference should be reasonably designed in a "double closed-loop" system and reference position should be chosen.

The compensation mechanism should be appropriately introduced, the detection signal should be solved and the measurand should be given.

1.9 The Characteristics and Advantages of the Resonator Sensors

Compared with other types of sensors, the essential features and unique advantages of the resonant sensor are as follows:

① The output signal is periodic and the measurand can be calculated by detecting the cycle signal. This feature determines that the resonant sensor can be expediently connected with the computer and the signal can be transmitted in a long distance.

② The sensor system is a closed-loop system in a resonant state. This feature determines that the output of the sensor system can automatically track the input.

③ Sensitive components of the resonator sensor which can also be described as the natural resonant characteristics of the harmonic oscillator decide the high sensitivity and resolution of it.

④ With respect to the vibrational energy of harmonic oscillator and the power consumption of system is a very small amount. This feature determines a strong anti-interference performance and good stability of the sensor system.

Questions and Exercises

1.1 Establish the second-order kinetic equation to the system that consists of mass, spring and dampers, and then explain the resonance phenomenon and its basic characteristics.

1.2 Resonant measurement principle is usually required to constitute a self-exciting closed-loop system based on the core harmonic oscillator (resonant sensitive components). Which parts is the composition of the self-exciting closed-loop system? What are they used for?

1.3 Why is the Q factor of the harmonic oscillator the key parameter of the resonant sensor? How to determine Q factor in a high precision and how to improve Q factor?

1.4 Illustrate the reason why the higher Q factor is better in terms of the self-exciting closed-loop conditions of the resonant sensor.

1.5 According to the underdamping system, $1 \gg \zeta_n > 0$, try to prove the Eqs. (1.24) and (1.25).

1.6 What are the main advantages of the resonant sensor? What are the shortage points?

1.7 Discuss the conditions of the closed-loop system implementation of the resonant sensor.

1.8 In the resonant sensors of frequency output, which methods are mainly used to measure the frequency? What are the respective characteristics?

1.9 For the resonator sensors constituted by the resonance phenomenon, despite of the sensing mechanism of the frequency detection, what are the other sensing mechanisms? What problems should be noticed?

1.10 Discuss the measurement errors with the referable Eq. (1.24), while using the normalized magnitude characteristic peak A_{max} of the harmonic oscillator to measure the Q factor.

Chapter 2
Resonant Pressure Sensor

Basic contents:

- Vibrating string
- Resonant film
- Resonant tube
- Resonant beam
- Quartz vibrating beam
- Silicon micromachined
- Electromagnetic excitation
- Piezoelectric excitation
- Electrostatic excitation
- Thermal excitation
- Open-loop excitation
- The self-exciting closed-loop system
- Differential output detection

2.1 Vibrating String Type Pressure Sensor

2.1.1 Structure and Principle

Fig. 2.1 shows the sketch of vibrating string type pressure sensor. It consists of the vibrating string type, magnet coil assembly and vibrating string clamping mechanism, and the other components.

Vibrating string is a string wire or string band. Its upper end is clamped with a clamping mechanism, and rigidly attached with the housing; while its lower end with a clamping

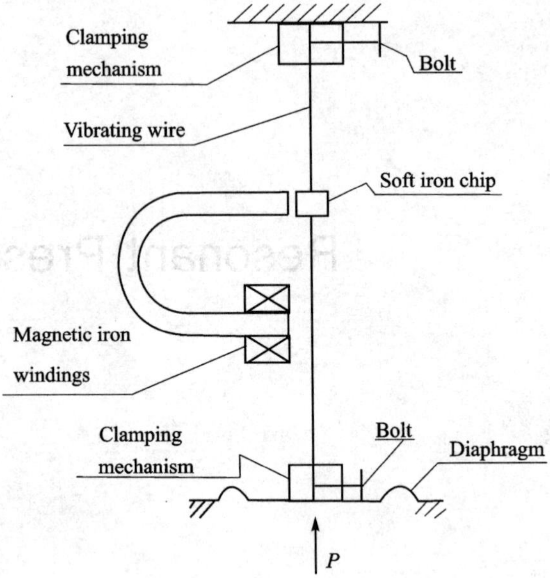

Fig. 2.1 Sketch of the vibrating string type pressure sensor

mechanism, and rigidly attached with the hard center of the diaphragm. A fixed preload is added when the vibrating string is clamped. When pressure p is applied the diagraph, there will initiate a relative displacement of the diagraph, which will result change of tensile force for the string wire. Correspondingly, resonant frequency of the string will change. By measuring the change of the resonant frequency, p can be calculated then.

The coil assembly of the magnet is used to generate the exciting force and to detect the vibration frequency, and the magnet could be a permanent magnet or a DC electromagnet. According to different excitation modes , coil assembly of the magnet may be one or two. When a coil assembly of the magnet is used, the coil is not only excitation coil but also the vibration pickup coil. When the pulse current passes the coil, the soft-iron plate fixed on the vibrating string magnet is attracted by the magnet and applies the excitation force to the vibrating string. The soft-iron plate will be released without the pulse current, and the vibrating string will vibrate freely on a natural frequency, so that an induction potential whose frequency is the same as the vibrating string will be induced in the magnet coil assembly. Due to the impact of the air damper, the free vibration of the vibrating string is gradually attenuating, so the pulse current whose frequency is the same as the vibrating string will be added to the excitation coil to keep the vibrating string vibrating.

If the measured pressure is different, the tension added to the vibrating string and the equivalent stiffness of the vibrating string will be different, so does the natural frequency of the vibrating string. Therefore, we can work out the pressure by measuring the natural frequency of the vibrating string.

2.1.2 Characteristic Equation

When the tension is added to the vibrating string, the lowest order natural frequency is

$$f_{\mathrm{TR1}}(p)=\frac{1}{2L}\sqrt{\frac{T_0+T_p}{\rho_0}} \tag{2.1}$$

where

$f_{\mathrm{TR1}}(p)$ is the lowest order natural frequency of the vibrating string under the pressure p (in Hz);

T_p is convert the measured pressure p to the tension acting on the vibrating string (in N);

T_0 is the initial tension of the vibrating string(in N);

L is the length of the vibrating string working segment (in m);

ρ_0 is mass per unit length of vibrating string (in kg/m).

From the Eq. (2.1), the relationship between the natural frequency of the vibrating string and the tension is nonlinear function. If the measured pressure is different, the tension added to the vibrating string and the equivalent stiffness of the vibrating string will be different. Therefore, we can work out the pressure by measuring the natural frequency of the vibrating string, and the frequency of the induced electrical potential in pick-up coil is related to the measured pressure.

2.1.3 Excitation Mode

Fig. 2.2 shows two different excitation modes of vibrating string type pressure sensor, while (a) shows intermittent excitation mode and (b) shows continuous excitation mode.

In the continuous excitation mode, there are two magnet coils assembly: one is an excitation coil and the other is a pick-up coil. The induced electromotive force of the coil 2 can not only be used as an output signal, but also feedback to the excitation coil 1 when it is amplified. As long as the signal after amplified can meet the requirement of amplitude and phase of the vibrating string system oscillation, the vibrating string will maintain vibrating.

Vibrating string pressure sensor has many advantages, such as high sensitivity, high accuracy, simple structure, small size, low power consumption and small inertia, so it is wide-

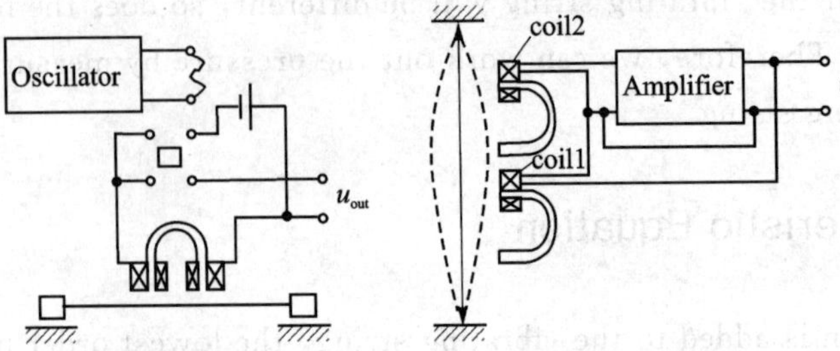

Fig. 2.2 Exciting method of the vibrating string

ly used in pressure measurement.

2.2 Resonant Diaphragm Pressure Sensor

2.2.1 Structure and Principle

Fig. 2.3 is a sketch of the resonant diaphragm pressure sensor. The circular flat diaphragm with clamped edges is elastic sensitive element, and an excitation electromagnetic coil is installed at the center of the diaphragm. The semiconductor strain element is affixed to the edge of the diaphragm to pick up the vibration. The element is placed on the edge of the diaphragm because that position can initiate maximum displacement to confirm the biggest signal output. A cited pressure nozzle is mounted on the base of the sensor. The reference pressure chamber and measured pressure chamber of the sensor are separated by the diaphragm.

The working principle of diaphragm pressure sensor, which is the same as vibrating cylinder pressure sensor, is to measure the pressure by that the natural frequency of the diaphragm changes with the measured pressure.

When the round diaphragm is given an exciting force, it will vibrate by natural frequency. When the measured pressure changes, the stiffness of the circular diaphragm will change, and will lead the natural frequency to change. In the meanwhile, the vibration of the circular diaphragm makes the stress at the edge change periodically, and thus we can detect the vibration signal of circular flat diaphragm through semiconductor strain gauge, and send it to the amplifier circuit through bridge output signal. On one hand the signal is fed back to the excitation coil to maintain the vibration of the diaphragm, on the other hand, it sends square wave signals to the subsequent measurement circuit after shaped. The diaphragm

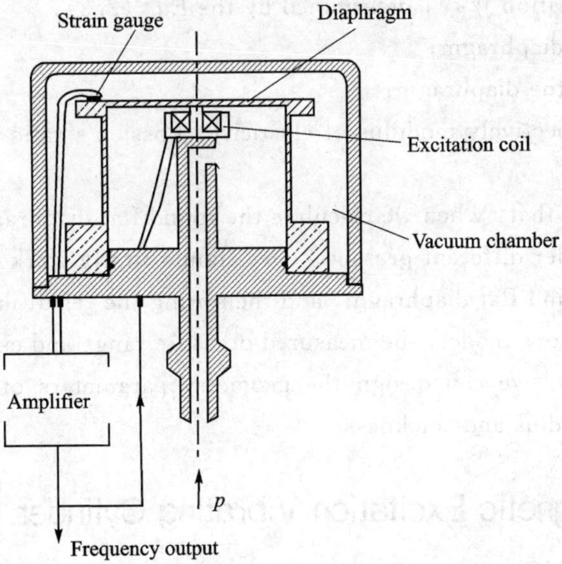

Fig. 2.3 Working principle of the vibration film-type pressure sensor

pressure sensor also has high accuracy, and is applied to the high-performance supersonic aircraft as a key sensor. Compared with the vibrating cylinder pressure sensor, the elastic sensitive element of vibrating membrane is of high stability in frequency and pressure characteristics, high measurement sensitivity, small size, small masses, and simple structure, but the processing difficulty is slightly increased.

2.2.2 Characteristic Equation

Under the uniform pressure, the approximation formula of the circular flat diaphragm's lowest order natural frequency can be described as

$$f_{R,B1}(p) = \frac{0.469H}{R^2}\sqrt{\frac{E}{\rho(1-\mu)^2}}\sqrt{1+\frac{(1+\mu)(173-73\mu)}{120}\left(\frac{W_{R,C}}{H}\right)^2} \tag{2.2}$$

$$p = \frac{16E}{3(1-\mu^2)}\left(\frac{H}{R}\right)^4\left[\frac{W_{R,C}}{H}+\frac{(1+\mu)(173-73\mu)}{360}\left(\frac{W_{R,C}}{H}\right)^3\right] \tag{2.3}$$

where

$f_{R,B1}(p)$ is the circular flat diaphragm's lowest order natural frequency under pressure p (in Hz);

$W_{R,C}$ is the largest normal displacement at the center of the circular flat diaphragm under great deflection deformation $W_{R,C}$, determined by the Eq. (2.3);

R is radius of the diaphragm;

H is thickness of the diaphragm;

E, μ and ρ are respectively modulus of elasticity, Possion's ratio and density of passion material.

It should be noted that: when we calculate the round flat diaphragm's natural frequency of the lowest order under different pressures, we should firstly work out the largest normal displacement of the round flat diaphragm, and then bring the result in the Eq. (2.2) to calculate. Based on the above model, the measured pressure range and circular flat diaphragm's frequency change ratio, we can design the geometry parameters of the circular flat diaphragm, namely the radius and thickness.

2.3 Electromagnetic Excitation Vibrating Cylinder Pressure Sensor

2.3.1 Structure and Principle

Fig. 2.4 shows the sketch of vibrating cylinder pressure sensor, it consists of the resonant cylinder unit and the driver amplifier.

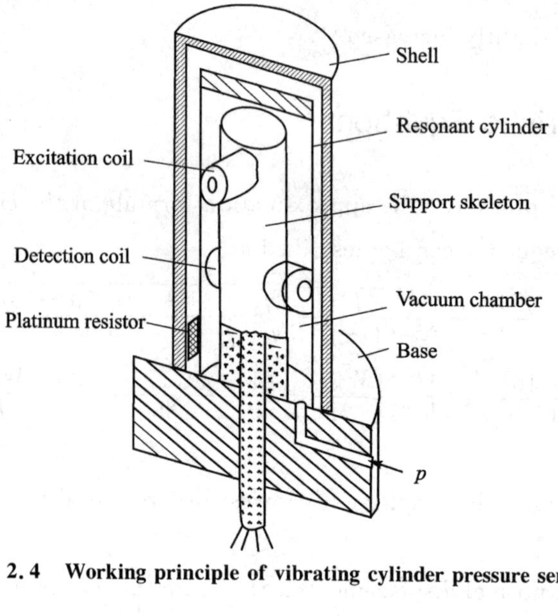

Fig. 2.4 Working principle of vibrating cylinder pressure sensor

The body of the sensor consists of vibrating cylinder, pick-up coil and the excitation coil. The sensor is the absolute pressure sensor, so there is the vacuum between the vibrating cylinder and the housing. Vibrating cylinder is molded by turning or spinning stretch and made through strict heat treatment process. The material is usually 3J53 or 3J58—constant elastic alloy (which is called Ni-Span-C in foreign). The typical dimensions of vibrating cylinder includes: diameter of 16～18 mm, wall thickness of 0.07～0.08 mm and the effective length of 45～60 mm. General requirements for the quality factor are greater than 5 000.

According to the structural characteristics and parameters of the resonant cylinder, Fig. 2.5 shows the possible vibration mode it may have. (a) shows the vibration mode of the circumferential direction of the vibrating cylinder and (b) shows the vibration mode of the generatrix direction of the vibrating cylinder. In the Fig. 2.5, m means half-wave number of vibration mode along the generatrix direction of the vibrating cylinder, and n means wave number of vibration mode along the circumferential direction of the vibrating cylinder.

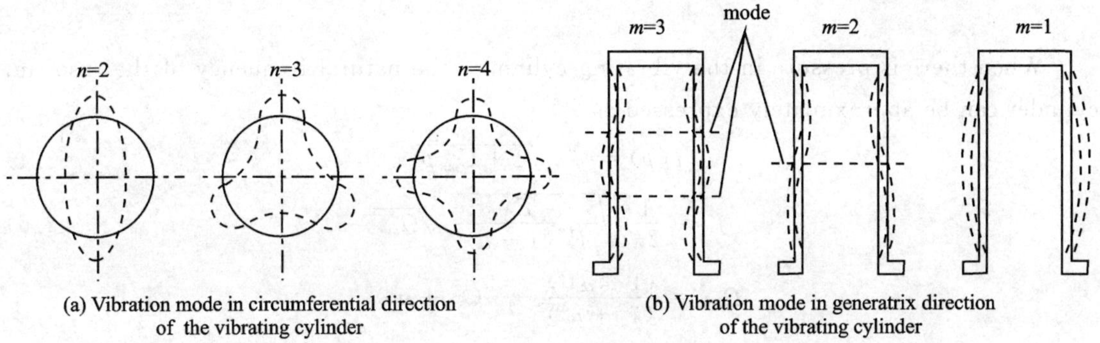

Fig. 2.5 Possible vibration mode for the vibrating cylinder

Fig. 2.6 is the sketch of the relationship between the vibration mode shapes and strain energy. Where in, (a) shows the curve of the lowest-order natural frequency's changing with the circumferential wave number n; (b) shows the curve of the relationship between tensile strain, bending strain energy and n. As can be seen from Fig. 2.6, When $m=1$, the desired strain of n from 3 to 4 can be minimized. Therefore, when designing the vibrating cylinder pressure sensor, we generally choose $m=1$, $n=4$.

When the measured pressure passing into the vibrating cylinder is different, the equivalent stiffness of the vibrating cylinder will be diverse. Therefore, the natural frequency of the vibrating cylinder will be different, too. The measured pressure can be measured by measuring the natural frequency of the vibrating cylinder.

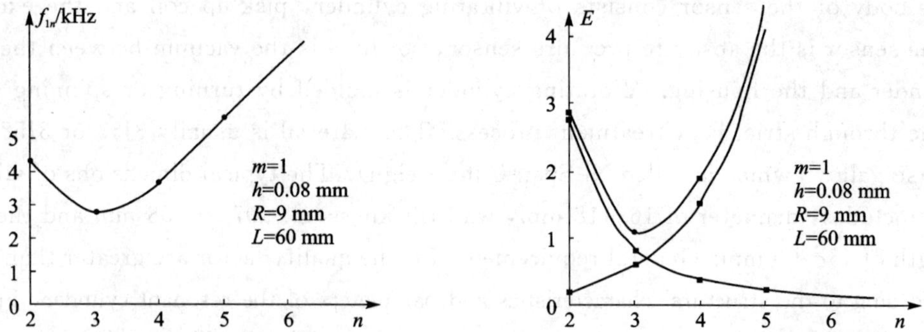

(a) curve of the lowest-order natural frequency's changing with the circumferential wave number n

(b) curve of the relationship between tensile strain energy and n

Fig. 2.6 Relationship between the vibrating mode and the strain energy

2.3.2 Characteristic Equation

When there is pressure in the vibrating cylinder, the natural frequency of the resonant cylinder can be approximately expressed as

$$f(p) = f_0 \sqrt{1 + C_{mn} p} \tag{2.4}$$

$$f_0 = \frac{1}{2\pi} \sqrt{\frac{E}{\rho R^2 (1-\mu)^2}} \sqrt{\Omega_{mn}} \tag{2.5}$$

$$\Omega_{mn} = \frac{(1-\mu)^2 \lambda^4}{(\lambda^2 + n^2)^2} + \alpha (\lambda^2 + n^2)^2$$

$$\lambda = \frac{\pi R m}{L}$$

$$\alpha = \frac{h^2}{12 R^2}$$

where

f_0 is the natural frequency of the vibrating cylinder when the pressure is zero (in Hz);

p is measured pressure (in Pa);

m is half-wave number of mode along the generatrix direction of the vibrating cylinder ($m \geqslant 1$);

n is wave number of mode along the circumferential direction of the vibrating cylinder ($n \geqslant 2$);

C_{nm} is the coefficient related to cylindrical shell material, physical parameters and vibration modes wave number (in Pa^{-1});

ρ is density of the material vibration cylinder(in kg/m^3);

μ is Poisson's ratio of the vibrating tube material;

E is modulus of elasticity of the vibration tube material (in Pa);

R is cylinder radius of vibrating cylinder (in m);

L is length of the portion of the vibrating cylinder (in m);

h is thickness of the wall of the cylinder (in m).

2.3.3 Excitation Mode

Pick-up and the exciting coil both consist of core and coil. In order to minimize the electromagnetic coupling between them , they are designed a certain distance apart and perpendicular to each other in the core. The core of the pick-up coil is magnet, and the core of the excitation coil is soft iron. Output voltage of the vibrating cylinder is proportional to the vibration speed of the pick-up coil, while the exciting force of the excitation coil is proportional to the square of current flowing through the coil.

Therefore, if AC current passing into the coil, we can see

$$i(t) = I_m \sin \omega t \tag{2.6}$$

Centrifugal force is

$$f_B(t) = K_f i^2(t) = K_f I_m^2 \sin^2 \omega t = \frac{1}{2} K_f I_m^2 (1 - \cos 2\omega t) \tag{2.7}$$

where

K_f is conversion coefficient (in N/A^2).

Therefore, the angular frequency of alternating force among excitation force is twice of the angular frequency of excitation current, which is not allowed in the resonant sensor. In order to keep them in the same frequency relationship, DC current should be passed through in the coil, excitation current should be

$$i(t) = I_0 + I_m \sin \omega t \tag{2.8}$$

and

$$f_B(t) = K_f (I_0 + I_m \sin \omega t)^2 = K_f \left(I_0^2 + \frac{1}{2} I_m^2 + 2 I_0 I_m \sin \omega t - \frac{1}{2} I_m^2 \cos 2\omega t \right) \tag{2.9}$$

When $I_0 \gg I_m$ is satisfied, it can be seen from the Eq. (2.8): among the exciting force generated by the excitation coil, the main component of the alternating force has the same frequency with the excitation current. Thus, to make the vibrating cylinder sensors of the electromagnetic excitation work properly, a certain amount of DC current (I_0) should be passed into the excitation coil, and should be guaranteed more than the amplitude I_m of the AC component.

For electromagnetic excitation, in order to prevent the interference of the external magnetic field on the sensor, the electromagnetic device maintaining oscillation should be shielded. Typically coaxial outer tube can be made of a high permeability alloy in order to be shielded.

Besides the electromagnetic excitation, we can also use the way of the piezoelectric actuator. We can use N piezoelectric characteristics of the piezoelectric transducer element to detect the vibration of the vibrating cylinder, use the characteristics of the inverse piezoelectric to produce excitation and constitute a self-exciting closed-loop circuit using a charge amplifier. The electromagnetic vibrating cylinder pressure sensor of piezoelectric actuators is better than electromagnetic excitation mode in the structure, volume, power, anti-interference ability and production costs, but maybe slightly higher in the hysteresis.

2.3.4 Characteristics Linearization and Error Compensation

Through the excitation amplifying circuit, the sensor output is a quasi-digital frequency signal, with high stability and without the influence of passing information.

It can be read by the general digital frequency meter, but can't display the pressure value directly. That's because the pressure to be measured does not have a linear relationship with the output frequency(see Eq. (2. 4)). It generally owns characteristics in Fig. 2. 7. When the pressure is zero, there is a large initial frequency, frequency increases along with the increase in the measured pressure.

The nonlinear error between the measured pressure and the output displayed value is too large to be easily interpreted. In the end, the processing of the "linear" should be added to the output of the sensor.

Linearization of vibrating cylinder pressure sensor, used MOS IC TTL logic component hardware to constitute compensation nonlinear error or temperature correction circuit at early stage, but because of its complex structure, high price and poor process, it is now rarely used. We have already started using the software compensation package, with the use of computer and microprocessor in sensor more and more. The package usually has two kinds. One uses computers of measurement and control system to directly convert the sensor output into the engineering units amended through solving, and directly shows the measured values or recorded by an external device, the other is to use a dedicated microprocessor to store test data in memory through a programmable memory, and find the measured pressure value by the table lookup method and interpolation formula.

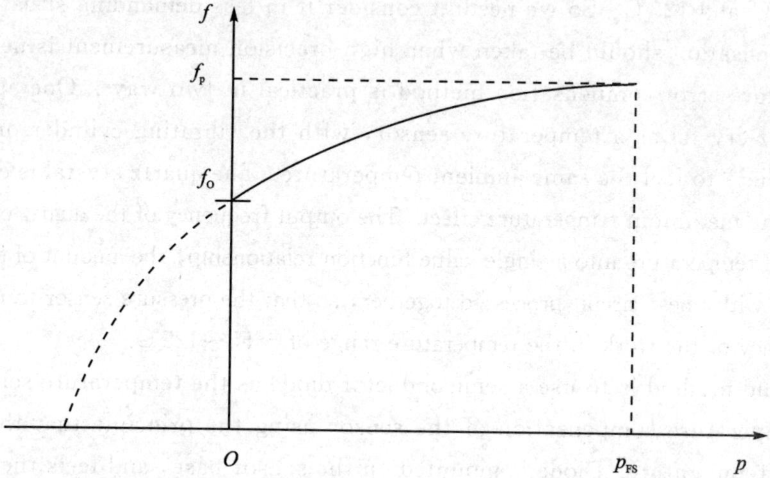

Fig. 2.7 Relationship between the frequency and pressure for the vibrating cylinder pressure sensor

Temperature effect will effect the measuring precision of the vibrating cylinder pressure sensor in two means:

① The elastic modulus of the vibrating cylinder metal material varies with temperature. Other dimensions, such as length, thickness and radius also changes slightly with temperature, but these effects are relatively small because using constant elastic material.

② Influence of temperature on the density of the measured gas. Although constant elastic material is used to make resonant sensitive element, quality of the gas in the barrel changes with the gas pressure and temperature. During the measurement, the measured gas is filled in the cylinder space, Therefore, when the cylinder vibrates, its internal gas will vibrates together with it, and gas mass must be attached to the quality of the barrel. The change in density of the gas causes the measurement error. The gas density (ρ_{gas}) can be represented by the formula following:

$$\rho_{gas} = K_{gas} \frac{p}{T} \tag{2.10}$$

where

p is the measured pressure (in Pa);

T is absolute temperature (in K);

K_{gas} is coefficient depending on gas components.

Therefore, in the pressure sensor of the vibrating cylinder, the impact of the gas density acts as the temperature error. Actual test shows that the change of output frequency is approximately 2% (the temperature error is about 0.01% /℃), when the temperature ranges between −55℃ and 125℃. So we needn't consider it in less demanding situations and temperature compensation should be taken when high-precision measurement is needed.

Temperature error compensation method is practical in two ways. One approach is the use of a quartz crystal as a temperature sensor, with the vibrating cylinder pressure sensor package together, to feel the same ambient temperature. The quartz crystal is cut into the direction with the maximum temperature effect. The output frequency of the quartz crystal temperature sensor and temperature into a single value function relationship, the amount of output frequency can be used with linear circuit processed together, so that the pressure sensor to reach 0.01% in the total accuracy of the work in the temperature range of −55～125℃.

The second method is to use a semiconductor diode as the temperature sensing element, and make temperature compensation of the sensor using the principle that the bias voltage changes with temperature. Diode is mounted on the sensor base, and feels the same ambient temperature with pressure sensors. Diode bias voltage sensitivity of it is up to 2 mV /℃, the sensitivity of the temperature sensing is 30 to 40 times higher than that of thermocouple, and the relationship between voltage variation and temperature is approximately linear. Of course, we can also use platinum resistance thermometer to measure temperature and make temperature compensation (see Fig. 2.4).

Through the test of vibrating cylinder pressure sensors under different temperature and pressure, the temperature error characteristic of the sensor under different pressures can be obtained. With this characteristic, the sensor temperature error can be corrected in order to achieve the desired measurement accuracy.

The accuracy of vibrating cylinder sensor is 1 to 2 orders of magnitude higher than that of the pressure sensor with average analog output, the work is extremely reliable, long-term stability, high repeatability, especially suitable for relatively harsh environmental conditions of the test. The test shows the error is only 0.004 5% FS when the 100 vibration acceleration effect is added to the sensor; and the error is only 0.001 5% FS when the power supply voltage fluctuates 20%. Because of a series of unique advantages, the high-performance supersonic aircraft is equipped with the vibrating cylinder pressure sensor through which we can obtain the correct height and speed in flight, and make parameter measurements of atmospheric data by computer's direct solving. It also can be used as the standard instrument for stress testing, and used to replace the non-mercury manometer.

2.4 Piezoelectric Exciting Resonant Cylinder Pressure Sensor

2.4.1 Structure and Principle

The electromagnetic exciting resonant cylinder pressure sensor with a cylindrical shell as its resonant sensing component is discussed in section 2.4. The advantage of this proposal lies in that the excitation unit and pick-up unit hardly have influence on the sensing structure of the resonant cylinder since it adopts the non-contact method to realize the energy supply to and gaining of the vibration characteristic of the resonant sensing structure. However, the electromagnetic-exciting-and-pick-up energy transformation has its own problem that needs to be seriously considered. Firstly, the exciting and pick-up units are both made of iron core and coils. In order to reduce the electromagnetic coupling between them, they need to be placed orthogonally in space, fixed by epoxy resin framework. Secondly, to prevent the interference of outer electromagnetic field, the electromagnetic device with a sustaining vibration state should be isolated, which demands coaxial outer cylinder make up of high permeability alloy materials to achieve the isolation goal. Furthermore, based on the working principles of the resonant cylinder pressure sensor, the space between the cylindrical shell resonant sensing component and the outer shell should be vacuumed, and the measured pressure will be led into the lumen of the cylindrical shell. Nonetheless, the exciting and pick-up units occupy large space in the lumen cylindrical shell. The measured gas media has an enormous influence on the vibrating characteristic of the resonant sensing component. It is even more obvious when the measured pressure is big, which forms a large squeeze film damping. This will affect the performance of the pressure, even stop its vibrating. All in all, electromagnetic exciting resonant cylinder pressure sensor has its own short comings like low energy conversion efficiency, larger volume, high energy consuming, poor anti-interference, low yield etc. Thus, one designing proposal came up to use piezoelectric element as the exciting and pick-up unit of resonant cylinder pressure sensor, as shown in Fig 2.8. In the diagram, both the piezoelectric exciting and pick-up unit are planted at the wave node locating at end of the cylindrical shell sensing structure, the lumen is completely empty. This proposal uses positive piezoelectric properties of piezoelectric transducer elements to detect the vibration of the cylinder, transforming mechanical vibration signal to electric signal; while negative piezoelectric properties are used to generate exciting force to the cylindrical shell resonant sensing component, realizing the transformation from electric exciting signal to mechanical exciting force. A charge amplifier is adopted to form a closed-loop self-exciting circuit. Apparently, when compared to electromagnetic excitation, piezoelectric exciting resonant cylinder pressure sensor overcomes some flaws existing in electromagnetic exciting resonant cylinder pressure sensor, with advantages of simpler structure, higher electromechanical conversion efficiency, easy to miniaturize, low energy consumption, convenience to form difference ways of closed-loop system and, however, bigger hysteresis error comparing to electromagnetic method. As for further problems concerning the vibrating cylinder pressure sensor with this particular structure, they will be thoroughly discussed in the following part.

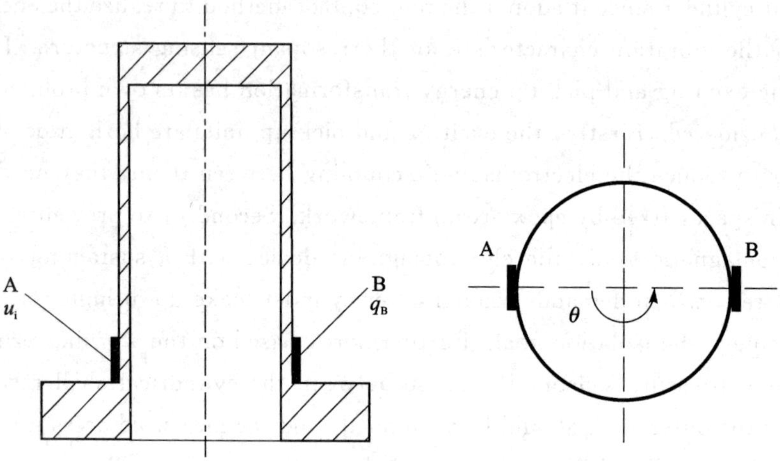

Fig. 2.8 Sketch of the piezoelectric exciting proposal

2.4.2 Characteristics of Piezoelectric Excitation

In Fig. 2.8, assuming A is piezoelectric exciting component and B is piezoelectric pick-up element. Based on inverse piezoelectric effect, the stress on exciting component A generated from exciting voltage (u_i) is

$$\left.\begin{aligned} T_1 &= \frac{E_t}{1-\mu_t} \cdot \frac{u_i}{\delta} d_{31} \\ T_2 &= T_1 \end{aligned}\right\} \tag{2.11}$$

where

E_t, μ_t and δ are respectively modulus of elasticity (in Pa), Poisson's ratio and thickness of the piezoelectric ceramics component;

d_{31} is the piezoelectric constant(in C/N).

T_1 and T_2 are the mechanical stress exerted on the cylindrical shell at the point A.

Under its influence, the transfer function of the displacement on the cylindrical shell u, v and w (unified as d) with regard to external forces is

$$\frac{d(s)}{T_j}=\sum_{m=1}^{\infty}\sum_{n=0}^{\infty}\frac{k_{nmd}}{\left(\frac{s}{\omega_{nm}}\right)^2+2\zeta_{nm}\left(\frac{s}{\omega_{nm}}\right)+1} \tag{2.12}$$

where

ζ_{nm}, ω_{nm} and k_{nmd} are respectively the equivalent damping ratio coefficient of the vibration modal, the natural frequency and the gain for the (n, m) modal (namely the wave number along the circumferential direction is n and the half wave number of the busbar is m).

s is the Laplace operator.

Eq. (2.12) written in the sum form is based on the orthogonality of the cylindrical shell vibration mode. d respectively expresses u, v and w. Thus, as for the nth-order symmetric vibration mode of the shell, it can be described as

$$\left.\begin{aligned}u&=u(s_1)\cos n\theta\\ v&=v(s_1)\sin n\theta\\ w&=w(s_1)\cos n\theta\end{aligned}\right\} \tag{2.13}$$

What needs to be paid extra attention is that θ starts calculating at point A, and s_1 is the coordinate on the axial direction.

$$\left.\begin{aligned}\varepsilon_{s_1}&=\frac{\partial u}{\partial s_1}\\ \varepsilon_\theta&=\frac{\partial v}{r\partial\theta}+\frac{w}{r}\end{aligned}\right\} \tag{2.14}$$

$$\left.\begin{aligned}\sigma_{s_1}&=\frac{E}{1-\mu^2}(\varepsilon_{s_1}+\mu\varepsilon_\theta)\\ \sigma_\theta&=\frac{E}{1-\mu^2}(\mu\varepsilon_{s_1}+\varepsilon_\theta)\end{aligned}\right\} \tag{2.15}$$

where

E and μ are respectively modulus of elasticity (in Pa), and Poisson's ratio of the cylindrical shell material;

r is radius of the middle cylinder section;

σ_{s_1} and σ_θ are stress suffered by pick-up component B.

According to positive piezoelectric effect:

$$q_B=d_{31}(\sigma_{s_1}+\sigma_\theta)A_0 \tag{2.16}$$

where

q_B is the amount of charge generated by pick-up component B;

A_0 is area distributed with charge (in m^2).

With the help of Eq. (2.12) to Eq. (2.16),

$$\frac{q_B}{u_i} = \sum_{m=1}^{\infty}\sum_{n=0}^{\infty} \frac{k_{nm}\cos n\theta}{\left(\frac{s}{\omega_{nm}}\right)^2 + 2\zeta_{nm}\left(\frac{s}{\omega_{nm}}\right)+1} \tag{2.17}$$

$$k_{nm} = 2k_t P_t\left[\frac{\mathrm{d}}{\mathrm{d}s_1}k_{nmu} + \frac{1}{r}(nk_{nmv} + k_{nmw})\right]$$

$$k_t = \frac{A_0 d_{31} E}{1-\mu}$$

$$P_t = \frac{d_{31}E_t}{(1-\mu_t)\delta}$$

Obviously, k_{nm} is related to the cylindrical shell's features (including boundary condition) and the characteristics of the piezoelectric ceramics component.

According to the Eq. (2.17), the phase shift of q_B with regard to u_i is determined by the mechanical features of the cylindrical itself. It remains the same within a given circumferential area, and in different area it shifts π or 0. Besides, the circumferential area is given by the wave number n on the circumferential direction and the position of the exciting point.

Fig. 2.9 gives the section division and related illustration when $n = 3$. When excited at point A, identical displacement takes place if picking-up at arc a_1b_1, arc a_2b_2 and arc a_3b_3. Considering arc a_1b_1 contains point A, this area is called "area with same phase", while area of arc a_2b_2 and arc a_3b_3 oppose to the above area, which is named "area with opposite phase".

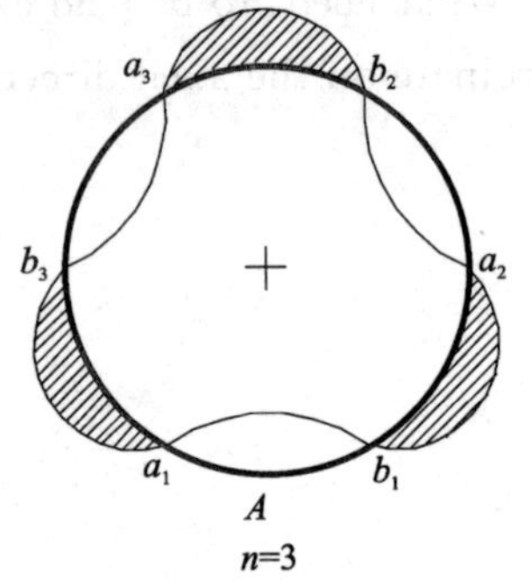

Fig. 2.9 Sketch of the phase distribution

The conclusion is obtained based on the piezoelectric exciting method, which stays the same while pressure changes. It shows a great importance in guiding the choosing of vibration mode and the gluing position the piezoelectric component.

2.4.3 Transformation of Picking-up Signals

When utilizing the positive piezoelectric effect of piezoelectric component, considering the special working mechanism of the components, they function actually as an electrostatic generator or a capacitor, as shown in Fig. 2.10(a). In the diagram, C_0 is a static capacitance, and R_x, C_x, L_x are all high frequency dynamic parameters. Considering the resonator working at a low frequency, and the piezoelectric component is in a fastening state, therefore its effective circle can be described as shown in Fig. 2.10(b). In practical measuring, the impedance matching is a problem which needs further consideration, namely a transducer with high input impedance is in demand to transfer high impedance input into a low impedance signal.

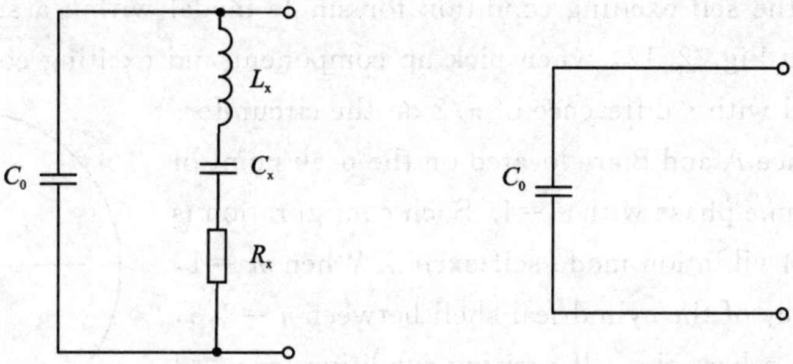

Fig. 2.10 Equivalent circuit of the vibrating picking element

Fig. 2.11 gives a proposal of a charge amplifier consisting of operational amplifier, and their relationship can be expressed as:

$$\frac{u_0}{u_B}=-\frac{R_f C_i s}{R_f C_f S+1} \tag{2.18}$$

$$C_i = C_0 + \Delta C \tag{2.19}$$

$$\frac{u_0}{q_B}=-\frac{R_f s}{R_f C_f + 1} \tag{2.20}$$

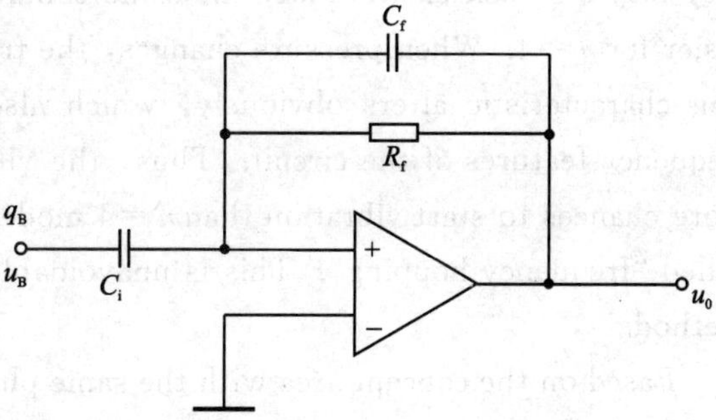

Fig. 2.11 Charge amplifier

where

R_f is the resistance;

C_i and C_f are the capacitance;

u_0 and u_B are output voltage and input voltage.

Thus, by the processing of the charge amplifier transducer, constant q_B is transferred into a low impedance input voltage signal u_0. The following work of setting the amplifier based on amplitude and phase condition becomes easy, and this problem will be no longer discussed.

2.4.4 The Realization of a Stable Single Modal Self-exciting System

According to the analysis of the pressure-frequency characteristics of vibrating cylinder, the (4,1) mode has a high sensitivity, therefore it is chosen as working modal. The next part will discuss how to form a stable single modal self-exciting system.

Single piece exciting and single pick-up modal system are vulnerable to interference. It is difficult to satisfy the self-exciting condition for single modal within a sufficient measuring range. As shown in Fig. 2.12, when pick-up component and exciting component are glued on the cylinder wall with a difference of $\pi/2$ on the circumferential direction, since A and B are located on the peak point of the area with the same phase with $n=4$. Such configuration is the easiest for $n=4$ vibration mode self-excite. When $m=1$, the natural frequency of the cylindrical shell between $n=2,3,4,5$ is close to each other, the self-exciting condition for $n=4$ also save its possibility for situation like $n=3$ or $n=5$ that they may also self-excite under the same condition. It is just easier for $n=4$. When pressure changes, the frequency selecting characteristic alters obviously, which also changes the frequency features of the circuit. Thus, the vibrating mode of $n=3$ or $n=5$ may occupy more chances to start vibration than $n=4$ mode. The transition from $n=4$ to $n\neq 4$, the so-called "frequency hopping". This is unavoidable for single piece exciting and single pick-up method.

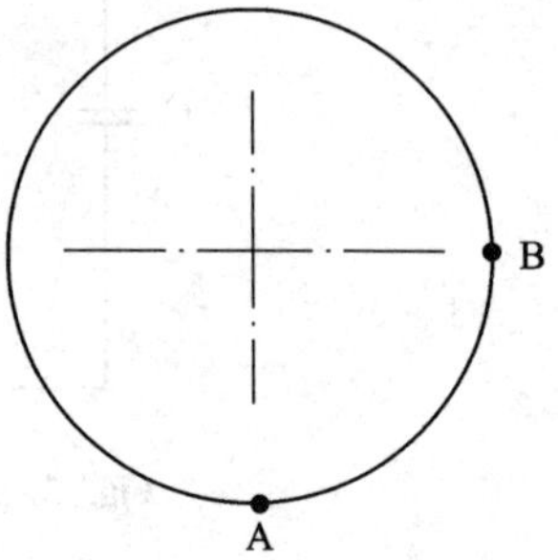

Fig. 2.12 Sketch of 90° phase difference for A and B

Based on the concept area with the same phase and area with opposite phase about vibrating mode, we can draw out these areas when n varies, according to the position of the exciting piece. If hoping for the mode when circumferential-direction wave number equals n_0, we should glue the exciting and pick-up component on the n_0 peak point. Since there is possibility that there are several other peak points, the principle of choosing to generate the expected vibrating mode n_0 is without other nearby mode.

Fig. 2.13 gives the position of the two excitation piece when $n=3,4,5$, as point A and B in the diagram, D is the position of the pick-up point. D with a "$-$" on top mean opposite phase compared to D, otherwise same phase.

We can illustrate why $n=4$ vibration mode stabilizes in accordance of Fig. 2.13. Assuming the exciting signals of point A and B have the influence on D, then only $\overline{A}$ and D can

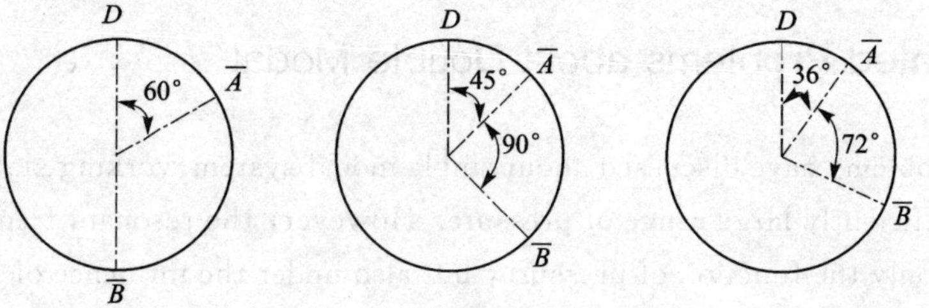

(a) $n=3$ (b) $n=4$ (c) $n=5$

Fig. 2.13 Position sketch of excitation and pick of the vibration

restrain the $n=5$ mode; only $\overline{B}$ and D can restrain the $n=3$ mode; since the included angle between $\overline{A}$ and $\overline{B}$ is 90°. And they are of the same phase, therefore $n=2$ mode will not appear. Thus, when $\overline{A}$ and $\overline{B}$ excite altogether, the $n=2,3,5$ will be restrained to make sure of the stability of $n=4$ mode. For the mode is of the same phase and the area with opposite phase will remain steady while pressure changes, therefore the vibration mode will sustain within a sufficiently large range of pressure to guarantee the system working stabilized and reliably.

Usually, the $n=4$ mode can utilize the method shown in Fig. 2.14 to make up a stabilized self-exciting closed-loop system. The diagram shows that R, R_1 and R_2 are all charge amplifiers, called accepting level; K is amplifying level. Theoretically, proposal (a) and (b) are the same. However, considering the error existing in actual shell processing, dual-excitation and single-pickup are less functional than single-excitation and dual-pickup.

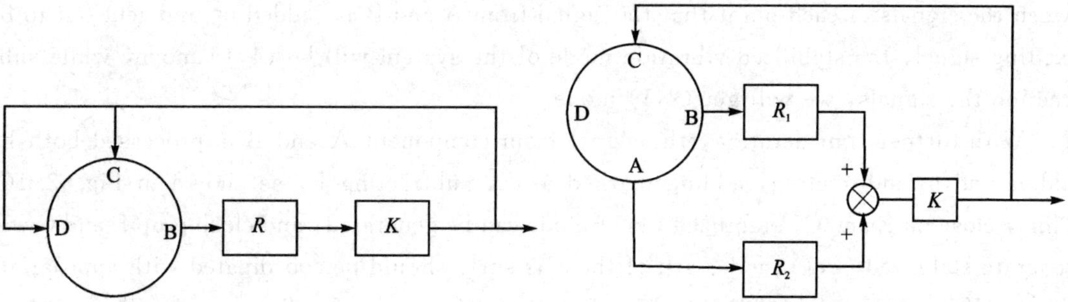

(a) Scheme of double excitation and single picking

(b) Scheme of single excitation and double picking

Fig. 2.14 Sketch of the closed-loop excitation

2.4.5 Related Problems about Double Modal

Related problems have discussed about single modal system working stabilized and reliably within a sufficiently large range of pressure. However, the resonant frequency of the cylindrical is not only the function of pressure, but also under the influence of factors like temperature, aging, environment pollution and gas density. In order to restrain the influence, based on difference detection, a proposal was brought up by realization two modal independently self-exciting on the cylindrical shell, utilizing the characteristic of sensing component itself to optimize. We will discuss the realization of two modal independently self-exciting using (4,1) mode and (2,1) mode.

As for stabilized single modal vibration, one pick-up component can be adopted. While in dual modal vibration, it usually needs two pick-up piezoelectric components. Here we will discuss the selection of modal with two pick-up component. As shown in Fig. 2. 15, assuming A and B are pick-up points. According to what we discuss above, when the system vibrates as (2,1) mode, the signal picked up by A and B have opposite phase; as to (4,1) mode, the phases are the same.

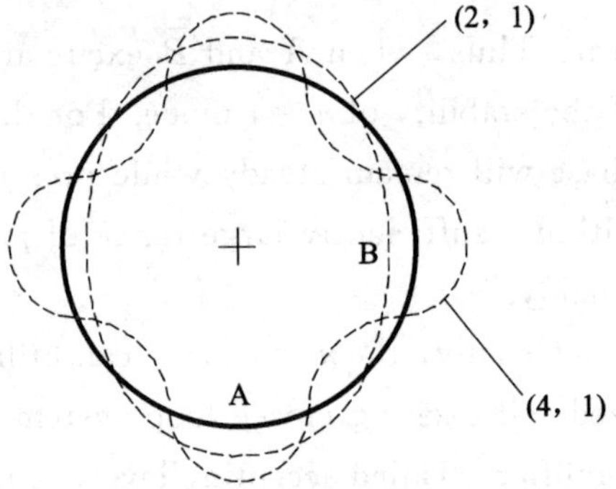

Fig. 2. 15 Distribution sketch of vibrating mode for (2,1) and (4,1)

It can be concluded from the phase relations between the signals of each point that the signal from A and B are added up and sent out to be exciting signal, the stabilized vibration mode of the system will be (4,1) mode, while subtracting the signals, we will get (2,1) mode.

With further consideration, the signal from component A and B is processed both by adding and by subtracting, adding marked as E, subtracting F, as shown in Fig. 2. 16. Thus, close up E and C, stabilized (4,1) mode can be generated; and closing up F and C will generate stabilized (2,1) mode. All of those is surly should be coordinated with appropriate circuit. If we add up E and F, marking P, it total equals to the signal at A. The result of these action lies that point P contains not only (4,1) mode signal but (2,1) mode signal as well. To close up P and C, (4,1) mode signal and (2,1) mode are both able to start vibration. It can also be achieved that when only A or B is attached, after one mode starts vibrating, by attaching the other pick-up component, the other mode can also be excited with the help of reasonable circuit configuration. It sufficiently guarantees that vibrating cylinder works stabilized at dual modal.

To further ensure its stability and reliability, and working more precisely, dual-pickup and dual-excitation are adopted, namely subtracting signal E and signal F and delivering the difference to D. The sketch of such dual modal vibrating system is shown in Fig. 2.17.

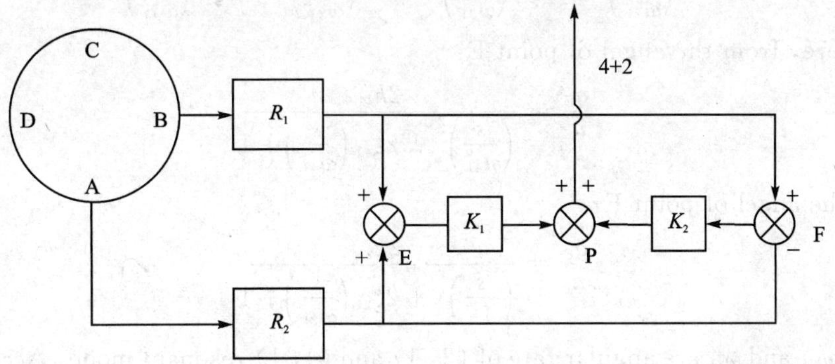

Fig. 2.16 Synthetic sketch of the signal

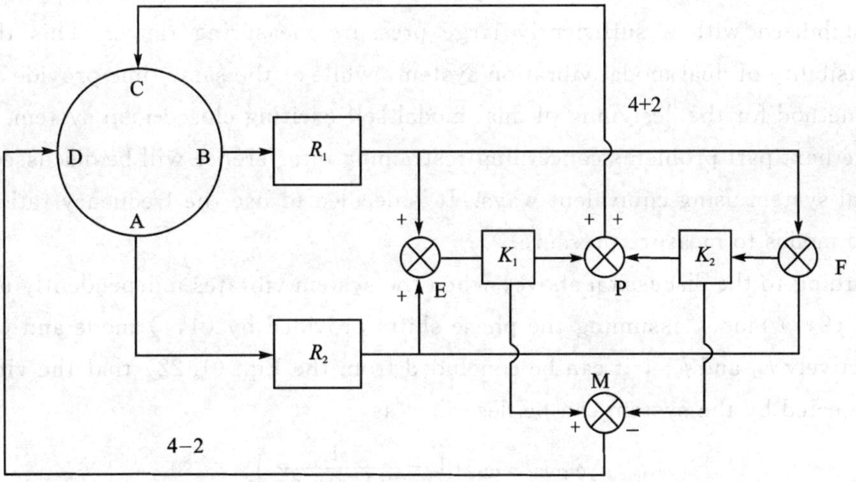

Fig. 2.17 Sketch of the dual-mode vibration system

As for the system in Fig. 2.17, considering from the angel of transfer function, Eq. (2.17) can be simplified as follows.

At point A:

$$\frac{q_A}{u_i}=\frac{k_{21}}{\left(\frac{s}{\omega_{21}}\right)^2+2\zeta_{21}\left(\frac{s}{\omega_{21}}\right)+1}+\frac{k_{41}}{\left(\frac{s}{\omega_{41}}\right)^2+2\zeta_{41}\left(\frac{s}{\omega_{41}}\right)+1} \tag{2.21}$$

At point B:

$$\frac{q_B}{u_i}=\frac{-k_{21}}{\left(\frac{s}{\omega_{21}}\right)^2+2\zeta_{21}\left(\frac{s}{\omega_{21}}\right)+1}+\frac{k_{41}}{\left(\frac{s}{\omega_{41}}\right)^2+2\zeta_{41}\left(\frac{s}{\omega_{41}}\right)+1} \tag{2.22}$$

Therefore, from the angel of point E:

$$\frac{q_E}{u_i}=\frac{2k_{41}}{\left(\frac{s}{\omega_{41}}\right)^2+2\zeta_{41}\left(\frac{s}{\omega_{41}}\right)+1} \tag{2.23}$$

From the angel of point F:

$$\frac{q_F}{u_i}=\frac{2k_{21}}{\left(\frac{s}{\omega_{21}}\right)^2+2\zeta_{21}\left(\frac{s}{\omega_{21}}\right)+1} \tag{2.24}$$

Where ω_{21} and ω_{41} are angular rate of (2, 1) and (4, 1) resonant mode. According to the Eq. (2.23), Eq. (2.24) and the analysis above, the system is now in a state of "decoupling", and is able to excite independently both (4, 1) mode and (2, 1) mode, and this work state is stabilized with a sufficiently large pressure measuring range. This theoretically proves feasibility of dual modal vibration system, while at the same time provide the foundation and method for the designing of dual modal self-exciting closed-loop system.

In the next part problems concerning restraining interference will be discussed about the dual modal system using equivalent ways. It is decided to use the frequency ratio (γ) of the above two modes to measure pressure.

According to the discussion above, when the system vibrates independently under (4,1) mode and (2,1) mode, assuming the phase shifts provided by (4,1) mode and (2,1) mode are respectively ϕ_{41} and ϕ_{21}, it can be concluded from the Eq. (1.22) that the vibrating frequency detected by the system can be described as

$$\omega_{41V}\approx\omega_{41}\left(1+\frac{1}{2Q_{41}\tan\phi_{41}}\right) \tag{2.25}$$

$$\omega_{21V}\approx\omega_{21}\left(1+\frac{1}{2Q_{21}\tan\phi_{21}}\right) \tag{2.26}$$

The detected frequency ratio is

$$\gamma=\frac{\omega_{21V}}{\omega_{41V}}=\frac{\omega_{21}\left(1+\frac{1}{2Q_{21}\tan\phi_{21}}\right)}{\omega_{41}\left(1+\frac{1}{2Q_{41}\tan\phi_{41}}\right)}=\gamma_n\beta \tag{2.27}$$

$$\gamma_n = \frac{\omega_{21}}{\omega_{41}}$$

$$\beta = \frac{1 + \dfrac{1}{2Q_{21}\tan\phi_{21}}}{1 + \dfrac{1}{2Q_{41}\tan\phi_{41}}} \approx 1 + \frac{1}{2Q_{21}\tan\phi_{21}} - \frac{1}{2Q_{41}\tan\phi_{41}}$$

According to Eqs. (2.25) and (2.26), the resonant frequency of the closed-loop system consists of two parts, namely the cylindrical shell related natural frequency ω_{21} and ω_{41} measurement concerning equivalent damping ratio (or quality factor) and working position ϕ_{21} and ϕ_{41}. The equivalent way is to regard the environment factor causing resonant frequency departure as the changes in physical parameter (modulus of elasticity, Poisson's ratio and density) of the cylindrical shell and in the equivalent damping ratio of the system.

Concluded from the previous analysis, the resonant frequency is proportional to $\sqrt{E/\rho}$. Therefore, when ρ or E changes, both ω_{41} and ω_{21} alter enormously, which also affects ω_{41V} and ω_{21V}. However ρ has no influence on γ_n. As for the error caused by the changes of Poisson's ratio, γ_n has no restraining towards it. Thus there exists such relation as follows:

$$\left|\frac{\Delta\omega_{41V}}{\omega_{41V}}\right| > \left|\frac{\Delta\gamma_n}{\gamma_n}\right| \tag{2.28}$$

Considering from the angel of closed-loop, the influences of environment interference can be equalized to the changes of Q_{41} and Q_{21}. According to the Eq. (2.27), when designing a dual modal system, it can always be achieved that $\tan\phi_{21} \cdot \tan\phi_{41} > 0$ and $(Q_{41}\tan\phi_{41})^{-1}$ is close to $(Q_{21}\tan\phi_{21})^{-1}$. Therefore,

$$\left|\frac{1}{2Q_{21}\tan\phi_{21}} - \frac{1}{2Q_{41}\tan\phi_{41}}\right| < \left|\frac{1}{2Q_{41}\tan\phi_{41}}\right| \tag{2.29}$$

The changing rate of γ is

$$\alpha_1 = \beta\frac{\Delta\gamma_n}{\gamma_n} + \Delta\beta \tag{2.30}$$

The changing rate of ω_{41V} is

$$\alpha_2 = \left(1 + \frac{1}{2Q_{41}\tan\phi_{41}}\right)\frac{\Delta\omega_{41}}{\omega_{41}} + \Delta\left(\frac{1}{2Q_{41}\tan\phi_{41}}\right) \tag{2.31}$$

According to the analysis above, it can be told that:

$$|\alpha_2| > |\alpha_1| \tag{2.32}$$

This indicates that the changing rate of measurand γ from the dual modal system consisting of (4,1) mode and (2,1) mode is much smaller than that of ω_{41V} from the single modal system consisting of only (4,1) mode.

What needs to be point out is that the sensitivity of vibration frequency of (4,1) mode of cylindrical shell with regard to pressure is much higher than that of (2,1) mode. Thus the sensitivity of frequency ratio γ is large enough.

Tab. 2.1 gives the measuring result of a dual modal resonant cylinder pressure sensor from Fig. 2.17. The material of the cylinder is 3J53. The effective length of the cylinder is 54 mm, diameter is 18 mm and the wall thickness is 0.08 mm.

Both theory and practice prove that the two modes making up the dual mode sensor had better have a larger difference of their resonant frequency range. These are the main basis to design the geometric parameter of the resonant cylinder.

When the vibrating is working under "dual mode" method, higher requirement is in demand for the aspects, such as the related processing technology, excitation and pick-up methods, amplifying circuit and signal process.

Tab. 2.1 Output period value of (4,1) mode and (2,1) mode

(n,m) / $10^{-1}\cdot p$/Pa	Positive travel		Negative travel	
	(2,1)	(4,1)	(2,1)	(4,1)
0	151.964	229.671	151.960	229.674
0.2	151.707	222.973	151.705	222.970
0.4	151.473	216.709	151.468	216.712
0.6	151.234	210.895	151.231	210.891
0.8	150.971	206.837	150.973	206.831
1.0	150.837	201.045	150.840	201.038
1.2	150.628	196.511	150.626	196.505
1.4	150.340	192.123	150.341	192.119
1.6	150.133	188.604	150.135	188.599
1.8	149.958	184.379	149.959	184.385
2.0	149.705	180.971	149.707	180.976
2.2	149.457	177.764	149.456	177.760

2.5 Quartz Vibration Beam Pressure Sensor

As for the three types of resonant pressure sensor, considering that their vibrating sensing component is made up of metal materials, the long term stability, aging problem and creeping of the material performance may cause frequency shifting. Besides, those components are easily under the influence of the electromagnetic interference and environment vibration. Therefore, the offset and sensitivity are highly unstable.

Quartz crystal has a stable natural vibrating frequency. When the frequency of the force vibration equals its natural vibrating frequency, it resonates. Via this feature quartz crystal resonator can be manufactured and by using different dimensions and different vibrating mode, we will be able to make quartz resonator with frequency from thousands of Hz to hundreds of MHz.

With the help of quartz resonator, we can research and manufacture quartz resonant pressure sensor. Considering that the quality factor of the quartz resonator is very high with a high natural frequency and a narrow bandwidth, therefore it has advantages on restraining interference and reducing the frequency error caused by phase angle difference. When it is made into a pressure sensor, the sensitivity and stability can reach a high level with a good dynamic response. Even though the processing of quartz has its difficulties, quartz resonant pressure sensor is still a promising pressure sensor.

2.5.1 Structure and Principle

Fig. 2.18 shows a cantilever pressure sensor consisting of quartz crystal resonator. Two opposing corrugated pipes are used to feel input pressures p_1 and p_2 . The pressure difference is exerted on the effective area to generate a resultant force, forming a moment of force rotating the fulcrum. This moment is balanced by the tensile or compression force of the quartz crystal resonant beam (referring to Fig. 2.19), which also changes the resonant frequency of the quartz crystal. The departure of the frequency is a single-valued function of the measured pressure, via which the measuring can be achieved.

Figure 2.19 shows the general view of the quartz resonant beam and its isolator. The quartz resonant beam is the sensing component of the pressure sensor, crossing over the right middle of the structure shown in Fig. 2.19. The intention of the isolator at the end of the beam is to prevent the energy loss of the substrate caused by the reaction force and moment of force, otherwise decreasing the quality factor; in the meanwhile, the isolating structure keeps the harmful interference from passing through to lower the stability and affect the performance of the resonator. The shape of the beam is chosen to be double-ended with a curved-manner vibrating method, which will achieve a high sensitivity toward force.

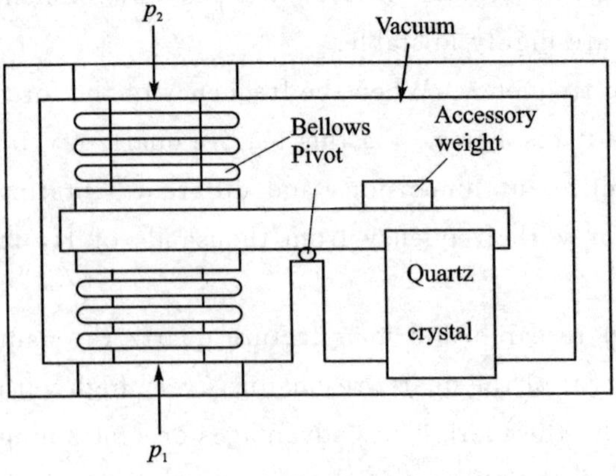

Fig. 2.18 Sketch of the quartz vibration beam pressure sensor

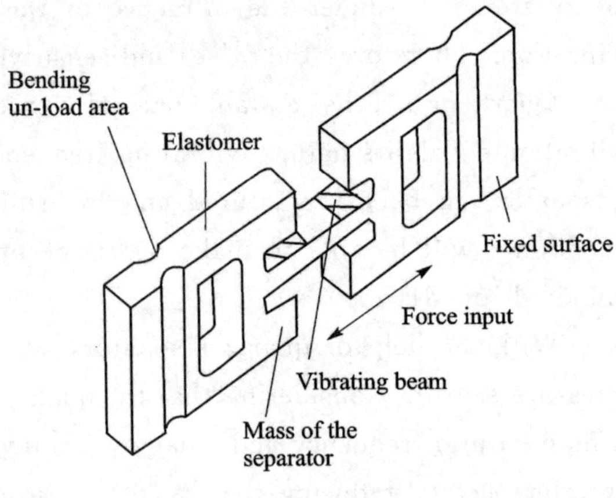

Fig. 2.19 Sketch of the quartz crystal resonator

Four electrodes are symmetrically placed on both side of the vibrating beam through evaporation deposition. Considering the piezoelectric effect the quartz crystal has itself, when the four electrodes are coupled with electric field, the beam will start vibration under the first-order bending vibration state. With no input pressure, its natural resonant frequency depends mostly on the geometric shape and structure of the beam. When the beam crystal is coupled with electric field, it turns into a parallelogram, as shown in Fig. 2.20, and how the beam crooks is determined by the polarity of the electric field. When one pair of electrodes opposes the polarity against the diagonal pair, the beam is in the first-order bending state; once the polarity of the electrodes reverses, the beam bends to the opposite direction. Just like that, using a sustaining vibrating circuit to replace the coupled electric field, the beam will resonate and remain vibrating with the help of the measuring circuit.

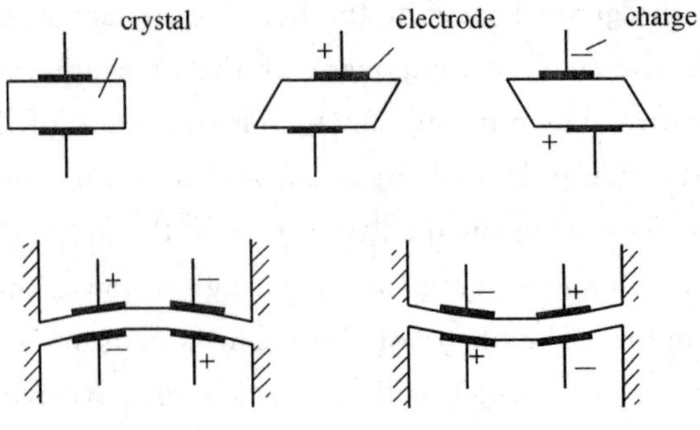

Fig. 2.20 Vibrating mode of the resonant beam

When exerting pressure and $p_1 > p_2$, the beam bears a tensile force with its stiffness increasing and resonant frequency rising; when $p_1 < p_2$, it does a compression force, stiffness decreases and resonant frequency drops. Therefore, the changes of the output frequency reflect the measurement of the input pressure.

The corrugated pipes are manufactured from highly purified material through special technology, with a function to transform the input pressure difference into the axial force on the vibrating beam (along the direction of the beam's length). To improve the measuring sensitivity, the delaying of the pipes needs to be small.

When the shape, geometric parameter and position of the quartz crystal resonator are fixed, the matching mass can be used to adjust the gravity center of the motion component to coincide with the fulcrum. With interfered by external acceleration, the matching mass has a function to compensate the acceleration. Since its lever arm is nearly zero, the resonator can only sense the moment generated by pressure, while staying a low sensitivity towards other external forces.

2.5.2 Characteristic Function

According to structure of Fig. 2.18, the relation from the input pressure p_1 and p_2 to the axial force exerted on the beam is shown as follows:

$$T_x = \frac{L_1}{L_2}(p_2 - p_1)A_E = \frac{L_1}{L_2}\Delta p A_E \tag{2.33}$$

where

A_E is effective area of the corrugated pipes (in m^2);

Δp is pressure difference (in Pa), and $\Delta p = p_2 - p_1$;

L_1 is distance between the corrugated pipe and the fulcrum (in m);

L_2 is distance between the vibrating beam and the fulcrum (in m).

The pressure difference generates axial force T_x on the beam. Under the influence of T_x, the relation between the lowest order (first-order) natural frequency and the pressure difference can be described as

$$f_{B1}(\Delta p) = f_{B1}\sqrt{1 + 0.2949\frac{T_x L^2}{Ebh^3}} =$$

$$f_{B1}\sqrt{1 + 0.2949\frac{L_1}{L_2}\cdot\frac{\Delta p A_E}{Ebh}\cdot\frac{L^2}{h^2}} \tag{2.34}$$

$$f_{B1} = \frac{4.730^2 h}{2\pi L^2}\sqrt{\frac{E}{12\rho}} \tag{2.35}$$

where

$f_{B1}(\Delta p)$ is the first-order bending natural frequency of the vibrating beam under the influence of the pressure difference (in Hz);

f_{B1} is the first-order bending natural frequency of the vibrating beam with zero pressure (in Hz);

ρ is density of the beam material (in kg/m^3);

μ is Poisson's ratio of the beam material;

E is modulus of elasticity of the beam material (in Pa);

L is length of the working part of the vibrating beam (in m);

b is width of the vibrating beam (in m);

h is thickness of the vibrating beam (in m).

This type of sensor has many advantages, with a low sensitivity towards external interferences such as temperature, vibration and acceleration. Some on-the-spot measuring data shows that it has a sensitivity temperature-drift of 4×10^{-5} % /℃, an acceleration sensitivity of 8×10^{-5} %(m · s^{-2}), outstanding stability, little volume (2.54), small mass (about 0.7 kg), high quality factor (reaching 40 000), high dynamic response (10^3 Hz). This type of sensor has been put into application in realms like atmosphere data system, jet engine experiments, digital program control and pressure secondary standard instrument.

2.6 Micromechanical Silicon Resonant Pressure Sensor

Since the mid-1990s, Micromechanical processing technology and resonant sensing technology are combined together with the development of a variety of micromechanical resonant sensor. The so-called "micro structure" is a miniaturized reproduction of common structure and its delicate new structure through micromechanical processing technology. The micro resonant sensor consisting of micro structure has a series of performances existing in conventional resonant sensor, as well as other characteristics, such as small mass, low power consumption, rapid responses and ease of integration. And as a result of these additional features, some new changes emerge in the related measuring mechanism. For example, thermal excitation can be a new working method.

Since the micromechanical silicon resonant sensor has so many advantages, it has now become a new direction of the sensor development. Currently, many research achievements will soon put into application. Especially in the realm of precision measuring, these applications will be more and more popular. This section holds a discussion based on a typical thermal exciting micromechanical silicon pressure sensor.

2.6.1 Sensing Structure and Mathematical Model

Fig. 2.21 shows a sensing structure of a typical thermal exciting micro resonant pressure sensor, consisting of square plain diaphragm, resonant beam oscillator and edge isolator. The square plain diaphragm acts as the first level sensing component to directly feel the measured pressure, transferring the pressure into the strain and stress; shallow groove is manufactured on the upper surface of the diaphragm, as well as the silicon beam, which is second level sensing component to sensing the stress of the diaphragm, namely feeling the pressure indirectly. The external pressure p changes the equivalent stiffness of the beam oscillator; therefore the natural frequency of the beam alters when pressure changes. By measuring the departure of the natural frequency of the oscillator, the difference of the external pressure can be told. In order to realize the self-exciting closed-loop of the micro sensor, methods such as resistance thermal exciting and piezoresistive pick-up are adopted. Considering the effect of excitation and pick-up and the process of signal transformation, the thermal exciting resistance is set at the middle of the oscillator beam and the piezoresistive pick-up resistance is located at the root of the oscillator beam.

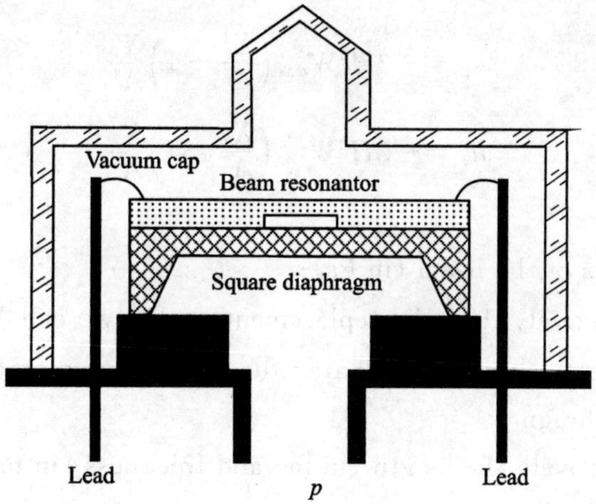

Fig. 2.21 Sketch of the resonant silicon MEMS pressure sensor

To build up a cartesian coordinates at the center of the diaphragm, as shown in Fig. 2.22. The Oxy plane coincides with the middle plane of the diaphragm and z axis points up. Under the influence of pressure p, the normal placement of the square plain diaphragm can be describes as

$$w(x,y) = \overline{W}_{S,\max} H\left(\frac{x^2}{A^2}-1\right)^2\left(\frac{y^2}{A^2}-1\right)^2 \tag{2.36}$$

$$\overline{W}_{S,\max} = \frac{49p(1-\mu^2)}{192E}\left(\frac{A}{H}\right)^4 \tag{2.37}$$

where

μ is Poisson's ratio of the beam material;

E is modulus of elasticity(in Pa);

A and H are respectively the side half-length and thickness of the diaphragm;

$\overline{W}_{S,\max}$ is the ratio between the maximum normal displacement of the diaphragm and its thickness under the influence of pressure p.

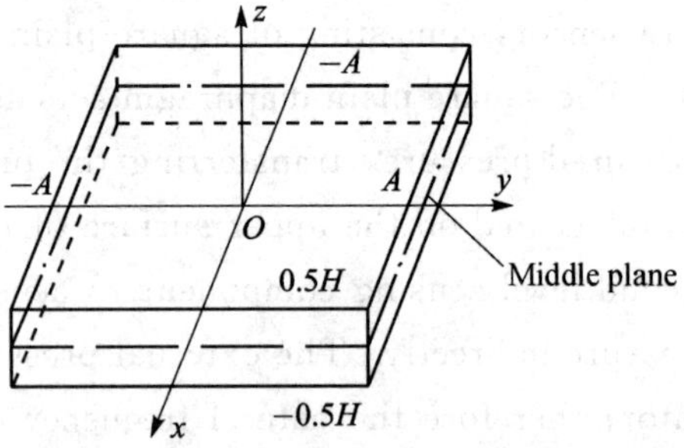

Fig. 2.22 Coordinate system of the square planar diaphragm (middle plane)

According to the practical working situation and mechanism of the sensing structure, and when the beam resonator is set up along the x axis at the point $x \in [X_1, X_2](X_2 > X_1)$, the initial stress under the influence of pressure p can be described as

$$\sigma_0 = E\frac{u_2 - u_1}{L} \tag{2.38}$$

$$u_1 = -2H^2 W_{\max}\left(\frac{X_1^2}{A^2}-1\right)\frac{X_1}{A^2} \tag{2.39}$$

$$u_2 = -2H^2 W_{\max}\left(\frac{X_2^2}{A^2}-1\right)\frac{X_2}{A^2} \tag{2.40}$$

where

σ_0 is the axial stress of the beam (in Pa);

u_1 and u_2 are respectively the axial replacement at the two end X_1 and X_2;

X_1 and X_2 are respectively the coordinate values of the beam on the cartesian coordinates of the square plain diaphragm;

L and H are respectively the length (in m) and thickness (in m) of the beam, and $L = X_2 - X_1$.

Under the influence of initial σ_0 (namely the pressure p), the first-order natural frequency of the double-ended beam can be described as

$$f_1(p) = f_1(0)\sqrt{1+0.294\,9\frac{K \cdot p \cdot L^2}{h^2}} \tag{2.41}$$

$$f_1(0) = \frac{4.730^2 h}{2\pi L^2}\sqrt{\frac{E}{12\rho_m}}$$

$$K = \frac{0.51(1-\mu^2)}{EH^2}(-L^2 - 3X_2^2 + 3X_2L + A^2)$$

where ρ_m is density of the beam material.

Eq. (2.38) to Eq. (2.41) give out the characteristic equation of the pressure and frequency of the micro silicon resonant pressure sensor. Using this model, a series of reference values were given here about the parameters of the micro sensor's sensing structure with a measuring range of a pressure from 0 to 0.1 MPa. The square plain diaphragm has a side length of 0.08 mm and a thickness of 0.007 mm; the beam resonator is located at the middle of the diaphragm along the x axis with a length of 1.3 mm, a width of 0.08 mm and a thickness of 0.07 mm. Moreover, the depth of the shallow groove is 0.002 mm. Considering the statics analyzing result of the diaphragm, the designing regulation for the structure parameter optimization is able to be given out here. Given some practical processing technology realization, the half side-length and the thickness of the diaphragm's edge isolation can be given about 1 mm and 1 mm.

When $E = 1.3\times10^{11}$ Pa, $\rho_m = 2.33\times10^3$ kg/m^3 and $\mu = 0.18$ under the measuring pressure range between 0 and 0.1 MPa, frequency of the beam calculated by the model is in the range of 31.81 and 45.10 kHz.

2.6.2 The Process of Signal Transformation

Fig. 2.23 is the sketch for the resistance thermal exciting and resistance thermal pick-up method of the beam oscillator component of the micro sensor. The thermal exciting resistance R_E is set up in the middle of the beam and the varistor R_D is put at the end of the beam. When the sensing component starts to work, an AC sinusoidal voltage ($U_{ac}\cos\omega t$) and a DC bias voltage are exerted on the exciting resistance (R_E) with heat generated.

$$P(t) = \frac{U_{dc}^2 + 0.5U_{ac}^2 + 2U_{dc}U_{ac}\cos\omega t + 0.5U_{ac}^2\cos 2\omega t}{R_E} \tag{2.42}$$

$P(t)$ consists of constant component P_s, alternating component $P_{d1}(t)$ and $P_{d2}(t)$ respectively with the same and twice the frequency of the excitation, as described below:

$$P_s = \frac{2U_{dc}^2 + U_{ac}^2}{2R_E} \tag{2.43}$$

$$P_{d1}(t) = \frac{2U_{dc}U_{ac}\cos\omega t}{R_E} \tag{2.44}$$

$$P_{d2}(t) = \frac{U_{ac}^2 \cos 2\omega t}{2R_E} \tag{2.45}$$

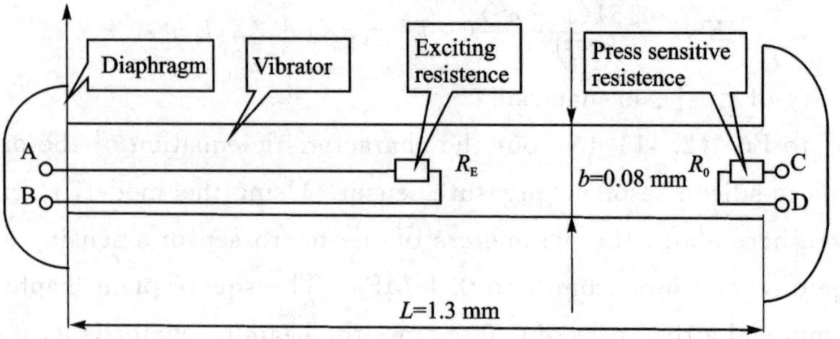

Fig. 2. 23 Sketch of the resonator

The alternating component $P_{d1}(t)$ will generate an alternating temperature difference distribution field $\Delta T(x,t)\cos(\omega t + \phi_1)$, thus an alternating thermal stress is generated on the beam resonator.

$$\sigma_{ther} = - E\alpha\Delta T(x,t)\cos(\omega t + \phi_1 + \phi_2) \tag{2.46}$$

where

α is the thermal stress coefficient of the silicon material;

x and t are respectively the axial position and time of the beam resonator;

ϕ_1 is the phase shift from thermal power to temperature distribution field;

ϕ_2 is the phase shift from temperature distribution field to thermal stress.

Apparently, ϕ_1 and ϕ_2 are related to exciting resistance's position on the beam resonator, the parameter of the exciting resistance itself, and the structure and material parameter of the beam.

The piezoresistive pick-up resistance planted at the end of the beam will feel the alternating thermal stress. According to the piezoresistive effect, the resistance difference is

$$\Delta R_D = \beta R_D \sigma_{axial} = \beta R_D E\alpha\Delta T(x_0,t)\cos(\omega t + \phi_1 + \phi_2) \tag{2.47}$$

where

σ_{axial} is stress of the end of the beam felt by the resistance;

β is sensitivity coefficient of the varistor;

x_0 is coordinate of the beam's end (in m).

Through the bridge, the difference of the pick-up resistance can be transformed into the changes of the AC voltage signal $\Delta u(t)$, which can be described as

$$\Delta u(t) = K_B \frac{\Delta R_D}{R_D} = K_B\beta E\alpha\Delta T(x_0,t)\cos(\omega t + \phi_1 + \phi_2) \tag{2.48}$$

where K_B is the sensitivity of the bridge (in V).

When the frequency (ω) of $\Delta u(t)$ coincides with the natural frequency of the natural frequency of the beam resonator, it resonates. Therefore, $P_{d1}(t)$ is the needed alternating signal to realize the electronic-heat-mechanical transformation.

2.6.3 The Temperature Field Model and Thermal Characteristic Analysis of the Beam Resonator

The constant component will generate a constant temperature difference distribution field ΔT_{av} on the beam resonator, causing initial thermal stress, therefore influence its resonant frequency.

The initial thermal stress caused by the resonator's temperature field is

$$\varepsilon_T = -\alpha \Delta T_{av} \tag{2.49}$$

where ΔT_{av} is the average temperature rising on the beam resonator, which is proportional to P_s.

Thus, with a thorough considering of the measured pressure and the distribution situation of exciting resistance's temperature field, the first-order natural frequency of the resonator is described as

$$f_{B1}(p, \Delta T_{av}) = \frac{4.730^2 h}{2\pi L^2}\left[\frac{E}{12\rho}\left(1 + 0.2949\,\frac{(Kp + \varepsilon_T)L^2}{h^2}\right)\right]^{0.5} \tag{2.50}$$

Based on the analysis above and Eq. (2.50), the temperature field caused by the exciting resistance will decrease the equivalent stiffness of the resonator, therefore the reduced stiffness should be limited to guarantee the stable working condition of the resonator. Usually, the constant power exerted on the resonator is determined by the formula as below:

$$0.2949\alpha \Delta T_{av}\frac{L^2}{h^2} \leqslant \frac{1}{K_s} \tag{2.51}$$

where K_s is safety coefficient, usually chosen between 5 and 7.

From the Eq. (2.50), the regulation of the influence the temperature field having on the pressure and frequency characteristic of the resonator is as follows:

① When taking into consideration the thermal power of the exciting resistance, the frequency of the resonator will decrease, the frequency shift changes monotonically with the exciting thermal power P_s;

② When the thermal power of the exciting resistance stays constant, the influence the temperature has on the pressure and frequency characteristic is fixed.

Based on the beam resonator's exciting and pick-up structure of the micro sensor and its relevant signal transforming regulation, when the AC sinusoidal voltage ($U_{ac}\cos\omega t$) and DC bias voltage are loaded on the exciting resistance, the problem needed to be solved the interfering signal from the double frequency alternating component $P_{d2}(t)$. Usually, what is done U_{dc} is to choose appropriate DC component, ensuring $U_{dc}\gg U_{ac}$ or to filter in the conditioning circuit. Therefore it can be given the mechanism block diagram of the sensor's self-exciting closed-loop system as is shown in Fig. 2. 24. In the diagram, the alternating signal $\Delta u(t)$ measured by the pick-up bridge is pre-amplified by the difference amplifier. Signals beyond the range are removed by the band-pass filter. The overall phase shift of the other links besides the closed-loop circuit is adjusted through phase shifter.

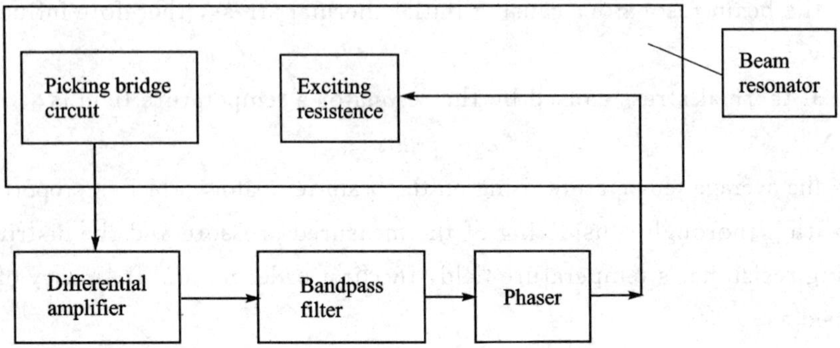

Fig. 2. 24 Sketch of the close-loop self-exciting system with biased direct current

The amplitude and phase condition (Eqs. (1. 15) and (1. 16)) can be used to design and calculate the parameter of the amplifier to ensure that the micro resonant pressure sensor is able to self- excite within the whole working frequency range, to maintain a stable and reliable working condition. Nonetheless, this proposal is easily affected by the influence the temperature difference distribution field ΔT_{av} have on the sensor performance.

To decrease the influence of ΔT_{av} on the resonator frequency to the greatest extend, the proposal of simple alternating exciting can be taken into consideration. With the help of Eq. (2. 52), the thermal exciting power under such circumstance can be described as

$$P(t)=\frac{U_{ac}^2+U_{ac}^2\cos 2\omega t}{2R_E} \tag{2.52}$$

Considering that the mechanical quality factor of beam resonator is very high, the exciting signal can be very small, therefore the current constant power $P_s=U_{ac}^2/2R_E$ stays low, with a capability to neglect its influence on the resonant frequency of the resonator. And the alternating component consists no longer one-time frequency component but the doubled one $P_{d2}(t)=\frac{U_{ac}^2\cos 2\omega t}{2R_E}$. A simple alternating exciting closed-loop self-exciting system needs a solution about the frequency-dividing problem. One practical proposal is adopting phase-locked frequency-dividing technology in the circuit, which is to attach a frequency multiplier in the feedback branch of the designed basic phase-locked loop to realize the frequency dividing. The principle is shown in Fig. 2.25.

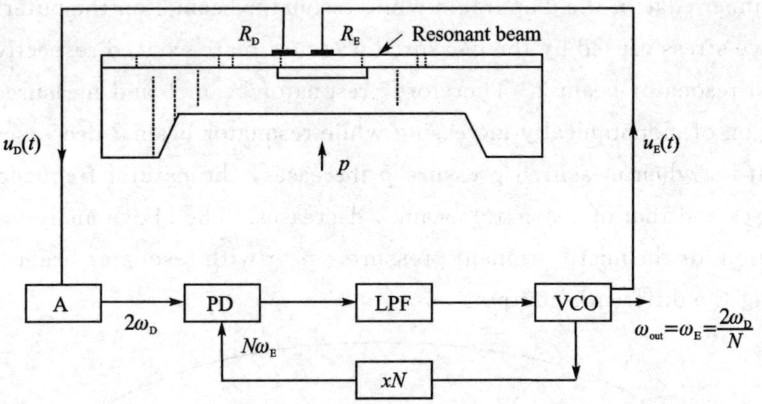

Fig. 2.25 Sketch of the close-loop self-exciting system with pure AC excitation

Assuming the two signal frequencies compared in the pick-up resistance phase comparator are $2\omega_D$ and $N\omega_E$, when the loop locks up, $2\omega_D = N\omega_E$ is a certainly , namely $\omega_E = \frac{2\omega_D}{N}$. Where N is the multiplying coefficient, determine how many times to divide frequency. When $N=2$, the output frequency ω_{out} of the voltage-controlling oscillator is equal to the detected beam resonator natural frequency ω_{out} of the voltage-controlling oscillator can realize the measuring of the pressure; in the meanwhile, to give feedback with a exciting signal frequency of the ω_D . For the frequency is modulated by the measured pressure, therefore by detecting the output frequency ω_{out} of $\omega_E = \omega_{out}$ to the exciting resistance, forming the self-exciting closed-loop system of the micro sensor.

2.7 Micro Mechanical Silicon Resonant Pressure Sensor with Differential Output

2.7.1 Sensing Structure and Working Mechanism

Fig. 2.26 is the sketch of the micro resonant pressure sensor with differential output. It is a delicately complex sensing structure using the silicon micro-machining process with the measured pressure p directly exerted to the lower surface of circle diaphragm with an E form. While the upper surface of the diaphragm are planted with a pair of silicon beam resonator performing differentially, sealed in the vacuum. Considering resonator beam 1 is installed on the inner edge of the diaphragm while resonator beam 2 on the outer, tensile stress and compressive stress caused by the measured pressure p are exerted respectively on resonator beam 1 and resonator beam 2. Therefore, resonator beam 1 and measured pressure follows a regulation of monotonically increasing while resonator beam 2 decreases alongside the pressure. That is, when measured pressure p increases, the natural frequency of resonator beam 1 increases and that of resonator beam 2 decreases. The above analysis lays the theoretical foundation for the micro resonant pressure sensor with resonator beam 1 and resonator beam 2 forming the differential output.

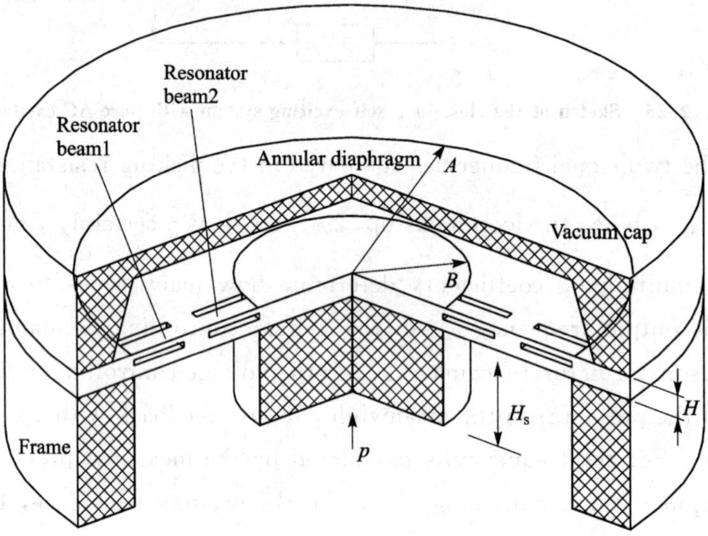

Fig. 2.26 Sketch of the resonant MEMS pressure sensor with differential output

2.7.2 Characteristic Equation

When measured pressure p is exerted on the lower surface of the E-type circle diaphragm, the initial stress generated by the resonator will cause the departure of the first-order natural frequency. When resonator beam 1 and 2 are planted on the inner and outer edge of the upper surface of the diaphragm, the characteristic equation of frequency of the two resonators can be described respectively as

$$f_1(p) = \frac{4.730^2 h}{2\pi L^2}\left[\frac{E}{12\rho}\left(1 + 0.295\frac{K_1^p p \cdot L^2}{h^2}\right)\right]^{0.5} \tag{2.53}$$

$$K_1^p = \frac{-3(1-\mu^2)R_2^3}{8EH^2 L}\left[\frac{L^3}{R_2^3} - \frac{L}{R_2} - \frac{L}{R_2}K^2 + \frac{R_2}{L}K^2\right]$$

$$f_2(p) = \frac{4.730^2 h}{2\pi L^2}\left[\frac{E}{12\rho}\left(1 + 0.295\frac{K_2^p p \cdot L^2}{h^2}\right)\right]^{0.5} \tag{2.54}$$

$$K_2^p = \frac{3(1-\mu^2)R_2^3}{8EH^2 L}\left[\left(\frac{R_2 - L}{R_2}\right)^3 - \frac{R_2 - L}{R_2} - \frac{R_2 - L}{R_2}K^2 + \frac{R_2}{R_2 - L}K^2\right]$$

where

R_1 and R_2 are inner radius and outer radius of the E-type diaphragm;

H is the thickness;

L is the length of the resonator beam;

h is height of the resonator beam;

K is ratio of R_1 and R_2.

As for the specific micro sensor, the system implementation is the same as the micro resonant pressure sensor shown in Fig. 2.21, whose output signal is the frequency difference $f_1(p) - f_2(p)$ between the two resonators.

This micro resonant pressure sensor with differential output has advantage of improving the measuring sensitivity, as well as the compensation for the common mode interference, such as temperature, environmental oscillation and overload. Therefore its performance indicators get enhanced effectively.

Questions and Exercises

2.1 In a resonant pressure sensor, what sensing components can be employed as the resonator?

2.2 Briefly introduce the working mechanism and characteristics of the resonant string pressure sensor.

2.3 Why is it necessary to exert preload on the resonant string of the resonant string pressure sensor?

2.4 Illustrate the characteristic of the quartz vibration beam pressure sensor.

2.5 Give out the sketch of the diaphragm pressure sensor and briefly introduce its working mechanism and characteristic.

2.6 Give out the sketch of the vibrating cylinder pressure sensor and briefly introduce its working mechanism and characteristic.

2.7 Draw the structure block diagram of the piezoelectric exciting vibrating cylinder pressure sensor.

2.8 Briefly explain the reason why choosing a vibrating cylinder of $m = 1, n = 4$ in the vibrating cylinder pressure sensor.

2.9 Compare the application features between the vibrating cylinder pressure sensors of electromagnetic excitation and piezoelectric excitation.

2.10 How to conduct temperature compensation in a vibrating cylinder pressure sensor?

2.11 One engineer planted the piezoelectric transducer of the piezoelectric exciting vibrating cylinder pressure sensor at the middle of the cylinder. Is this proposal feasible? and why?

2.12 The plane diaphragm sensing component of a vibrating diaphragm pressure sensor is developed through 3J53 process and its thickness is designed as 0.15 mm according to the process condition. If the measured pressure varies between 0 and 1.35×10^5 Pa, try to work out the range of the radius of the circle plain diaphragm with the requirement for the sensor of a relative frequency variation between 25% and 30%.

2.13 A vibrating cylinder is processed through 3J53 with a geometric structure parameter of $R = 9$ mm, $L = 55$ mm, $h = 0.078$ mm. Try to solve the natural frequency under zero pressure according to Eq. (2.4) when the vibrating mode has 1 (or 2) half-waves along the axial direction on the surface of the cylinder, and 2 (or 3) whole-waves on the circumferential direction.

2.14 Compared with resonant tube or diaphragm pressure sensor, what are the application features of the micro silicon resonant pressure sensor?

2.15 As for micro silicon resonant pressure sensor, besides thermal excitation, what are other methods of excitation? Refer to related literature to learn about other methods and compare to thermal excitation.

Chapter 3
Resonant Inertial Sensor

Basic contents:

- Acceleration sensor
- Angular velocity
- Inertial technology
- Quartz vibrating beam
- Resonator gyro
- Frequency measurement
- Amplitude ratio measurement
- Dual self-exciting closed-loop system
- Direct output frequency

3.1 Introduction

Resonant inertial sensors include resonant acceleration sensors and resonant angular rate sensors which is also known as resonant gyroscope. Before the 1970s, because the sensitive components of such sensor use vibrating wire or tuning fork and so on, it is of small practical value. After the 1970s, resonator angular rate sensor gained great development and successful application because of the use of cylindrical shell or hemispherical shell as the sensing element. In recent years, MEMS(Micro-Electro Mechanical Systems) resonant gyroscope has also gained great development with the rapid development of micromechanical technology, especially showing significant advantages in the low-cost, low-power, fast response, batch production, and is attached importance by the relevant application areas.

The resonant acceleration sensor is a typical resonant sensor of sensitive frequency, while the resonant angular velocity sensor, particular the current practical one, is a typical resonant sensor of sensitive amplitude ratio. In recent years; the MEMS resonant angular velocity sensor based on working principle of complex resonance is developed which can output frequency value. Although it is not put into practice, it will demonstrate unique advantage.

3.2 Quartz Vibrating Beam Acceleration Sensor

The quartz vibrating beam accelerate sensor is a typical micromechanical inertia device, which include the quartz resonator, the flexible support, the sensitive mass, frequency measurement circuit and so on. As is shown in Fig. 3.1, the sensitive mass is constrained by the precision flexible support, so that it could have a single degree of freedom. It uses squeeze film damping gap as further constraints to the mass when over range, it also used as a mechanical shock limit to protect the crystal from overvoltage damage. The open-loop structure is a typical second-order mechanical system. The quartz vibrating beam accelerometer sensor uses the double-end fixed tuning fork structure as a resonator. The main advantages are: the torque and stress generated by the two fork arm in its binding premises can be mutual offset, so that the entire resonator could have characteristics of decoupling when vibrating, and have no apparent reaction to the structure around. The consumption of the resonator is negligible. In order to produce large axial force of inertia of the limited mass, the mechanical structure should be chosen legitimately so that the inertia force could be enlarged several times, or even a hundred times.

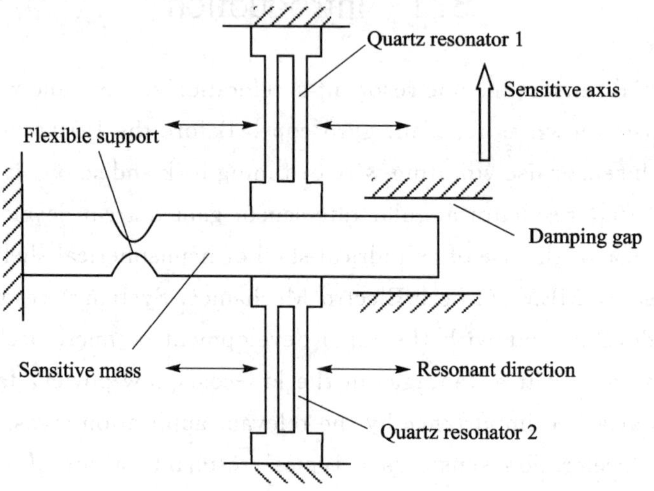

Fig. 3.1 Working principle of the quartz beam accelerometer

The working principle of quartz vibrating beam acceleration sensor can be described as follows:

$$\Delta f = K_f f_0^2 F_C \tag{3.1}$$

$$F_C = -ma$$

where

f_0 is the initial frequency of quartz resonator is related to harmonic frequency, resonator material, structural parameters, the shell material, structure parameters and so on (in Hz);

Δf is the change amount of the frequency of the quartz resonator (in Hz);

K_f is related to harmonic frequency, resonant material, structural parameters, the shell material, structure parameters and so on;

F_C is Coriolis force caused by the measured acceleration and sensitive mass, and it is in the opposite direction to the measured acceleration.

As a value for on the quartz resonator, the Coriolis force is the quartz resonator under tension, the resonant frequency increases; quartz resonator is compressed when the negative value of the Coriolis force, the resonant frequency decreases.

As is shown in Fig. 3.1, the quartz vibrating beam accelerate sensor has the differential detector structure, therefore, the resonant acceleration sensor has a good inhibition of conjugated interference, such as temperature, random vibration, to the sensor.

The resonant beam of quartz vibrating beam acceleration sensor, which is shown in Fig. 3.1, will initiate resonance under the drive of oscillator electronic circuits. When an acceleration is put in, an inertia force is generated on the sensitive mass, which can be enlarged N times in accordance with the principle of the lever in mechanics. The enlarged inertial force is applied to the axial direction of the resonant beam, so that the frequency of the resonant beam changes. When the axial tension is applied to a quartz resonator, the resonant frequency will increase; while the axial pressure is applied to another quartz resonator, the resonant frequency will decrease. The two output signals will be compensated and calculated in the frequency measurement circuit, so that we can obtain the measured acceleration.

3.3 Resonant Silicon Micromachined Acceleration Sensor

Fig. 3.2 shows the working principle of the resonant silicon MEMS accelerometer, which is composed by support beam, the proof mass, the resonant beam, the vibration exciting unit, and vibration pick-up unit, and converts acceleration measurement to the resonant frequency measurement through the two-stage sensitive structure. Sensitive structure of the first stage is constituted by the support beam and the mass, while the mass converts the acceleration to the inertial force and output it. Sensitive structure of the second stage is a resonant beam, the inertia force is applied to the axial direction of the resonant beam and cause changes in resonant frequency. Resonator in the acceleration sensor works in resonant state, and achieves tracking of the natural frequency of the resonator through the self-exciting closed-loop. Similar to the closed-loop of the resonant silicon microstructures pressure sensors, the closed-loop mainly includes the exciting unit, resonator, pick-up unit, amplitude modulation link and phase shift link, the excitation signal transforms the resonant excitation force to the resonator through the exciting unit, and the pick-up unit converts vibration signal into an electrical signal and output it, amplitude modulation and phase shifting links are used to adjust the amplitude gain and phase shift of the entire closed-loop, in order to meet the amplitude conditions and phase conditions of the self-exciting closed-loop.

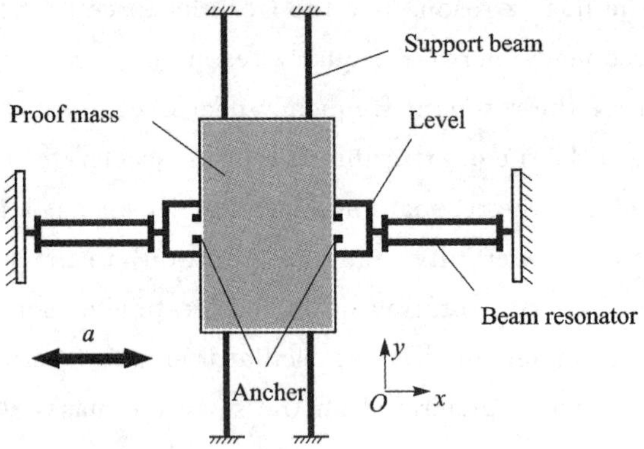

Fig. 3.2 Working principle of the resonant silicon MEMS accelerometer

The working principle when double-clamped beam harmonic oscillator work at baseband can be described as

$$f(F_C) = f_0\left(1 + 0.1475\,\frac{F_C L^2}{Ebh^3}\right)^{0.5} \tag{3.2}$$

$$f_0 = \frac{4.730^2 h}{2\pi L^2}\left(\frac{E}{12\rho}\right)^{0.5}$$

$$F_C = -ma$$

where

f_0 is the initial frequency of quartz resonator (in Hz);

L, b and h are respectively length, width and thickness of harmonic oscillator of the beam, in order to fully reflect the structural characteristics of the beam, $L \gg b \gg h$;

E and ρ are respectively modulus of elasticity and density of the beam material of resonant oscillator;

F_C is Coriolis force caused by the measured acceleration and sensitive mass.

As is shown in Fig. 3.2, silicon micro mechanical resonant acceleration sensor has a good inhibition to conjugate interference because it's also a differential detection structure, such as the influence temperature, random vibration on the sensor.

3.4 Piezoelectric Excitation Resonant Cylindrical Shell Angular Rate Sensor

Figs. 3.3 and 3.4 respectively shows structure sketch and closed-loop structure sketch of the cylindrical shell angular speed sensor. It uses the cylindrical shell whose top opens as the sensitive element; A, D, C, B, A′, D′, B′ and C′ are the 8 piezoelectric transducer element, which are uniformly distributed around the ring at the open end. Fig. 3.4 shows two independent loops. The signal measured by B travels from the phase-locked loop G_1 and a low-pass filter G_2 to the loop A, which is known as the excitation circuit maintaining the harmonic oscillator vibrating; the signal measured by C′ travels from the band pass filter circuit G_3 to the loop D′, which can be called measuring damping circuit. The two signals obtained by B and C′ can calculate the rotation angular rate around the central axis of cylindrical shell through the discriminator phase detector G_4, and ensure the direction.

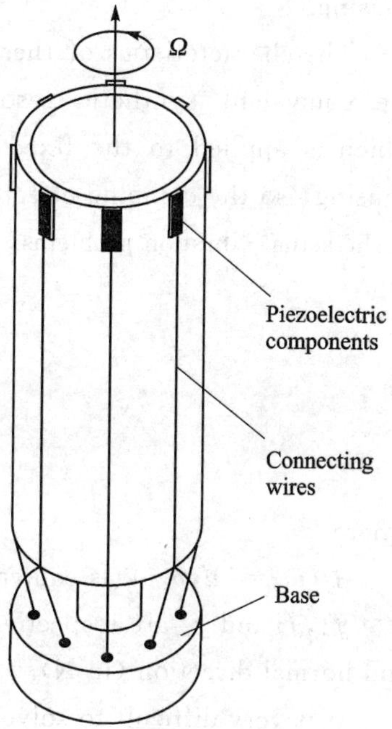

Fig. 3.3 Structure sketch of the resonant cylindrical shell angular speed sensor

The free vibration equation the cylindrical shell whose top opens can be abbreviated

$$\boldsymbol{LV} = \boldsymbol{O} \tag{3.3}$$

$$\boldsymbol{V} = [u \quad v \quad w]^{\mathrm{T}}$$

Where $\boldsymbol{L}$ is the operator matrices of 3×3.

For the symmetric mode whose circumferential wave number is n, that is

$$\left.\begin{aligned} u &= A\cos n\theta \sin \omega t \\ v &= B\sin n\theta \sin \omega t \\ w &= C\cos n\theta \sin \omega t \end{aligned}\right\} \tag{3.4}$$

where

u, v and w are respectively displacement of cylindrical shell in the axial direction, the tangential direction and the normal direction;

A, B and C are respectively the amplitude of the displacement of cylindrical shell in the axial direction, the tangential direction and the normal direction;

ω is vibration frequency of the related vibration modes of the casing (in rad / s), which is the inherent physical property of housing.

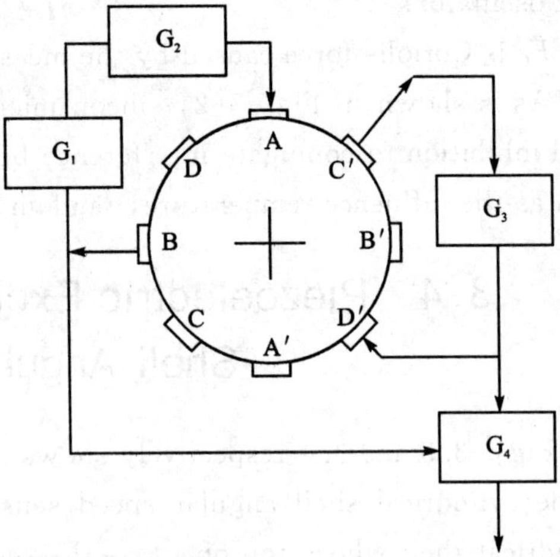

Fig. 3.4 Structure sketch of the close-loop method for the cylindrical shell angular speed sensor

The characteristics of these sensors are the equivalent synthetic resonance force, which is applied to the fixed point of the housing, so the damping effect and the influence of the equivalent excitation force are considered in the actual vibration problems, the Eq. (3.3) can be written as

$$(\boldsymbol{L}+\boldsymbol{L}_{\mathrm{d}})\boldsymbol{V}=\boldsymbol{F} \tag{3.5}$$

$$\boldsymbol{F}=[f_u \quad f_v \quad f_w]^{\mathrm{T}}$$

$$\boldsymbol{L}_{\mathrm{d}}=\begin{bmatrix}\beta_{uu} & \beta_{uv} & \beta_{uw}\\ \beta_{vu} & \beta_{vv} & \beta_{vw}\\ \beta_{wu} & \beta_{wv} & \beta_{ww}\end{bmatrix}\frac{\partial}{\partial t}$$

where

β_{ij} $(i, j=u, v, w)$ is equivalent damping ratio;

f_u, f_v and f_w are respectively equivalent exciting force in axial direction, radial direction and normal direction (in N).

It is very difficult to solve the Eq. (3.5), considering the the physical meaning of practical problems, the steady state solution of the Eq. (3.5) in time-domain is

$$\left.\begin{aligned} u_0 &= A\cos n\theta \sin \omega_v t\\ v_0 &= B\sin n\theta \sin \omega_v t\\ w_0 &= C\cos n\theta \sin \omega_v t\end{aligned}\right\} \tag{3.6}$$

where ω_v is resonant frequency of the housing when constituting the system (in rad / s). it is different from the vibration frequency in Eq. (3.4).

Thus Eq. (3.5) can be written in the equivalent form of time domain decoupling

$$\boldsymbol{L}_e \boldsymbol{V} = \boldsymbol{F} \tag{3.7}$$

$$\boldsymbol{L}_e = G(t)\boldsymbol{I}_{3\times 3}$$

$$G(t) = \frac{\partial^2}{\partial t^2} + \frac{\omega_v}{Q}\frac{\partial}{\partial t} + \omega_v^2$$

where

$\boldsymbol{I}_{3\times 3}$ is 3×3 unit matrix;

Q is equivalent quality factor of the bending vibration, the equivalent values in u, v, w three directions are the same.

Solution of Eq. (3.7) in the frequency domain can be expressed as

$$\left.\begin{aligned} u_0(s) &= \frac{f_u(s)}{G(s)} \\ v_0(s) &= \frac{f_v(s)}{G(s)} \\ w_0(s) &= \frac{f_w(s)}{G(s)} \end{aligned}\right\} \tag{3.8}$$

$$G(s) = s^2 + \frac{\omega_v}{Q}s + \omega_v^2$$

where s is Laplace operator.

Obviously, the expression of steady-state solution of Eq. (3.8) in the time-domain is Eq. (3.6).

When the housing rotates with an arbitrary rotational angular velocity $\boldsymbol{\Omega} = \boldsymbol{\Omega}_x + \boldsymbol{\Omega}_{yz}$, which is shown in Fig. 3.5. In the moving coordinate system whose rotating angular is $\boldsymbol{\Omega}$, we can create a kinetic equation and introduce the inertia force generated by $\boldsymbol{\Omega}$.

$$\boldsymbol{F} = \boldsymbol{F}_0 + \boldsymbol{F}_v \tag{3.9}$$

$$\boldsymbol{F}_0 = \rho_m hr\,\mathrm{d}s_1\,\mathrm{d}\theta\left[r\left(\boldsymbol{\Omega}_x^2 + \frac{1}{2}\boldsymbol{\Omega}_{yz}^2\right)\boldsymbol{e}_\rho - r\dot{\boldsymbol{\Omega}}_x\boldsymbol{e}_\theta - s_1\boldsymbol{\Omega}_{yz}^2\boldsymbol{e}_s\right] \tag{3.10}$$

$$\boldsymbol{F}_v = \rho_m hr\,\mathrm{d}s_1\,\mathrm{d}\theta\left\{\left[w\left(\boldsymbol{\Omega}_x^2 + \frac{1}{2}\boldsymbol{\Omega}_{yz}^2\right) + v\dot{\boldsymbol{\Omega}}_x + 2\boldsymbol{\Omega}_x\frac{\partial v}{\partial t}\right]\boldsymbol{e}_\rho + \left[-2\boldsymbol{\Omega}_x\frac{\partial w}{\partial t} - w\dot{\boldsymbol{\Omega}}_x + v\left(\boldsymbol{\Omega}_x^2 + \frac{1}{2}\boldsymbol{\Omega}_{yz}^2\right)\right]\boldsymbol{e}_\theta + u\boldsymbol{\Omega}_{yz}^2\boldsymbol{e}_s\right\} \tag{3.11}$$

where

$\boldsymbol{F}_0$ is item in the inertial force generated with $\boldsymbol{\Omega}$, which has nothing to do with the displacement;

$\boldsymbol{F}_v$ is item in the inertial force generated with $\boldsymbol{\Omega}$, which is not related to the displacement;

r and h are respectively radius and wall thickness of the middle cylinder for the cylindal shell harmonic oscillator;

s_1 and θ are respectively the coordinates in the axial direction and the circumferential direction of the round case;

E and ρ_m are respectively modulus of elasticity (in Pa) and mass density of the round shell material (in kg/m^3);

$\boldsymbol{e}_s$, $\boldsymbol{e}_\theta$ and $\boldsymbol{e}_\rho$ are respectively unit motion vector of circular shell in axial direction, the tangential direction and the normal direction;

$\dot{\boldsymbol{\Omega}}_x$ is angular acceleration of the circular housing in the axial direction.

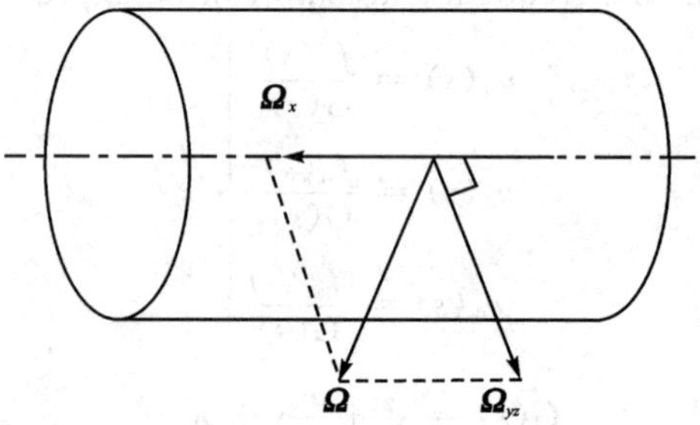

Fig. 3.5 Rotating sketch for the cylindrical shell with arbitrary angular speed $\boldsymbol{\Omega}=\boldsymbol{\Omega}_x+\boldsymbol{\Omega}_{yz}$

F_0 which has nothing to do with u, v and w will cause the initial strain of the housing and affect the resonant frequency ω of the housing, so it can be omitted, and the kinetic equation of vibration of the housing can be written as

$$\left.\begin{aligned}&\frac{\partial^2 u}{\partial t^2}+\frac{\omega}{Q}\frac{\partial u}{\partial t}+(\omega^2-\Omega_{yz}^2)u=f_u\\&\frac{\partial^2 v}{\partial t^2}+\frac{\omega}{Q}\frac{\partial v}{\partial t}+\left(\omega^2-\Omega_x^2-\frac{1}{2}\Omega_{yz}^2\right)v+2\Omega_x\frac{\partial w}{\partial t}+w\dot{\Omega}_x=f_v\\&\frac{\partial^2 w}{\partial t^2}+\frac{\omega}{Q}\frac{\partial w}{\partial t}+\left(\omega^2-\Omega_x^2-\frac{1}{2}\Omega_{yz}^2\right)w-2\Omega_x\frac{\partial v}{\partial t}-v\dot{\Omega}_x=f_w\end{aligned}\right\}\tag{3.12}$$

Examine the first formula of Eq. (3.12), it still stays decoupled form, so we should focus on the 2th and 3rd formula of Eq. (3.12), which stay mutual coupling.

$$\begin{bmatrix}v(s)\\w(s)\end{bmatrix}=\frac{1}{D}\begin{bmatrix}s^2+\frac{\omega}{Q}s+\omega_0^2 & -2\Omega_x s-\dot{\Omega}_x\\2\Omega_x s+\dot{\Omega}_x & s^2+\frac{\omega}{Q}s+\omega_0^2\end{bmatrix}\begin{bmatrix}f_v(s)\\f_w(s)\end{bmatrix}\tag{3.13}$$

$$D=\left(s^2+\frac{\omega}{Q}s+\omega_0^2\right)^2+(2\Omega_x s+\dot{\Omega}_x)^2\tag{3.14}$$

$$\omega_0^2=\omega^2-\Omega_x^2-\frac{1}{2}\Omega_{yz}^2\tag{3.15}$$

Because $\omega^2 \gg \Omega_x^2 + \frac{1}{2}\Omega_{yz}^2$, $\omega^4 \gg \dot{\Omega}_x{}^2$, the constant term of the Eq. (3.14) determines the resonance frequency of the casing, if we set the resonance frequency to be $\omega(\Omega)$, then

$$\omega^4(\Omega) = \omega_0^4 + \dot{\Omega}_x^2 \tag{3.16}$$

From the Eq. (3.15), we can know the rate of change initiated by Ω_x , Ω_{yz} and $\dot{\Omega}_x$ to the resonant frequency $\omega(\Omega)$ can be expressed as

$$\alpha(\Omega_x) = \frac{\Omega_x}{\omega} \tag{3.17}$$

$$\alpha(\Omega_{yz}) = \frac{\Omega_{yz}}{2\omega} \tag{3.18}$$

$$\alpha(\dot{\Omega}_x) = \frac{\dot{\Omega}_x}{2\omega^3} \tag{3.19}$$

In the general sense, $\omega \geqslant 1\ 000$ Hz, $\Omega \leqslant 1$ Hz , and $\dot{\Omega}_x$ is also very small, so we can know $\alpha(\Omega_x)$, $\alpha(\Omega_{yz})$ and $\alpha(\dot{\Omega}_x)$ are also very small from Eqs. (3.17), (3.18) and (3.19). The Eq. (3.13) can be written as

$$\begin{bmatrix} v(s) \\ w(s) \end{bmatrix} = \frac{1}{G^2(s)} \begin{bmatrix} s^2 + \frac{\omega}{Q}s + \omega^2 & -2\Omega_x s - \dot{\Omega}_x \\ 2\Omega_x s + \dot{\Omega}_x & s^2 + \frac{\omega}{Q}s + \omega^2 \end{bmatrix} \begin{bmatrix} f_v(s) \\ f_w(s) \end{bmatrix} \tag{3.20}$$

The first formula of Eq. (3.20) can be written as

$$v(s) = v_0(s) - P_v(s) \tag{3.21}$$

where

$$P_v(s) = \frac{2\Omega_x s + \dot{\Omega}_x}{s^2 + \frac{\omega}{Q}s + \omega^2} w_0(s) = \frac{2\Omega_x s + \dot{\Omega}_x}{s^2 + \frac{\omega}{Q}s + \omega^2} \cdot \frac{\omega_v^2}{s^2 + \omega_v^2} C\cos n\theta =$$
$$P_{vs}(s) + P_{vi}(s) \tag{3.22}$$

$P_{vs}(s)$ and $P_{vi}(s)$ are respectively the steady-state solution and the transient solution of $P_v(s)$.

Through the derivation we can get

$$P_{vs}(s) = \frac{\omega_v^2(L_1 s + L_2)}{s^2 + \omega_v^2} C\cos n\theta \tag{3.23}$$

$$P_{vi}(s) = \frac{\omega_v^2(L_3 s + L_4)}{s^2 + \frac{\omega}{Q}s + \omega^2} C\cos n\theta \tag{3.24}$$

$$\left.\begin{aligned}
L_1 &= \frac{2\Omega_x(\omega^2-\omega_v^2)-\dot{\Omega}_x\frac{\omega}{Q}}{(\omega^2-\omega_v^2)^2+\frac{\omega^2\omega_v^2}{Q^2}} \\
L_2 &= \frac{2\Omega_x\omega\frac{\omega_v^2}{Q}+\dot{\Omega}_x(\omega^2-\omega_v^2)}{(\omega^2-\omega_v^2)^2+\frac{\omega^2\omega_v^2}{Q^2}} \\
L_3 &= -L_1 \\
L_4 &= \frac{1}{\omega_v^2}(\dot{\Omega}_x-L_2\omega^2)
\end{aligned}\right\} \tag{3.25}$$

So the steady-state solution in the time-domain is

$$P_{vs}(t) = KC\cos n\theta\sin(\omega_v t+\phi) \tag{3.26}$$

$$K = \frac{2\Omega_x Q}{\omega}\sqrt{\frac{1+\frac{\dot{\Omega}_x^2}{4\Omega_x^2\omega_v^2}}{1+\frac{\omega^2-\omega_v^2}{\omega^2\omega_v^2}}} \tag{3.27}$$

$$\left.\begin{aligned}
\phi &= \arctan\frac{2\Omega_x\omega_v}{\dot{\Omega}_x}-\phi_1 \\
\phi_1 &= \begin{cases}\arctan\dfrac{\omega\omega_v}{(\omega^2-\omega_v^2)Q} & (\omega\geqslant\omega_v) \\ \pi-\arctan\dfrac{\omega\omega_v}{(\omega_v^2-\omega^2)Q} & (\omega<\omega_v)\end{cases}
\end{aligned}\right\} \tag{3.28}$$

According to the above analysis, the steady-state solution in two directions of v and w can be obtained

$$\left.\begin{aligned}
v &= B\sin n\theta\sin\omega_v t-KC\cos n\theta\sin(\omega_v t+\phi) \\
w &= C\cos n\theta\sin\omega_v t+KB\sin n\theta\sin(\omega_v t+\phi)
\end{aligned}\right\} \tag{3.29}$$

From the Eqs. (3.12) and (3.29), we can see: in such angular rate sensor, the vibration mode of cylindrical shell axis's direction is basically unchanged, the vibration mode of the the link direction and the tangential direction produces "antisymmetric vibration modes" caused by the Coriolis effects which is based on the original symmetric vibration modes. Amount of "antisymmetric vibration modes" is substantially proportional to Ω_x, the follower coordinate system point of view, the vibration mode only has a small offset, and will not appear continuous precession, because the equivalent excitation force is applied to the fixed point of the housing. When using piezoelectric ceramics as the transducing element, we can detect at the wave section of vibration mode of the case as following:

$$q = \frac{KA_0 d_{31} E(nC + B)}{[r(1 - \mu)]\sin(\omega_v t + \phi)} \tag{3.30}$$

where

A_0 and d_{31} are respectively the charge distribution area of piezoelectric ceramic element (in m^2) and the piezoelectric constant (in C/N);

E and μ are respectively modulus of elasticity (in Pa) and Poisson's ratio of vibration cylinder material;

r is radius of the vibrating cylinder in middle cylindrical surface (in m).

We can know from the Eqs. (3.27) and (3.30):

① The detection signal q is proportional to Ω_x and to the amplitude of the harmonic oscillator, so Ω_x can be obtained through detecting q directly. In order to eliminate the impact of amplitude change caused by the closed-loop system excitation energy changes on measurement results, Ω_x can be work out through the ratio of amplitude at wave section and wave antinode.

② The detection signal q is related to the change rate $\dot{\Omega}_x$ of the measured angular velocity (angular acceleration), so for such a resonant angular rate sensor, the measurement error should be considered during the process of dynamic measurement.

3.5 Static Exciting Hemispherical Resonant Angular Velocity Sensor

Fig. 3.6 shows the structure sketch for the hemispherical resonant gyroscope (HRG), whose sensing component is the hemispherical shell with an opening made up of fused quartz.

The main components of the hemispherical resonant gyroscope are vacuum sealed base, capacitive sensor (namely annunciator), gas suction chamber, hemispherical resonator, ring force electrode and discrete electrode, vacuum housing, etc.. The function of the gas getter is to absorb the residue gas molecule within the vacuum shell. On the vacuum sealed base is planted the sealing insulator connecting the inner and outer wire. The purpose of vacuum sealing is to reduce the air damp, raising the quality factor, therefore lifting its working timing constant (currently reaching over 27 minutes). The annunciator has 8 capacitive signal picking-up components used to pick up the signal and determine the position of the vibrating diagram of the resonator. By this means, the rotated angle around the core axis can be determined and, therefore the angular information of rotating shell can be determined. The hemi-

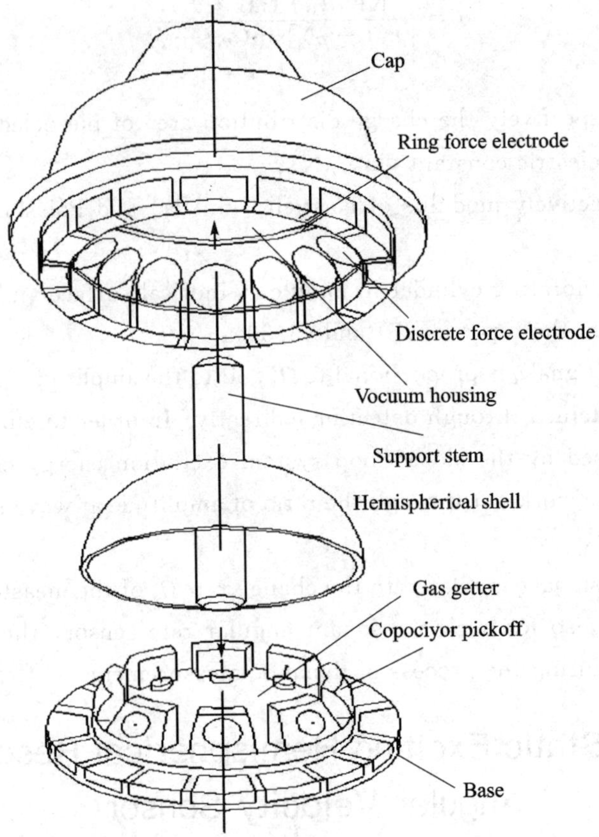

Fig. 3.6 Structure sketch for the hemisphere resonant gyroscope

spherical shell. The hemispherical resonator is the core component of the gyroscope. It hangs on and is supporting by the center supporting beam, which is clamped tightly by the force generator and annunciator on both sides to reduce the harmful coupling of the supporting structure. In addition, it is necessary to precisely fix the notch on the periphery of the hemispherical shell to help the resonator reaching dynamic balance, having equal amplitude vibration on each direction while remaining insensitive to external interference. The force generator consists of the core generator made of ring force electrode, generating square-wave voltage to maintain the resonator's amplitude a constant, compensating the energy loss caused by damping, as well as 16 discrete electrodes, that equidistantly distributed, controlling the vibration diagram and restraining the inapplicable vibrating mode within the four antinodes (mostly orthogonal vibration). In order to lift the quality factor of the resonator and ensuring its insensitivity to temperature alteration, the resonator, force generator and the annun-

ciator are all made up of fused quartz and are attached to each other with indium. A thin film of chromium is plated on the resonator and gold on the surface of the force generator and annunciator.

Figs. 3.7 and 3.8 give respectively the description of the structure sketch for the hemisphere shell and the sketch of the spherical coordinate for the hemisphere shell.

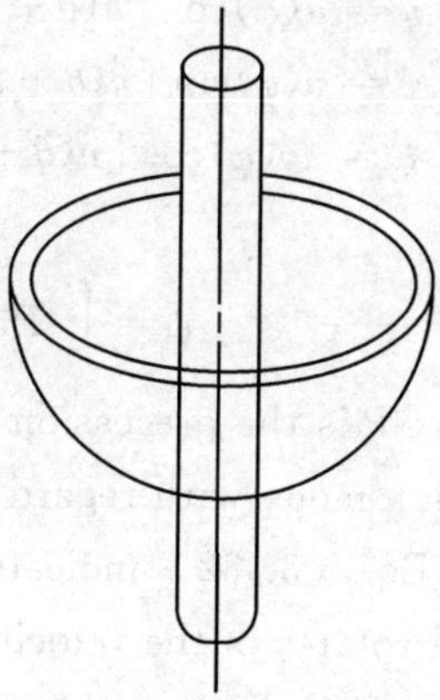

Fig. 3.7 Structure sketch for the hemisphere shell

When the hemispherical shell remains steady, the logarithmic modes of its circumferential wave number n are

$$\left.\begin{aligned} u &= u(\varphi)\cos n\theta\cos \omega t \\ v &= v(\varphi)\sin n\theta\cos \omega t \\ w &= w(\varphi)\cos n\theta\cos \omega t \end{aligned}\right\} \tag{3.31}$$

where

$u(\varphi)$, $v(\varphi)$ and $w(\varphi)$ are the vibrating mode along the axial direction;

ω is the corresponding natural frequency (in rad/s).

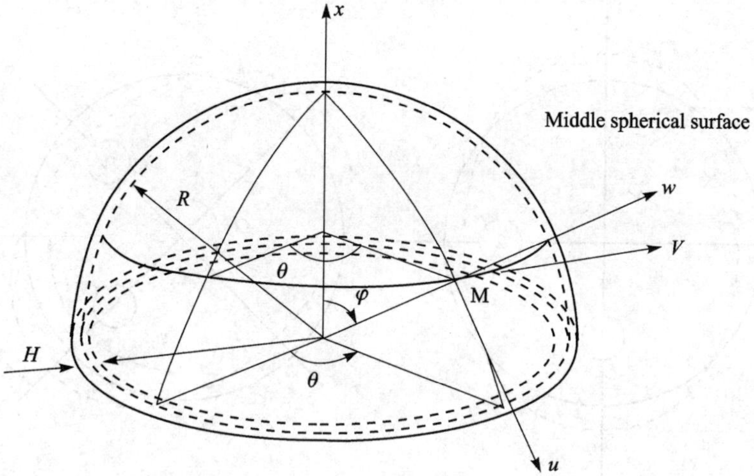

Fig. 3.8 Description sketch of the spherical coordinate for the hemisphere shell

The Coriolis effect happens when the hemispherical rotates around the center axis causes displacement on the circumferential direction (namely the precession of the vibration mode), which is also the mechanism of realizing the measurement of angular information. When the shell rotates in the inertial space with a velocity of $\Omega = \Omega_x + \Omega_{yz}$ (see Fig. 3.9), in this rotating space (see Fig. 3.10), its vibration mode can be described as

$$\left.\begin{aligned} u &= u(\varphi)\cos\left[n(\theta+\psi)\right]\cos\omega t \\ v &= v(\varphi)\sin\left[n(\theta+\psi)\right]\cos\omega t \\ w &= w(\varphi)\cos\left[n(\theta+\psi)\right]\cos\omega t \end{aligned}\right\} \tag{3.32}$$

$$\psi = \int_{t_0}^{t} P\mathrm{d}t$$

where P is the precession velocity of the vibration mode with regard to the shell.

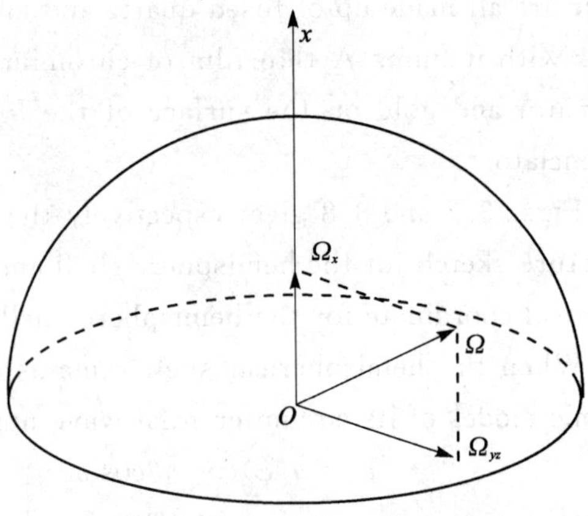

Fig. 3.9 Rotating sketch of the shell with angular speed $\Omega = \Omega_x + \Omega_{yz}$ around the inertial space

Eq. (3.32) indicates that when the shell rotates in the velocity of Ω_x around the center axis passing the angle of $\psi_1 = \int_{t_0}^{t} \Omega_x \mathrm{d}t$ viewing from the shell the circumferetial direction vibration mode, the shell rotates along the reversed direction of Ω_x in the velocity of P, passing the angle of $\psi = \int_{t_0}^{t} P\mathrm{d}t$.

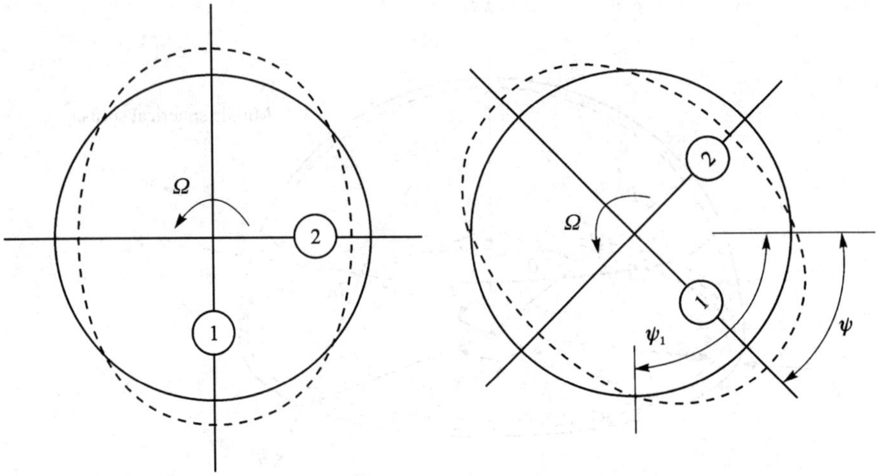

(a) In perpendicular direction (b) At 45° angle direction

Fig. 3.10 Moving sketch of the hemisphere shell in radial direction

According to the principle of measuring the hemispherical resonant gyroscope as mentioned above, in order to form the closed-loop system of the hemispherical resonant gyroscope, what needs to be done first is to maintain the hemispherical resonator in the state of "free resonance" with equal amplitude. As a matter of fact, there always exists damping when resonator vibrates. To maintain its constant vibration, energy needs to be absorbed from outer world as compensation. When exciting force equivalently exerts on the "instantaneous" antinodes of the resonator's vibration mode and the energy compensation coincide with the vibration, the above resonator's "free vibration" can be realized. Of course, this is not the free resonance under typical physics concept, which is more precisely called "nearly free vibration state".

As is discussed above, the schematic diagram can be given here for the sketch of closed loop system for hemisphere resonant gyroscope as shown in Fig. 3.11. In the sketch, C_1 and S_1 are the displacement sensors to detect the vibration mode of the resonator, with the gain of G_d. C_2 and S_2 are the exciting resources acting on the resonator, the equivalent gains for the exciting forces sharing the same frequency generated by the resonator are all G_f. Assuming the resonators are even symmetric, the amplitude is equal along each circumferential direction. K_C and K_S are the amplification element of the circuit, with the same amplitude gain G_k. C_1 and C_2 locate on the same spot θ_C along the circumferential direction on the shell, S_1 and S_2 on the same spot θ_S along the circumferential direction on the resonator. θ_C and θ_S has a difference of one fourth wave number along the circumferential direction, namely

$$\theta_S - \theta_C = \frac{\pi}{2n} \tag{3.33}$$

where n is the wave number on the circumferential direction.

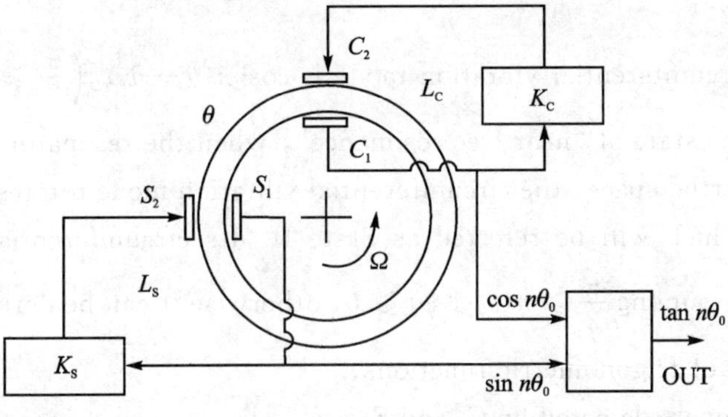

Fig. 3.11 Sketch of closed-loop system for hemisphere resonant gyroscope

As for the resonator under the self-exciting state, its vibration mode on the normal direction with the circumferential wave number n is

$$w(\theta,t) = W_0 \cos[n(\theta - \theta_0)]\cos \omega t \tag{3.34}$$

Based on the C_1 and S_1 with the assumption mentioned above, the detected displacement is

$$\left.\begin{aligned} x_C(t) &= G_d W_0 \cos(\theta_C - \theta_0)\cos\omega t \\ x_S(t) &= G_d W_0 \cos(\theta_S - \theta_0)\cos\omega t \end{aligned}\right\} \tag{3.35}$$

Signal $x_C(t)$ and $x_S(t)$ are sent to the exciting sources C_2 and S_2 via the amplification element $\overline{K}_C$ and $\overline{K}_S$, the generated exciting force are respectively as follows:

$$\left.\begin{aligned} F_C(t) &= G_K G_f x_C(t)\delta(\theta - \theta_C) \\ F_S(t) &= G_K G_f x_S(t)\delta(\theta - \theta_S) \end{aligned}\right\} \tag{3.36}$$

According to the superposition principle, under the influence of $F_C(t)$ and $F_S(t)$, the vibration mode generated by the resonator is proportional to

$$\begin{aligned} \overline{w}(\theta,t) = G\{&[\cos^2 n(\theta_C - \theta_0) + \cos^2 n(\theta_S - \theta_0)]\cos[n(\theta - \theta_0)] + \\ &\sin[n(\theta_S + \theta_C - 2\theta_0)]\cos[n(\theta_S - \theta_0)]\sin[n(\theta_C - \theta_0)]\}\cos\omega t \\ G &= G_K G_f G_d W_0 \end{aligned} \tag{3.37}$$

Insert Eq. (3.33) into Eq. (3.37), it can be concluded that

$$\overline{w}(\theta,t) = G\cos(\theta - \theta_0)\cos\omega t \tag{3.38}$$

Under the control of the closed-loop mentioned above, the system is able to track the resonator's original vibration mode $\cos(\theta - \theta_0)$, therefore the resonator can reach the "near free resonance state".

As for the closed-loop system shown in Fig. 3.11, to take into general consideration, take $\theta_C = 0, \theta_S = \frac{\pi}{2n}$, and consider two different vibration state.

State I: The circumferential vibration mode is $\cos[n(\theta - \theta_0)]\left(\frac{\pi}{2} \geqslant n\theta_0 \geqslant 0\right)$. Because the shell is under the state of "near free resonance", when the resonator rotates over the angle ψ_1 around the inertial space, the circumferential vibration mode rotates angle ψ with regard to the resonator, which will be referred as State II. Its circumferential vibration mode is $\cos n(\theta - \theta_0 - \psi)$ (assuming $\frac{\pi}{2} \geqslant n(\theta_0 + \psi) \geqslant 0$, otherwise it can be derived according to the properties and logic of trigonometric functions).

State II: the signals detected by C_1 and S_1 are

$$\left.\begin{aligned} x_C(t) &= G_d W_0 \cos n\theta_0 \cos\omega t = D_C \cos\omega t \\ x_S(t) &= G_d W_0 \cos n\left(\frac{\pi}{2n} - \theta_0\right)\cos\omega t = D_S \cos\omega t \end{aligned}\right\} \tag{3.39}$$

where D_C and D_S are respectively the amplitude of signal $x_C(t)$ and $x_S(t)$.

According to Eq. (3.39), by sending the signals detected by C_1 and S_1 to the logical comparator and divider, it can be obtained that

$$\left.\begin{aligned} \tan n\theta_0 &= \frac{D_S}{D_C} \quad (D_S \leqslant D_C) \\ \cot n\theta_0 &= \frac{D_C}{D_S} \quad (D_S > D_C) \end{aligned}\right\} \tag{3.40}$$

Therefore, by detecting the amplitude ratio between signal $x_S(t)$ and $x_C(t)$, $n\theta_0$ can be solved out and consequently determines the circumferential position θ_0 of State I. Similarly, the circumferential position of State II can also be determined as $\theta_0 + \psi$. Thus, it can be told that State II has influence on the circumferential vibration mode of State I, causing angular displacement ψ. According to the precession regulation of vibration with regard to resonator, the shell (resonator) rotates over angular displacement $\psi_1 = \frac{\psi}{K}$ around the inertial space. So by the use of closed-loop system, measuring angle is realized, and by differentiating ψ_1 can be measured the angular velocity.

The closed-loop system shown in Fig. 3.11 is realized by utilizing two independent annunciators and two independent exciting sources. Practically, on the one hand, to improve the measuring precision, we can deploy multiple annunciators to pick up the vibration signal, while on the other hand, to maintain the resonator under an ideal vibration state, generating the vibration mode with needed circumferential wave number n, multiple exciting resources is can be deployed. For instance, as for common four-antinode vibration with $n=2$, 8 independent annunciators and 16 independent exciting resources are deployed.

The sketch of signal measurement system for hemisphere resonant gyroscope is given in Fig. 3.12.

The designing thought is: first of all, rectify $x_S(t)$ and $x_C(t)$, and generate half-wave sine burst series; Then, distribute these bursts to integrator, maintain the integrator close to balance. When the given computer sampling period ends, the ratio between the number of bursts with bigger amplitude and that with smaller amplitude can be roughly regarded as the signal amplitude ratio; Meanwhile, the off-balance information of the integrator at the end of sampling period provides some extra information for precise calculation.

Signal $x_S(t)$ corresponding to θ_S is sending into the integrator after full-wave rectifying, as shown in Fig. 3.13(a). Assuming the integrating starts at the point $t=0$, when the wave exactly passes the zero point. When the integrating stops at t_1, the number of complete half-wave is N_S, and the time gap of the last incomplete half-wave is $M_S = t_1 - \frac{T}{2}N_S$. Therefore, the integral value is

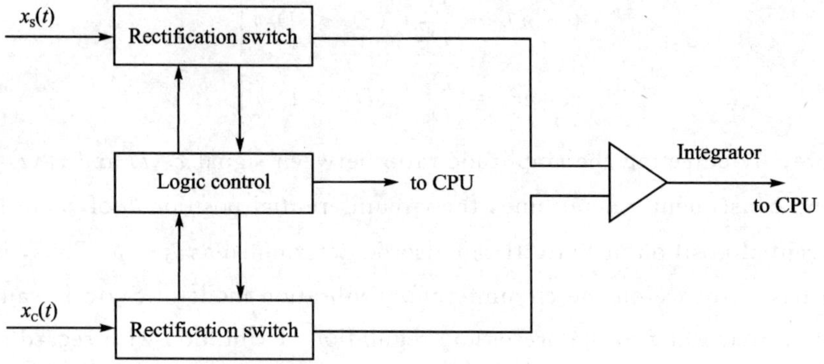

Fig. 3.12 Sketch of signal measurement system for hemisphere resonant gyroscope

$$A_S = \frac{1}{\tau}\int_0^{t_1} |x_S(t)| \, dt = \frac{1}{\tau}\left(D_S N_S \int_0^{\frac{T}{2}} \sin \omega t \, dt + D_S \int_0^{M_S} \sin \omega t \, dt\right) = \frac{2D_S}{\pi\tau}\left[\frac{T}{2}N_S + B(T, M_S)\right] \tag{3.41}$$

$$B(T, M_S) = \frac{T}{4}\left(1 - \cos\frac{2\pi M_S}{T}\right) \tag{3.42}$$

where

τ is the time constant of the integrator(in s);

T is the period of the signal(in s).

According to Eq. (3.41), the integral value A_S is mainly related to the time of former N_S half-waves, while another small amount of item $B(T, M_S)$ is the extra information when losing balance as mentioned before. In practical calculation, considering that the period T for vibration signal is a determined constant, $B(T, M_S)$ can be obtained by piecewise interpolation, that is, with a given M_S, a corresponding $B(T, M_S)$ can be "looked up".

Similarly, we can have $x_C(t)$ after rectification and integrating (see Fig. 3.13(b)):

$$A_C = \frac{2D_C}{\pi\tau}\left[\frac{T}{2}N_C + B(T, M_C)\right] \tag{3.43}$$

Based on Eqs. (3.41) and (3.43), it can be concluded that:

$$\frac{D_S}{D_C} = \frac{A_S\left[\frac{T}{2}N_C + B(T, M_S)\right]}{A_C\left[\frac{T}{2}N_S + B(T, M_C)\right]} \tag{3.44}$$

Eq. (3.44) is the mathematical model of detecting signal amplitude ratio of dual channel as shown in Fig. 3.12. By measuring the seven parameters A_S, A_C, N_S, N_C, M_S, M_C and T can the amplitude ratio between the two channels' signal be obtained. Among them, A_S and A_C

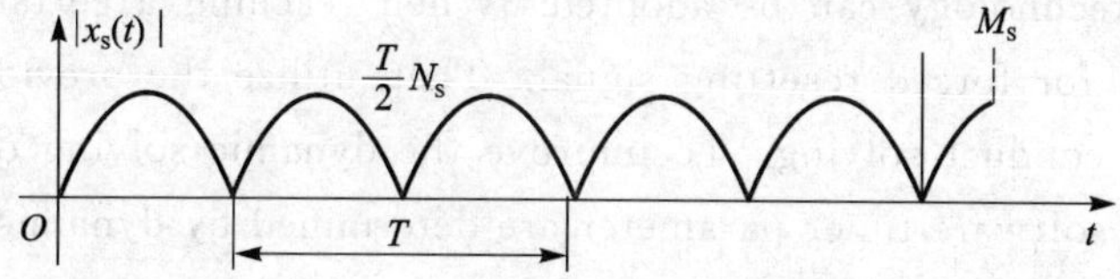

(a) $x_s(t)$

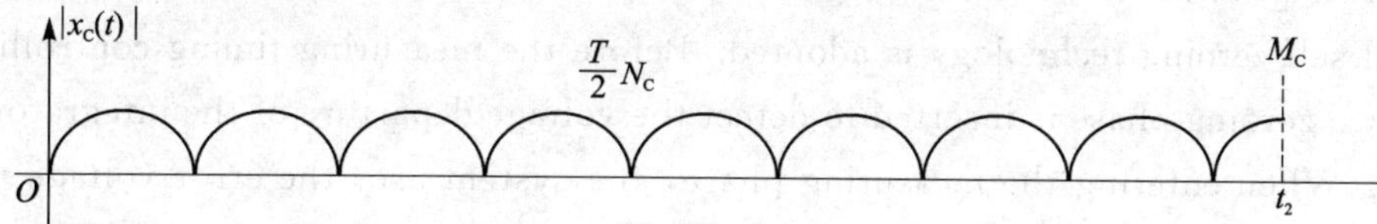

(b) $x_c(t)$

Fig. 3.13 Solved sketch of the signal

can be transferred into digitals via A/D converting, while the other 5 themselves are digitals. Therefore, by measuring the digitals, we can get the measured values of the amplitude ratio.

The advantages of this proposal are as follows:

① The measurement of amplitude is transferred indirectly into the measuring of time gap and two DC electronic signals' A/D converting.

② Based on the theoretical analysis above, such method does not need total precisely same phase of the two channel signals.

③ As for some non-strict sine wave, phase error and random interference, it has certain suppression over those influences.

④ What's more is continuous measurement and real-time.

It needs to be pointed out that several practical problem as follows should be taken into consideration when designing hardware and software.

(1) The comparison of two channels' signal

The purpose is to improve the precision when solving angle ψ, which is reflected in Eq. (3.40).

(2) The quadrant problem of angle 2 ψ

This problem can be solved by telling whether $x_S(t)$ and $x_C(t)$ are inverting or non-inverting and by the previous sampling of $x_S(t)$ and $x_C(t)$.

(3) The signal processing problem when it approaches 0° and 45° $\left(0, \frac{\pi}{4}\right)$

At the time the channel with smaller amplitude may have problem reaching scheduled value after integrating for a long time. In order to maintain the system's real-time and precision, software timer interrupt technology can be adopted. When reaching a certain time point, the system stops waiting for forced resetting signal. Then utilize the previous and present sampling information to conduct solving. To improve the dynamic solving quality, the scheduled integral value and software timer parameter are determined by dynamic determining method, namely within each measuring period, these two parameters can be given to CPU according to the practical changing situation of the signal.

(4) Error problem of the measuring zero point

Digital self-zeroing technology is adopted. Before the measuring timing controlling signal is sent, a zeroing phase is inserted to detect the voltage departure of the integrator's analog output. When entering the measuring phase, the system uses the error voltage to compensate the error factor that is affecting to make sure that the final result does not contain the value of zero point departure.

3.6 Micro Silicon Resonant Angular Velocity Sensor

3.6.1 Silicon Capacitive Surface Micromechanical Gyroscope

Fig. 3.14 shows the structure diaphragm of a symmetric surface micro gyroscope with a characteristic of decoupling. Its sensing structure has supporting "anchor" at each of four outside corners and it is connected sufficiently with driving and sensing electrodes through beams. Since the natural frequency of the two vibrating mode has no influence on each other, therefore such linking methods prevent from mechanical coupling. Besides, aside from regular realization of directly attached to the "anchors", it takes full use of a symmetric structure to put the sensing mass onto the linking beam, and via the beam to connect with the driving and sensing electrodes.

The working mechanism of micromechanical gyroscope is based on Coriolis effect. When working, a DC bias voltage is exerted on the sensing mass and a proper AC exciting voltage between the moving and fixed interdigital, therefore a natural vibration is generated on the sensing mass along the y axis. When the gyroscope sense the angular velocity surrounding the z axis, the mass will generate additional vibration along the x axis due to Coriolis effect. By measuring the amplitude of the additional vibration, the measured angular velocity can be obtained.

Usually, the driving mode and detecting mode of the vibrating gyroscope is coupling

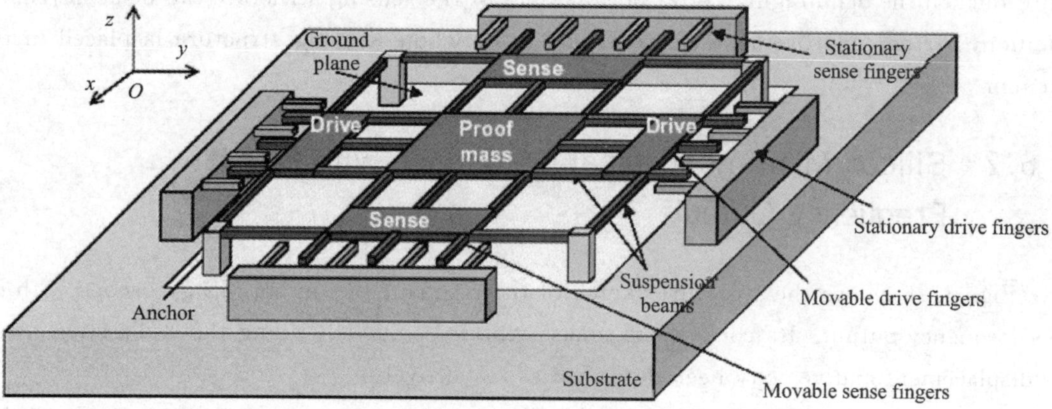

Note: The whole designed structure is symmetric with driving and detecting mode decoupling with each other and x and y axis sharing the same resonant frequency.

Fig. 3.14 Geometrical sketch for MEMS gyroscope

with each other. Due to the designing method of mutual decoupling spring, this structure helps solving the problem mentioned above. The designing still remains the overall symmetry and the sensitivity of the micromechanical gyroscope remains a high level.

Fig. 3.15 is the SEM (scaning electronic microscopy) diagram of the manufactured micromechanical gyroscope, with an outer plain structure parameter of 1 mm^2 and thickness of 22 μm. Because of its extreme thin structure with a capacitance of fF order of the driving and detecting electrodes, there exists a limitation for its performance. However, due to the symmetry of the overall structure, its performance is relatively ideal.

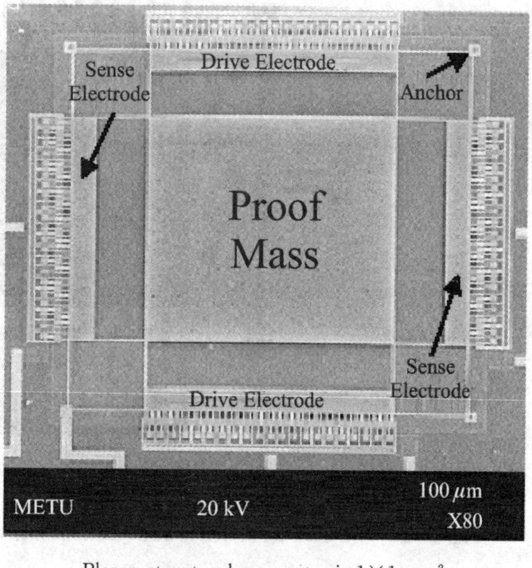

Planar structural parameter is 1×1 mm^2

Fig. 3.15 SEM sketch of MEMS gyroscope

Results of practical measurement indicate that under regular atmosphere condition, the sensing structure has a resolution better than 0.37(°)/s. With a few measures to reduce parasite capacitance and air damping, the resolution can be improved furthermore. If more advanced technology is adopted to thicken the diaphragm, better performance of the sensing structure can be achieved. In addition, further improvement is promising if the whole sensing structure is placed in the vacuum.

3.6.2 Silicon Micromechanical Gyroscope with Direct Frequency Output

Fig. 3.16 shows the structure sketch of the resonant silicon MEMS gyroscope with direct frequency output. Its core sensing mass vibrates sinuously along the x direction, with its displacement and velocity been described as respectively:

$$x(t) = X_0 \sin \omega_x t \tag{3.45}$$

$$v_x(t) = \omega_x X_0 \cos \omega_x t \tag{3.46}$$

where

X_0 is the amplitude of the sensing mass vibrating at the x direction;

ω_x is the vibrating frequency of the sensing mass on the x direction.

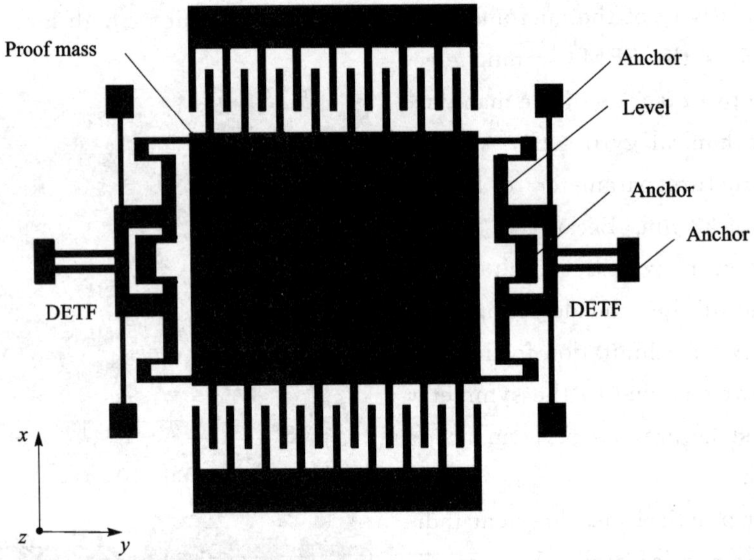

Fig. 3.16 Structure sketch of the resonant silicon MEMS gyroscope with direct frequency output

When there is an angular velocity surrounding the z direction, the sinuous vibration on the x direction will sense the Goriolis acceleration, generating inertial force on y direction, which can be described as

$$f_C(t) = 2m\Omega_z v_x = 2m\Omega_z \omega_x X_0 \cos \omega_x t \tag{3.47}$$

where

Ω_z is the angular velocity surrounding z direction of the sensing mass;

m is the value of the sensing mass.

Via the outer frame structure and leverage, the inertial force is exerted on both ends of the resonant fork on the axial direction, thus changing the resonant frequency of the double ended tunneling fork(DETF) combining with tuning. The difference of the resonant frequency reflects the value of the angular velocity.

What needs to be pointed out is that due to the fact the inertial force exerted on the DETF with tuning function is caused by the sinuous vibration of the core sensing mass along the x direction. Such inertial force is a sinuous force sharing the same frequency with the sinuous vibration mentioned above. It differs enormously compared to other usual sensors using tuning fork as the resonant sensing component accomplishing measurement. The resonant state of the tuning forking itself is under a modulation state, whose modulating frequency is the frequency of the core sensing mass sinuously vibrating along the x direction. To precisely solve the resonant frequency of the tuning fork and maintain the functional working status of the resonant gyroscope, it is required that the resonant frequency should exceed much above the frequency of the sinuous vibration mentioned above. Additionally, the DETFs with tuning function on both sides forms a differential working mode, which helps improving its sensitivity and the ability to resist interference.

Questions and Exercises

3.1 Draw a sensing structure diaphragm of a type of resonant acceleration sensor and briefly illustrate its measuring principle.

3.2 Briefly explain the basic principle of the quartz vibrating beam acceleration sensor shown in Fig. 3.1 with the ability to restrain the influence of temperature.

3.3 Draw the principle block of the resonant acceleration sensor shown in Fig. 3.2.

3.4 Briefly describe the main characteristics of micromechanical inertial sensor, compared to conventional inertial sensor.

3.5 Draw the diagram of the piezoelectric exciting resonant cylinder shell angular velocity sensor.

3.6 Illustrate the reason why the double closed-loop circuit is adopted based on the principle diagram of the closed-loop structure of the electronic exciting resonant angular velocity sensor shown in Fig. 3. 4.

3. 7 Draw the diagram of the static exciting hemispherical resonant angular velocity sensor.

3. 8 Compare the similarities and differences of the working mechanism between the piezoelectric exciting resonant cylinder shell angular velocity sensor and the static exciting hemispherical resonant angular velocity sensor.

3. 9 Explain the working mechanism of hemispherical resonant angular velocity sensor using diaphragm and formula.

3. 10 Why is it need to realize the "nearly free resonant state" for the Hemispherical resonant angular velocity sensor? and how?

3. 11 Use the principle diagram of closed-loop system of the hemispherical resonant gyroscope shown in Fig. 3. 11 to illustrate the mechanism of the readout of the signal.

3. 12 Briefly introduce the working mechanism of the micromechanical gyroscope shown in Figure 3. 14 and illustrate its application characteristics.

3. 13 Explain the working mechanism of the micromechanical gyroscope with direct frequency output, build up its mathematical model and illustrate its working characteristics.

3. 14 Compare the main difference between the micromechanical gyroscope shown in Figure 3. 14 and that with direct frequency output shown in Fig. 3. 16.

Chapter 4
Resonant Mass Flow Sensor

Basic contents:

- Mass flow
- U elastic bend
- Phase/delta-T measurement
- Amplitude ratio measurement
- The measurement of two-component fluid
- Detection circuit
- The self-exciting closed-loop system
- The classification of mass flow sensor
- Interference factor and its restrain

4.1 Introduction

Because the flowmeter is based on the Coriolis effect and its sensitive structure is working in the resonant state and the mass flow is measured directly, this flowmeter is called resonant Coriolis direct mass flowmeter or simply Coriolis mass flowmeter (CMF) for short. It includes the resonant Coriolis direct mass flow sensor and the corresponding detection circuit. The Coriolis mass flow sensor is short for the resonant Coriolis direct mass flow sensor and it is a key part of the Coriolis mass flowmeter. The detection circuit of Coriolis mass flowmeter, known as the secondary instrument, plays an important role in Coriolis mass flowmeter. It is the important technical platform to realize the multi-functions and intelligent of the sensor.

Coriolis mass flowmeter research and development began in the early 1950s. As the practical technical problem that fluid should be in a rotating coordinate while moving in a straight line has not been well solved, it has not been able to reach the industrial application stage. Until the mid-1970s, James E. Smith skillfully introduced fluid into the instrument tube being in a resonant state, and then he invented the resonant direct mass flowmeter in the combination of the two movements using the Coriolis effect. In 1977, U. S. corporation Rosemount successfully developed the first mass flowmeter based on the Coriolis effect principle all over the world. After this, many enterprises in the field of instrumentation technology has developed this kind of flowmeter.

Coriolis mass flowmeter technology has the high technical contents and difficulties, which refers to the multidisciplinary field involving materials, mechanics, mechanical engineering, electronic, magnetic, control engineering, computers, etc. Due to the excellent performance, high value-added market applications and broad market application, it has powerfully developed since its appearance. The United States, Germany, Japan and other countries have invested a lot of manpower, material and financial resources in the development and application. The most significant world-renowned representatives like Rosemount, Krohne, E+H, Foxboro, Schlumberger and ABB has already developed related products and acquired considerable economic benefits. Among these companies, Rosemount as a pioneer in Coriolis mass flowmeter development, launched the most representative products in the principles, techniques, and performance, which accounted for more than 60% of the market share in the domestic and foreign mass flowmeter market and represented the highest level and direction of mass flowmeter development.

Domestic study of the Coriolis mass flowmeter began in 1987. After nearly 20 years of development, many companies can produce Coriolis mass flowmeter in different models and specifications. Taiyuan aviation instrument company, China test technique institute and Beijing Shouke Petrochemical Automation Equipment Co. Ltd. the stronger production capacity among them. Compared with foreign ones, the main technical indicators of domestic Coriolis mass flowmeter products in commonly used models have reached the level of similar products abroad over the same period. But domestic products have a certain gap in the site adaptability, diversification of the tubes and specifications and are still lagging behind the foreign ones in general.

The annual direct domestic demand is more than 2 000 sets of sales of more than one hundred million RMB, of which domestic products account for about 20% of the market share.

In recent years, with the development and progress of science and technology, direct mass flowmeter based on the Coriolis effect have developed very fast and some new hot spots have been emerged, such as micro machined Coriolis mass flowmeter based on micro electro mechanical systems (MEMS) for the detection of micro-flow; intelligent flow test technology based on digital technology and Coriolis mass flowmeter based on the high-sensitivity detection of gases.

4.2　The Basic Working Principle

The basic measurement principle is explained below using the twin-tube structure. The resonant sensitive structure topology model of resonant Coriolis direct mass flow sensor is shown in Fig. 4.1(a). It consists of three parts: a pair of parallel elastic elbow T and T′; an elastic excitation unit K placed in the "center"; a pair of measuring elements B and B′ that are symmetrically distributed in the center. When the system is working, flexible incentive units are used to maintain the sensitive structure in the first bending resonance state. T and T′ as the main bending vibrations in reverse synchronization are centered on K (shown in Fig. 4.1(b)). When the mass flow is flowing through the vibration elbow, "Coriolis effect" that making T and T′ on the basis of the main bending vibration induces the second order

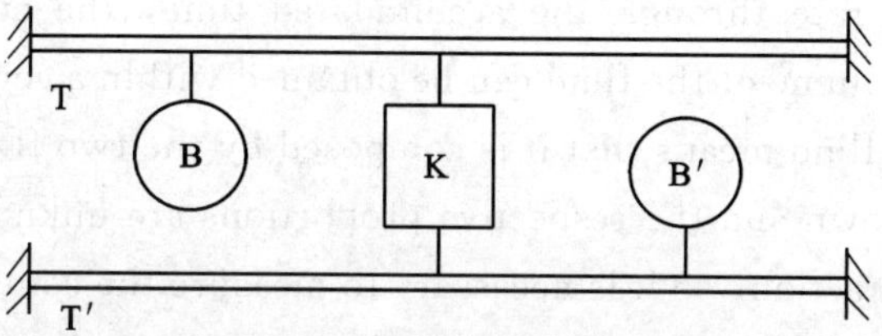

(a) Topological model of the resonant mass flow sensor

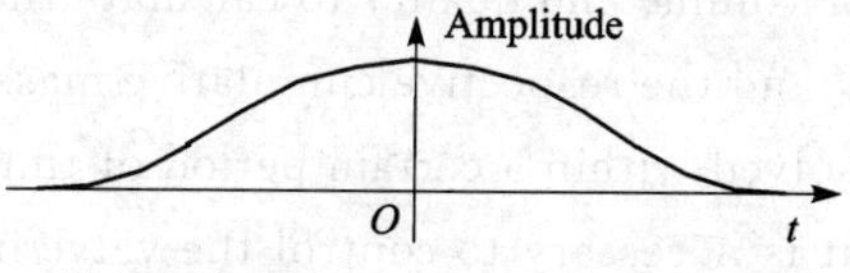

(b) Sketch of first order bending vibration

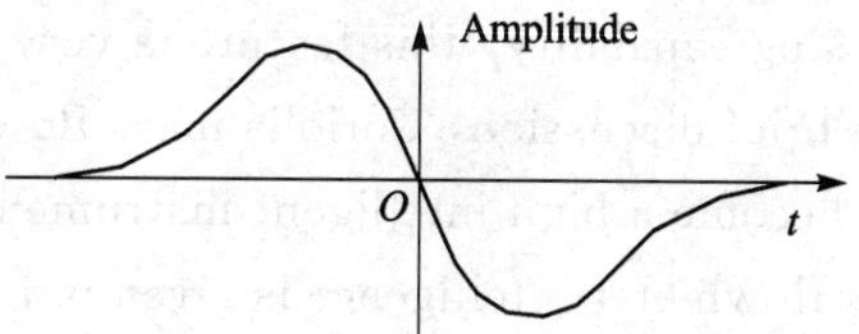

(c) Sketch of second order bending vibration

Fig. 4.1　Topological model and two orders bending vibration sketch of the resonant direct mass flow sensor

sub-vibration antisymmetrized on K (shown in Fig. 4.1(c)). As Coriolis effect is proportional to the mass flow in the measurement prime tube, the "secondary undulation" is directly in proportion to the mass flow (in kg/s) flowing past the tube and is the mass flow of the fluid can be obtained directly through the synthetic vibration of the detection bend of the measuring elements B and B′. This is the basic working principle of Coriolis mass flowmeter.

On the other hand, when the measuring tube is filled with the fluid, the overall quality of the resonant sensitive structure of the sensor and the mass of the fluid in the measuring tube are closely related, that is, the natural frequency of the resonant sensitive structure is closely tied to the fluid quality in the measuring tube. As the volume of the measuring tube in the measuring process is constant, the natural frequency of the resonant sensitive structure and the density of the fluid are closely related. The density of the measured fluid can be calculated by measuring the resonant frequency of the measurement tube through the measuring elements B and B′. Considering that the system has gained the mass flow rate and density of the fluid, therefore the volume of fluid flow can be obtained. By using the resulting mass and volume flow rate through the accumulated time, the cumulative mass and the amount of cumulative volume of the fluid can be obtained within a certain period of time. The certain two-component fluid means that it is composed by the two fluids of which the respective components are known and the respective proportions are unknown. For example, a lot of water exist in oil extraction, so it is necessary to measure the content of water in the crude oil. This belongs to the above situation. When they are physically incompatible, it is possible to use the conservation of volume and quality to calculate the respective mass and volume flow of the two components, and the respective cumulative mass and volume of the two-component fluid can further be solved within a certain period of time. In recent years, according to the fluid filling process, it is necessary to control the valve in accordance with the quality of the canned fluid to achieve the batch control functions. As Coriolis mass flowmeter has a powerful information processing capability, this feature is very appropriate to achieve.

According to the above brief discussion, Coriolis mass flowmeter is a kind of typical resonator sensors, which has become a high intelligent instrument. The basic working principle diagram of Coriolis mass flowmeter intelligence is given in Fig. 4.2. With the continuous development and progress of science and technology, the features of Coriolis mass flowmeter will be continually enhanced and the application mode for the personalization of the users will also be expanded.

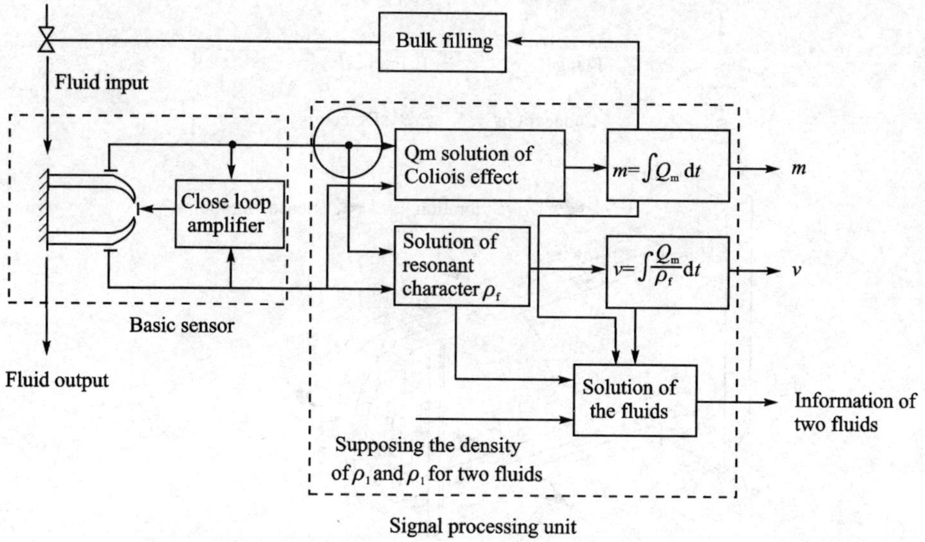

Fig. 4. 2 Basic working principle of the intelligentized Coriolis mass flow sensor

4.3 Mathematical Model of the Measurement

4.3.1 The Measurement of Mass Flow

The structure and operation schematic diagram of the typical U-shape tube which is used as the sensitive element of resonant direct mass flow sensor is shown in Fig. 4. 3. The excitation unit E causes the first order bending vibration of a pair of parallel U-shape tube, so that the working point can be built. When the mass flow flows past the tube, the U-shape tube can produce the first order torsional vibration on the center axis owing to Coriolis effect. The first order torsional sub-vibration is equivalent to the second order bending vibration (see Fig. 4. 1 (b), (c)). Meanwhile, the sub-vibration is directly in proportion to the flowing mass flow (in kg/s). Therefore, the mass flow can be directly obtained by detecting the synthetic vibration of the U-shape tube through the measuring elements B and B′.

Mathematical model of the U-shape tubule mass flow sensor is shown in Fig. 4. 4. When there is no fluid flowing in the tube, the harmonic oscillator can produce the main vibration around the CC' axis under the effect of the exciter. It can be written as

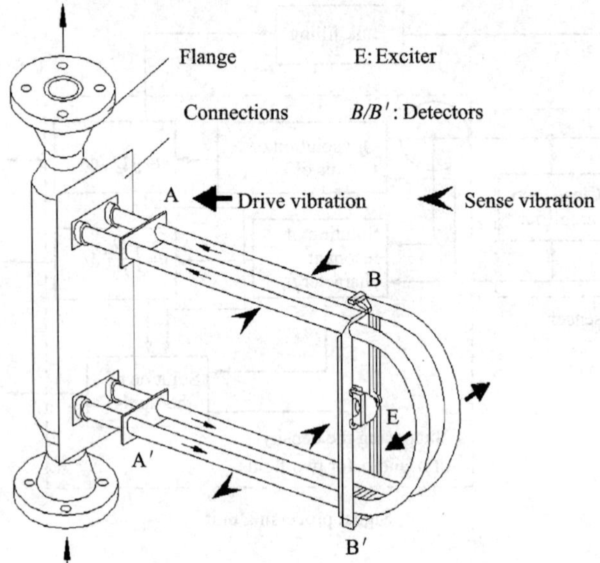

Fig. 4.3 Structure sketch of the U-shape resonant mass flow sensor

$$x(s,t) = A(s)\sin \omega t \tag{4.1}$$

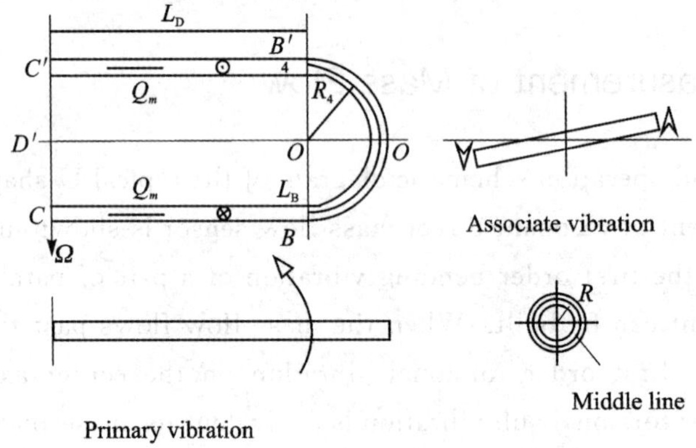

Fig. 4.4 Mathematical model the U-shape resonant mass flow sensor

where

ω is the main vibration frequency of the system, and is determined by the overall structure of the resonator including the elastic bend and the resilient support;

$A(s)$ is corresponding to the main vibration type of ω ;

s is the curvilinear coordinate along the tube axis direction.

The vibration of the tubes can be regarded as the rotation of CC' axis, and the equivalent rotation is cyclical. So the angular velocity of elastic tube around CC' axis can be written as

$$\Omega(s,t) = \frac{\mathrm{d}x(s,t)}{\mathrm{d}t} \cdot \frac{1}{x(s)} = \frac{A(s)}{x(s)}\omega\cos\omega t \tag{4.2}$$

where $x(s)$ is the distance from any point on the tube to CC' axis.

When the fluid is flowing in the tube with the velocity $\boldsymbol{v}$, it can be seen as the rotation of the coordinate system at the same time along with the relative linear motion and then the Coriolis acceleration is produced which can cause the "Coriolis force". When the elastic tube has the forward vibration, the Coriolis force suffered in the micro segment ds of the section CBD can be described as

$$\mathrm{d}\boldsymbol{F}_C = -\boldsymbol{a}_C\mathrm{d}m = -2\boldsymbol{\Omega}(s)\times\boldsymbol{v}\mathrm{d}m = -2Q_m\omega\cos\alpha\cos\omega t\frac{A(s)}{x(s)}\mathrm{d}s \tag{4.3}$$

Similarly, in the segment $C'B'D$, the Coriolis force on the symmetric point of the micro segment ds which is centered on DD' can be written as

$$\mathrm{d}\boldsymbol{F}'_C = -\mathrm{d}\boldsymbol{F}_C \tag{4.4}$$

Eqs. (4.3) and (4.4) have a difference of a negative sign, which indicates the opposite direction between them. When the fluid flows past the vibratory harmonic oscillator, the couple for DD' axis will be generated on the effect of $\mathrm{d}\boldsymbol{F}_C$ and $\mathrm{d}\boldsymbol{F}'_C$.

$$\boldsymbol{M} = \int 2\mathrm{d}\boldsymbol{F}_C \times \boldsymbol{r}(s) \tag{4.5}$$

where $\boldsymbol{r}(s)$ is the distance from the axis DD' to the representative elemental volume.

By the Eqs. (4.3) and (4.5), it can be expressed as

$$M = 2Q_m\omega\cos\alpha\cos\omega t\int\frac{A(s)r(s)}{x(s)}\mathrm{d}s \tag{4.6}$$

where

Q_m is the mass flow of the fluid that flows past the tube;

α is the angle between the direction of the fluid velocity and the x axis which is not given in Fig. 4.4.

The couple caused by Coriolis effect compels the harmonic oscillator to generate a twisting motion around the axis. With respect to the main vibration of the harmonic oscillator, it is called as the "sub-vibration", and the motion equation relative to the harmonic oscillator can be written as

$$x_1(t) = B_1(s)Q_m\omega\cos(\omega t + \varphi) \tag{4.7}$$

where

$B_1(s)$ is the sensitivity coefficient of the sub-vibration response (in m · s^2/kg), which is relative to the sensitive structure, parameters and the position of the detection point;

φ is the phase changes of the sub-vibration response for the torsional couple.

According to the above analysis, when the fluid is flowing past the tube, the vibration equation at the two points B and B' of the harmonic oscillator can be respectively written as

At the point B:

$$S_B = A(L_B)\sin\omega t - B_1(L_B)Q_m\omega\cos(\omega t + \varphi) = A_1\sin(\omega t + \varphi_1) \tag{4.8}$$

$$A_1 = [A^2(L_B) + Q_m^2\omega^2 B_1^2(L_B) + 2A(L_B)Q_m\omega B_1(L_B)\sin\varphi]^{0.5}$$

$$\varphi_1 = \arctan\frac{Q_m\omega B_1(L_B)\cos\varphi}{A(L_B) + Q_m\omega B_1(L_B)\sin\varphi}$$

At the point B':

$$S_{B'} = A(L_B)\sin\omega t + B_1(L_B)Q_m\omega\cos(\omega t + \varphi) = A_2\sin(\omega t + \varphi_2) \tag{4.9}$$

$$A_2 = [A^2(L_B) + Q_m^2\omega^2 B_1(L_B) - 2A(L_B)Q_m\omega B_1(L_B)\sin\varphi]^{0.5}$$

$$\varphi_2 = \arctan\frac{Q_m\omega B_0(L_B)\cos\varphi}{A(L_B) - Q_m\omega B_1(L_B)\sin\varphi}$$

where L_B is the coordinate value at the point B along the axis direction.

Therefore, the phase ($\varphi_{BB'} = \varphi_2 - \varphi_1$) difference can be generated at the two-points B and B' between the signal S_B and $S_{B'}$, as shown in Fig. 4.5. According to the Eqs. (4.8) and (4.9), we have

$$\tan\varphi_{BB'} = \frac{2A(L_B)Q_m B_1(L_B)\omega\cos\varphi}{A^2(L_B) - Q_m^2 B_1^2(L_B)\omega^2} \tag{4.10}$$

As the inequality $A^2(L_B) \gg Q_m^2 B_1^2(L_B)\omega^2$ is always established in the practical application, Eq. (4.10) can be described as

$$Q_m = \frac{A(L_B)\tan\varphi_{BB'}}{2B_1(L_B)b\omega\cos\varphi} \tag{4.11}$$

Eq. (4.11) is the basic equation based on the phase difference $\varphi_{BB'}$ of S_B and $S_{B'}$ which can directly solve the mass flow Q_m. If the inequality $\varphi_{BB'} \leqslant 5°$ is founded, we can obtain the following formula:

$$\tan\varphi_{BB'} \approx \varphi_{BB'} = \omega\Delta t_{BB'} \tag{4.12}$$

$$Q_m = \frac{A(L_B)\Delta t_{BB'}}{2B_1(L_B)\cos\varphi} \tag{4.13}$$

At this time, there is no relation between the mass flow Q_m and the vibration frequency of the elastic structure, and Q_m is only proportional to the time difference between the two signals S_B and $S_{B'}$. This is a virtue of such sensors. As it is relevant to cos φ, the actual measurement will bring a certain error and the real-time detection is not ideal. For this reason, the amplitude ratio detection method can be considered for application.

According to the Eqs. (4.8) and (4.9), we have

$$S_{B'} - S_B = 2B_1(L_B)Q_m\omega\cos(\omega t + \varphi) \tag{4.14}$$

$$S_{B'} + S_B = 2A(L_B)\sin \omega t \tag{4.15}$$

If the amplitude ratio between $S_{B'} - S_B$ and $S_{B'} + S_B$ is set as R_a, then we have

$$Q_m = \frac{R_a A(L_B)}{B_1(L_B)\omega} \tag{4.16}$$

Eq. (4.16) is the basic equation based on the amplitude ratio R_a of the difference and sum between the two signals S_B and $S_{B'}$, which can directly calculate the mass flow Q_m.

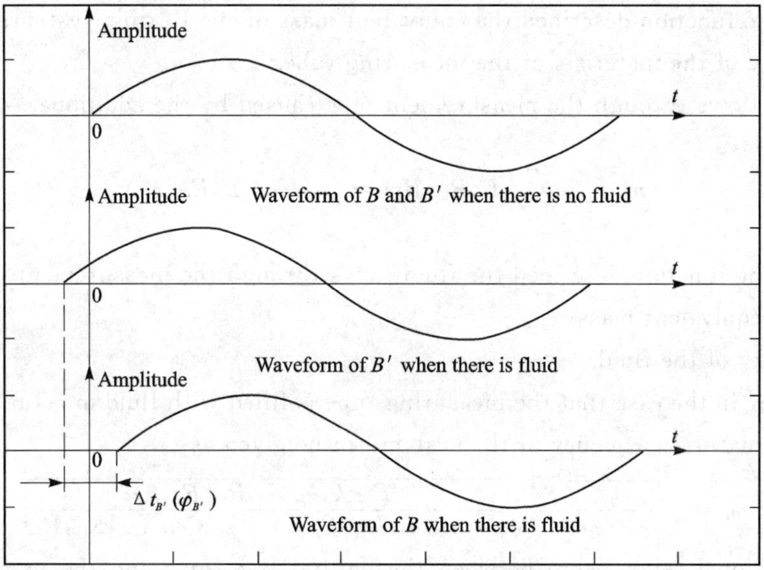

Fig. 4.5 Signal sketch of *B* and *B′*

4.3.2 Density Measurement

Based on the structure and working principle of the resonant direct mass flow sensor as shown in Fig. 4.3, equivalent stiffness of the flexible system of measuring tube can be written as

$$k = k(E, \mu, L, R_C, R_f, h) = E \cdot k_0(\mu, L, R_C, R_f, h) \tag{4.17}$$

where

$k(\cdot)$ is function described the equivalent stiffness of the flexible system;

E is modulus of elasticity of the tube;

μ is Poisson's ratio;

R_C is the axis radius of the arc portion of the U-shape measuring tube;

L is length of the straight section of the U-shape measuring tube at working part;

R_f is inner radius of the measuring tube;

h is the wall thickness of the measuring tube.

Equivalent mass of the flexible system of the measuring tube can be described as

$$m = m(\rho, L, R_C, R_f, h) = \rho \cdot m_0(L, R_C, R_f, h) \tag{4.18}$$

where

$m(\cdot)$ is the function described the equivalent mass of the flexible system;

ρ is density of the materials of the measuring tube.

The fluid flows through the measurement pipe caused by the additional equivalent mass can be written as

$$m_f = m_f(\rho_f, L, R_C, R_f) = \rho_f \cdot m_{f0}(L, R_C, R_f) \tag{4.19}$$

where

$m_f(\cdot)$ is the function described the fluid flows through the measurement pipe caused by the additional equivalent mass;

ρ_f is density of the fluid.

Therefore, in the case that the measuring tube is filled with fluid in actual measurement condition, the natural frequency of the system can be given as

$$\omega_f = \sqrt{\frac{k}{m + m_f}} = \sqrt{\frac{E \cdot k_0(\mu, L, R_C, R_f, h)}{\rho \cdot m_0(L, R_C, R_f, h) + \rho_f \cdot m_{f0}(L, R_C, R_f)}} \tag{4.20}$$

The functional relationship between the natural frequency and the parameters of the measuring tube of the structure and materials is given in the Eq. (4.20), in which the mechanism that the resonant direct mass flow sensor can simultaneously measure the density of the fluid is demonstrated.

According to the Eq. (4.20), when there is no fluid in the measuring tube, the following relationship can be described as

$$\omega_0^2 = \frac{k}{m} \tag{4.21}$$

Moreover, when the measuring tube is filled with the fluid, the relationship below also can be written as

$$\omega_f^2 = \frac{k}{m + m_f} \tag{4.22}$$

In conclusion, according to the Eq. (4.17) to Eq. (4.22), we have

$$\rho_f = K_D\left(\frac{\omega_0^2}{\omega_f^2} - 1\right) \tag{4.23}$$

$$K_D = \frac{\rho \cdot m_0(L, R_C, R_f, h)}{m_{f0}(L, R_C, R_f)}$$

where K_D is the coefficient relative to the material and geometric parameters of the measuring tube.

4.3.3 The Measurement of the Two-component Fluid

Generally, when the measured fluid is a mixture of two immiscible (oil and water), the respective mass flow rate and volume flow rate of the two-component fluid can be well measured.

Based on the volume conservation and mass conservation relation, considering the situation that the fluid is filled with the measuring tube.

$$V = V_1 + V_2 \tag{4.24}$$

$$V\rho_f = V_1\rho_1 + V_2\rho_2 \tag{4.25}$$

where

V_1 and V_2 are respectively the volumes of the fluid of which the densities are ρ_1 and ρ_2 ;

ρ_1 and ρ_2 as the set values are respectively the densities of component 1 and component 2;

ρ_f is density of the mixed liquid in actual measurement.

According to Eqs. (4.24) and (4.25), the respective proportion of component 1 in the density ρ_1 and component 2 in the density ρ_2 in the total volume of fluid can be calculated as

$$R_{V1} = \frac{V_1}{V} = \frac{\rho_f - \rho_2}{\rho_1 - \rho_2} \tag{4.26}$$

$$R_{V2} = \frac{V_2}{V} = \frac{\rho_f - \rho_1}{\rho_2 - \rho_1} \tag{4.27}$$

The respective proportion of component 1 and component 2 in the total mass can be described as

$$R_{m1} = \frac{V_1\rho_1}{V\rho_f} = \frac{\rho_f - \rho_2}{\rho_1 - \rho_2} \cdot \frac{\rho_1}{\rho_f} \tag{4.28}$$

$$R_{m2} = \frac{V_2\rho_2}{V\rho_f} = \frac{\rho_f - \rho_1}{\rho_2 - \rho_1} \cdot \frac{\rho_2}{\rho_f} \tag{4.29}$$

By Eqs. (4.28) and (4.29), the mass flow of the component 1 and component 2 can be respectively described as

$$Q_{m1} = \frac{\rho_f - \rho_2}{\rho_1 - \rho_2} \cdot \frac{\rho_1}{\rho_f} \cdot Q_m \tag{4.30}$$

$$Q_{m2} = \frac{\rho_f - \rho_1}{\rho_2 - \rho_1} \cdot \frac{\rho_2}{\rho_f} \cdot Q_m \tag{4.31}$$

where Q_m is the mass flow of the two-component fluid measured by the mass flow sensor in actual measurement.

The volume flow of the component 1 and component 2 can be respectively given as

$$Q_{V1} = \frac{\rho_f - \rho_2}{\rho_1 - \rho_2} \cdot \frac{1}{\rho_f} \cdot Q_m \tag{4.32}$$

$$Q_{V2} = \frac{\rho_f - \rho_1}{\rho_2 - \rho_1} \cdot \frac{1}{\rho_f} \cdot Q_m \tag{4.33}$$

The respective mass and volume of the two-component fluid flowing through the mass flowmeter in a certain period can be calculated by using the Eq. (4.30) to Eq. (4.33).

In some industrial production, although the measured two-component fluid does not react chemically, the physical miscible phenomenon occurs. This means the sum volume of the two components is greater than the volume of the mixture. At this time the above model is no longer valid, but the targeted engineering approach can be given through engineering practice.

4.4 Signal Detection Scheme and Its Realization

4.4.1 Phase Difference Detection of the Mass Flow

Based on the signals S_B and $S_{B'}$ measured by B and B', the mass flow Q_m can be directly calculated by detecting the phase difference $\phi_{BB'}$ between them or the amplitude ratio R_a of the "difference" signal and the "sum" signal. As for the density of the fluid, it can be obtained by the frequency of the detection signal S_B and $S_{B'}$. The following discussion is about the specific detection methods.

Analog detection principle is usually adopted on the measurement of the phase difference, which means analog comparators are used for the zero-crossing testing to achieve the phase difference detection. Taking into account the variety of vibration and electromagnetic interference in the actual use, various types of noise exist in the input signal of the detection circuit. These noise components will change the zero crossing position of the sine wave, and thereby affect the accuracy of phase difference detection. It is necessary to use the analog filter to filter out the noise. As the order of analog filter order is limited, the problems that eliminating the noise close to the useful signal frequency is difficult and there are two inconsistent filter characteristics as well as component parameters drift could cause the detecting error. Digital signal processing method can be effective to avoid the element parameter drift and other problems, by which the more effective noise suppression method becomes possible. Currently, based on the digital signal processing technology, there are two phase difference detection methods: one is using FFT in the frequency domain, and the other one is using the method of cross-correlation. As both algorithms require the entire cycle of sampling and the signal cycle of the measuring system is not fixed, a more complex set of measuring circuit which has the intricate calculation is needed to ensure that the integer multiple relationship between the sampling period and the signal period.

Based on the advanced digital processing technology to achieve the zero crossing digital phase difference detection principle, the above problems can be solved. DSP can be used in real time time-domain analysis of the signal waveform, and then the time difference and the phase difference of the zero-crossing point of the two signals can be calculated.

The phase difference of the two signals in the digital zero-crossing detection principle is shown in Fig. 4.6. After the sensing signals at the points B and B' are sampled by AD converter, a series of data points are obtained. The intersection point as the curve zero crossing point of the fitting curve and the abscissa axis can be obtained by fitting the data curve in the vicinity of zero. Therefore, the time difference of the curve zero crossing points of the two signals can also be gained, and then the phase difference of the signal can be calculated.

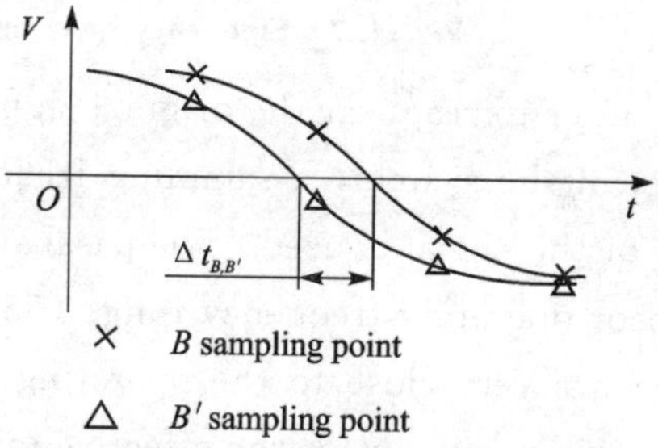

Fig. 4.6 Measuring principle of the digital zero crossing point for phase difference of two signal channels

A specific example is described as below.

The frequency range of a Coriolis mass flow sensor is between 65 Hz and 110 Hz, and the phase difference range is between 0.09° and 1.8°. The results of the spectral analysis of the output signal in a fluid mining is shown in Fig. 4.7. It is concluded from the sketch, that besides the working frequency $f_0 = 77.32$ Hz, it is not difficult to find there are the operating frequencies of $2f_0$, $3f_0$ and 50 Hz in addition. That is caused by the non-linearity of the sensor itself and is relative to the structural parameters of the sensor and the working condition. The working signal at 50 Hz is created by the power supply system of the surrounding environment of the sensor. These interference signals have a greater impact on the accuracy of the phase difference calculation. Therefore, the signal should be filtered out to improve the SNR before the phase difference calculation. Because of the shortcomings of the analog filter mentioned above, the digital band-pass filtering can be used in the DSP programs.

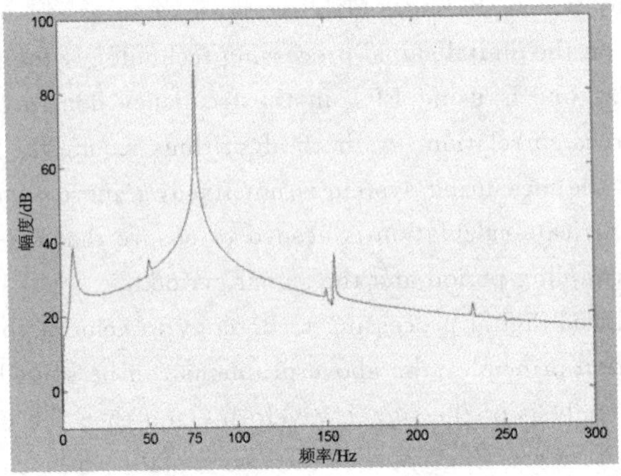

Fig. 4.7 Frequency spectrogram sketch before the solution of the data

In order to reappear the original signal better and improve the phase difference detection accuracy of the system, a sampling frequency $f_{\text{Sample}} = 19.2$ kHz much higher than the frequency of the signal is used. The passband of the selected digital filter is slightly larger than the sensor operating frequency range (55~120 Hz). As some of the interference signal frequencies are very close to the operating frequency of the sensor, the filter transition band must be steep enough for the effective inhibition of these interfering signals. To achieve this passband characteristics, 3 000-orders FIR filter is initially selected for band-pass filtering.

The conventional 3 000-order digital filter has a large amount of calculation and it is difficult to achieve in the practical application. The modified band-pass filter of finite impulse response is designed to achieve real-time filtering processing by existing mature digital filter analysis and computer simulation. The band-pass filter structure is shown in Fig. 4.8.

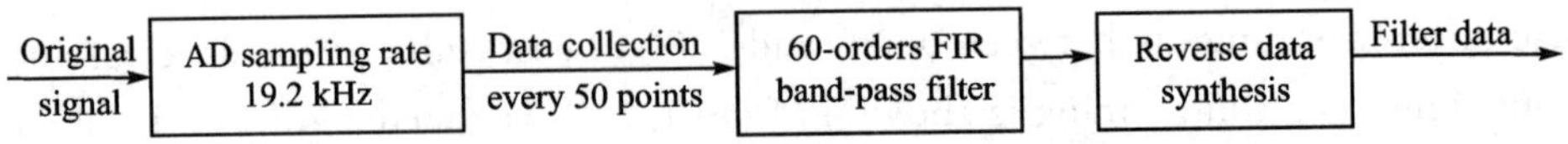

Fig. 4.8 Structure of the improved filtering arithmetic

50 subsequences can be obtained by the anthropogenic sub sampling of AD data. Each of the data subsequence is equivalent to the original signal after the sample in the sampling frequency 19 200 Hz/50=384 Hz. A standard 60-order FIR band-pass filter ($W_n=[W_1, W_2]=[0.1432, 0.3125]$) is used to filter each data subsequence and have the reverse composition of the 50 sets of data of the output of the filter to give the final filtered result. In every filter operation, the FIR filtering processes of the 50 sets of data are not at the same time. But the respective data subsequence 61 times multiplications and 60 times adder are needed only for the current sampling.

The FIR filter of this improvement retains the advantage linear phase shift of a conventional FIR filter. Meanwhile, in this real-time signal processing system at the same time, at every sampling time intervals, the filter calculation requires only 61 multiplications and 60 times the adder, while achieving the 3 000-orders FIR filter with the same filtering effect is required 3 001 multiplications arithmetic and 3 000 times adder. Obviously, the calculation is greatly reduced. Using the above band-pass method on DSP TMS320VC33, the results of the original data after filtering collected from the scene can be analyzed by Mat lab as shown in Fig. 4.9. The horizontal axis represents frequency, and the vertical axis is amplitude.

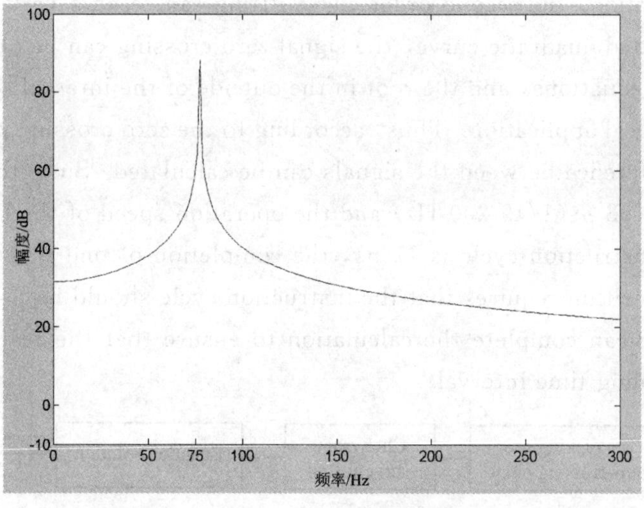

Fig. 4.9 Frequency spectrogram sketch of the filtered data

According to the comparison of Figs. 4.9 and 4.7, the filtering effect is quite obvious. This algorithm is very effective in inhibiting the signal interference and improves the signal-to-noise ratio, which provides a guarantee to the following extraction of the phase difference signal. The characteristics of the FIR filter shows that it also satisfies the characteristics of the linear phase shift. As for the mass flowmeter, due to the change of the fluid density, the sensor resonant frequency also changes. Therefore, the sampled value at different times represents the different frequency information. The characteristics of the digital filter is to use the front of the data of N points for filtering and the traditional nonlinear phase shift filter results in the presence of a calculation error. As long as the two signals of the flowmeter pass the FIR filter in the same coefficient, the phase shift can be linear. For this reason, the effect of the traditional filter on the phase difference of the two signals can be overcome.

The modified FIR band-pass filter improves the signal-to-noise ratio, and the phase shifts of the two signals are the same. Therefore, the accuracy of the above phase difference detection algorithm can be ensured. In order to meet the real-time nature of the system, the system must complete two-way data filtering, curve fitting and zero phase difference and frequency calculations in two sampling interval. The structure of the zero-crossing detection algorithm is shown in Fig. 4.10. The filtered data can be detected in real time by the software. If $x(n) > 0$ and $x(n+1) < 0$ or $x(n) < 0$ and $x(n+1) > 0$ occurs, the zero-crossing is regarded between $x(n)$ and $x(n+1)$. Therefore, the front and the back of five points are stored into the storage unit specified for providing the raw data to Chebyshev curve fitting. Through the simulation, the second order curve fitting can reach a very high computational accuracy. In the fitted quadratic curve, the signal zero crossing can be calculated by the traditional solution of equations, and the root in the outside of the interval $[x(n), x(n+1)]$ is discard in the practical application. Thus, according to the zero crossing point of the two signals, the phase difference between the signals can be calculated. Since the sampling interval of the system is 52.08 μs(1/19 200 Hz) and the operation speed of the DSP (TMS320VC33 example) of each instruction cycle is 17 ns, the completion of one time sampling, filtering and retardation algorithm requires that the instruction cycle should be 17 ns×2 000=34 μs. Consequently, DSP can complete the calculation to ensure that the real-time nature of the system in the sampling time interval.

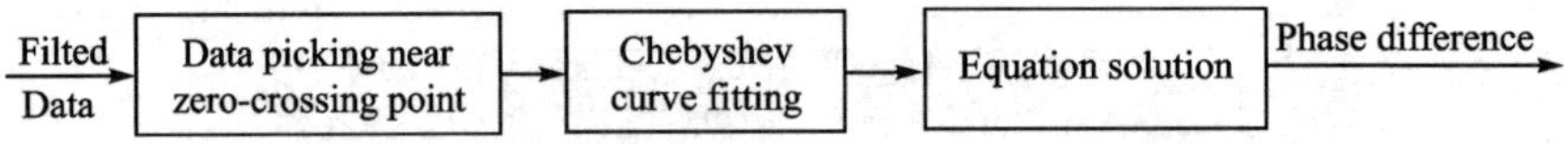

Fig. 4.10 Arithmetic structure sketch of the phase difference

4.4.2 The Amplitude Ratio Detection of the Mass Flow

The principle circuit of the amplitude detection is shown in Fig. 4.11. S_B and $S_{B'}$ are the two output signals of the mass flow sensor. According to the amplitude detection sensor by MCU, the amplitude ratio can be calculated and then the mass flow can be obtained.

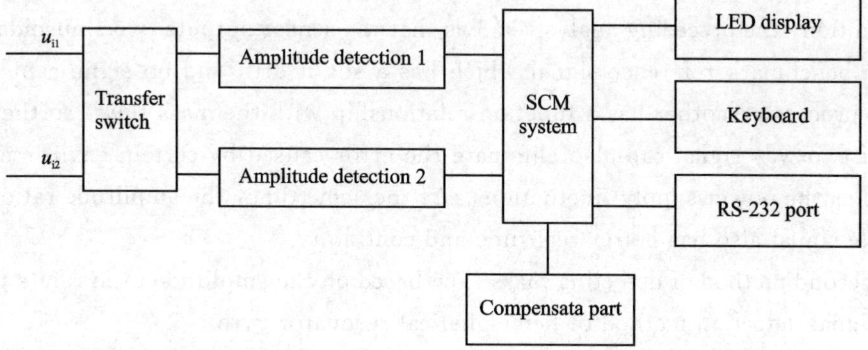

Fig. 4.11 A kind of theory circuits to measure the amplitude rate

The amplitude detection principle circuit of the periodic signal detection circuit is given in Fig. 4.12. Through the characteristics that the wizard passes and the reverse cuts off of diode, the AC signal can be rectified, And the maintenance characteristics of the capacitor can be used to obtain the signal amplitude.

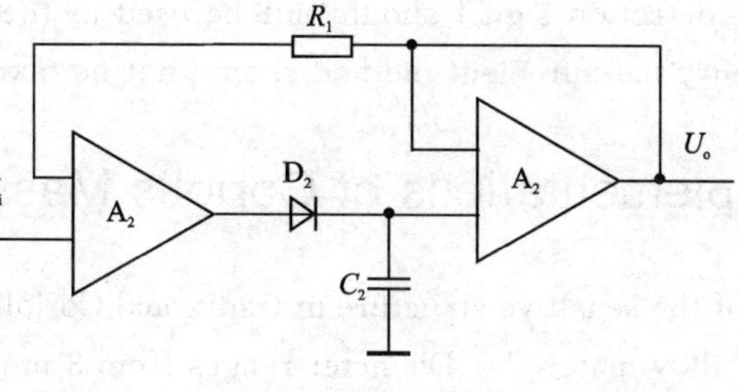

Fig. 4.12 A kind of amplitude measuring circuit for periodical signal

According to the circuit shown in Fig. 4.11, the better the symmetry of the two amplitude detection member is, the higher the accuracy of the system has. As the device may generate the asymmetric, the following steps should be carried out in the amplitude measurement and the amplitude ratio measurement process.

① The amplitude detection 1 is used to detect the amplitude of the input signal S_B which is recorded as A_{11}, and the amplitude detection 2 is used to detect the amplitude of the input signal $S_{B'}$ which is recorded as A_{22}.

② The amplitude detection 2 is used to detect the amplitude of the input signal S_B which is recorded as A_{12}, and the amplitude detection 1 is used to detect the amplitude of the input signal $S_{B'}$ which is recorded as A_{21}.

③ $B_1 = A_{11} + A_{12}$, $B_2 = A_{21} + A_{22}$. $C = B_1/B_2$ is used for detecting the amplitude ratio of the input signal.

In addition, the preceding analysis shows that the sensor outputs two sinusoidal signals. One is the benchmark reference signal which has a slight drift and no significant change in the whole process, Another has a function relationship with the mass flow, so the ratio calculation of two-way signal can also eliminate the error caused by certain environmental factors, such as the power supply fluctuations. At the same time, the amplitude ratio detection of the cycle signal also has better real-time and continuity.

The second method of detecting mass flow based on the amplitude ratio refers to the section 3.5 signal detection method of hemispherical resonator gyro.

4.4.3 Density Detection

Based on the working principle of Coriolis mass flowmeter, the measurement of the fluid density can be implemented through the sensitive structure of the main vibration frequency. For the conventional Coriolis mass flowmeter, taking into account the variation range of the output frequency of the detection signal should not be used in the audio 80 Hz to 1 kHz, which means the frequency measurement method should not be taken. See Fig. 1.7.

4.5 The Implementations of Coriolis Mass Flow Sensor

The flexible bend of the sensitive structure in traditional Coriolis mass flowmeter is constructed from precision alloy materials. Diameter ranges from 3 mm to 250 mm.

The elastic elbow structure in Coriolis mass flowmeter applications has many forms, they first bending vibration induced by the Coriolis effect is slightly different with the second bending vibration form, so the corresponding natural frequency also has the difference that reflects in the Coriolis mass flowmeter work mode, sensitivity, linearity, scope of work, the dynamic response characteristics, and application characteristics. In fact, the majority of the elastic bend structure of Coriolis mass flowmeter can be seen as an appropriate combination of straight pipe segment of the arc-shape bend, and any arbitrary structure can be approximated as a combination of certain straight and arc-type unit. Fig. 4.13 shows the structure of the basic unit established by the finite element model specification. Fig. 4.13(a) is a straight-shape unit and (b) is an arc-shape unit. Fig. 4.14 shows four kinds constantly used elastic elbow structures in mass flow sensor. Fig. 4.14(a) is a straight tube, (b) is the U-type tube, (c) is double loop tube, and (d) is the S-shape tube. It is real example of engineering application.

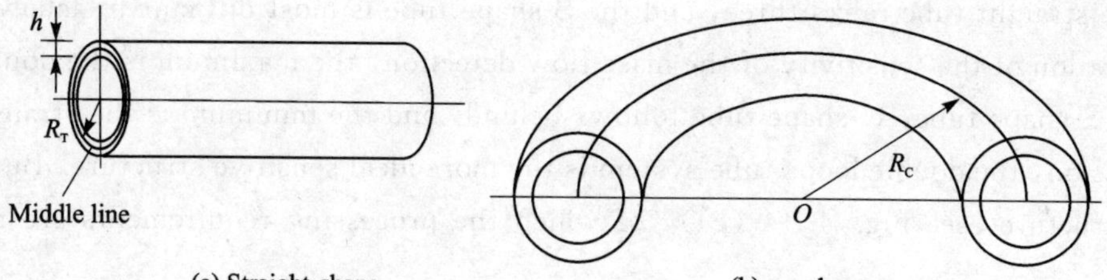

(a) Straight-shape　　(b) arc-shape

Fig. 4.13　Basic units for elastic bending tube: straight and arc

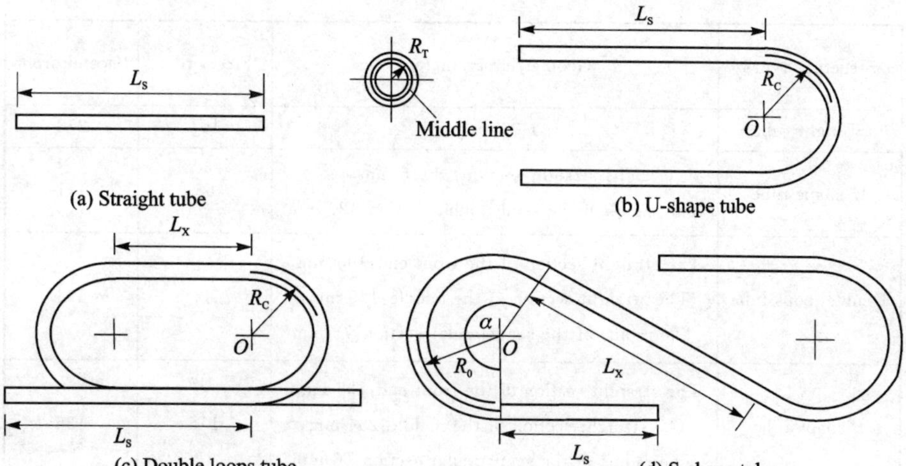

(a) Straight tube　　(b) U-shape tube　　(c) Double loops tube　　(d) S-shape tube

Fig. 4.14　Typical sensitive structure for four kinds of elastic bending tube

To facilitate the analysis and comparison, long flexible elbow structure is selected for calculation. Wherein the modulus of elasticity, the density, and Poisson's ratio of the tube are respectively 194.5 GPa, 785 050 kg/m^3 and 0.32. The average radius of the cross section of the tube is 8 mm, and the thickness of the wall is 2 mm. The tube has a total length of 1 000 mm. The above other four geometrical parameters of the structure and the corresponding natural frequency of the corresponding order bending vibration are given in Tab. 4.1.

According to the simulation results: for the four compared flexible elbow sensitive structure, the order of the main first order bending vibration stiffness from small to large is U-shape tube, double loops tube, straight tube, and the S-shape tube. The ascending order of the second order bending sub-vibration stiffness is double loops pipe, S-shape tube, U-shape tube and the straight tube. Therefore, the U-shape tube is most likely to achieve the self-excited system of closed-loop of the mass flow sensor, followed by the double loops tube, the straight tube ranks three, and the S-shape tube is most difficult to achieve. From the realization of the sensitivity of the mass flow detection, the maximum is the double loops tube and S-shape tube, U-shape tube follows behind, and the minimum is the straight tube. In conclusion, the double loops tube system is the more ideal sensitive structure. But it is the spatial structure (see Fig. 4. 14 (c)), of which the processing requirements are relatively high.

Tab. 4. 1 Several typical structures of the geometric parameters and the corresponding natural frequency (in Hz) of the first and second order bending vibration

Structure types	Geometric parameters	First order	Second order
Straight tube	1 000 mm	101	278
U-shape tube	The straight section: 311. 5 mm, The radius of the semicircular section: 120 mm	68	171
Double loops tube	The straight section of the front end: 190 mm, The straight section of the middle: 130 mm, The radius of the semicircular section: 78 mm	82	134
S-shape tube	The straight section of the front end: 135 mm, The straight section of the middle: 234 mm, The radius of the semicircular section: 68 mm	115	135

The simulation results show that: based on the physical mechanisms of the mass flow sensor, the ideal type is the double semicircle with the straight segment on the choice of the sensitive structure of the sensor. If the main first order bending vibration stiffness of the elastic elbow structure need to be improved, the geometry of the center point "antisymmetric" type can be selected, such as the S-shape tube. Conversely, the "symmetrical" type can be selected, such as a U-shape tube or a double loops tube. The second order bending stiffness of the sub-vibration is just opposite to the above conclusion.

In recent years, with the development of MEMS technology, scientists and engineers have developed the micro Coriolis mass flowmeter on silicon chip, of which the pipe cross-sectional area of the flow is about $(10 \times 100)\ \mu m^2$. A micro-mechanical mass flow sensor based on Coriolis effects is shown in Fig. 4.15. Where in, (a) is for the sensor dimensional view; (b) is the cross-sectional view for the sensor. Its working principle is the same as Coriolis mass flow sensor described in Figs. 4.8~4.12. It can achieve a direct measurement of the mass flow and measure the density of the fluid as well. Silicon micro-machined mass flow sensor not only has the advantage of small size, but also has some unique advantages that low cost, fast response and high resolution as well.

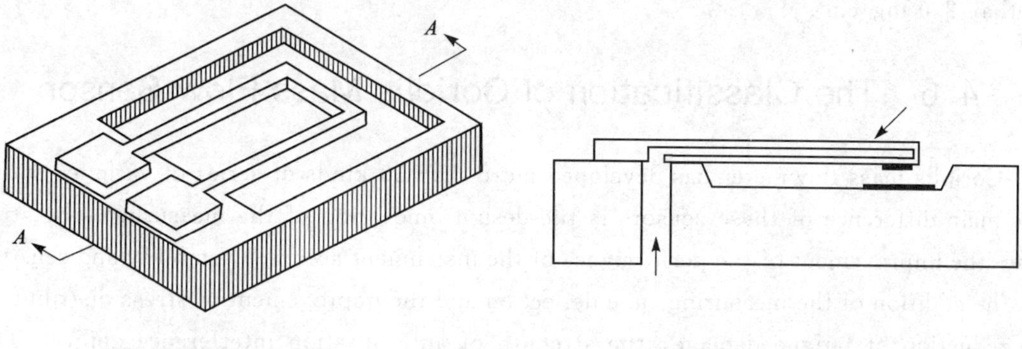

(a)Three dimensional sketch (b) Cross section sketch

Fig. 4.15 MEMS Coriolis mass flow sensor

Shown in Fig. 4.15, the basic structure of the micro-mechanical mass flow sensor comprises a micro-tube and a glass base. The root of the microtubules and the glass base combine together with a wafer encapsulated in vacuum. The U-shape micro-pipe is formed on a silicon substrate through the deep boron diffusion. Microtubule vibration is detected by capacitive, which is simple, practical and highly accurate.

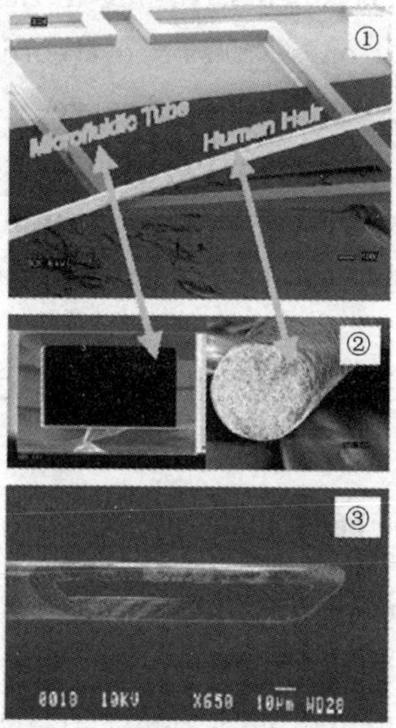

Note: ① is sketch of the overall structure. ② and ③ are cross section sketch of the micro measuring pipe.

Fig. 4.16 MEMS Coriolis mass flow sensor

Shown in Fig. 4.16, ① and ② are respectively the developed overall structural view and cross sectional view of the micro mechanical mass flow sensor. The cross section of micro-tubes can be made of a rectangular or trapezoidal shape, which is parallel to the silicon substrate. Micro-pipe can be easily achieved in different shapes and parameters. For example, the cross-sectional view of the micro-tube can be made into the size of one hair (100 μm in diameter), which can also be made into the 1/10 size of the cross section of one hair. Details are shown in ① and ② of Fig. 4.16.

For a specific micro mechanical mass flow sensor prototype, the measured vibration frequency of the micro tube is approximately 16 kHz and the mechanical quality factor can reach 1 000. It has an excellent mass flow resolution of 2 μg/s and the fluid density resolution better than 2.0 mg/cm^3.

4.6 The Classification of Coriolis Mass Flow Sensor

Coriolis mass flowmeter has developed more than 30 kinds of series of varieties so far. The main difference of these sensors is the design innovation of the measuring tube structure; the improvement of the performance of the instrument accuracy, stability and sensitivity; the addition of the measuring tube deflection and the improvement of stress distribution; the reduction of fatigue damage; the strength of anti-vibration interference ability. Thus there are a variety of shapes and structures of the measuring tube, shown in Fig. 4.17. For this reason, the classification and discussion are only from different angles.

CMF classified according to the shape of measuring tube, can be divided into the curved and straight ones; classified according to the number of pipe sections, can be divided into single-tube and double-tube type; classified according to the connection of the measuring tube with the double tube mode, can be divided into the parallel ones and the serial ones; classified according to the layouts of the flowing direction of the measuring tube and the process pipe, can be divided into parallel and vertical way.

4.6.1 The Classification According to the Shape of the Measuring Tube

1. The bend mode

The measuring tube instrument put on the market firstly were bent in a U-shape tube, and now the developed curved shapes are the Ω-shape, the B-shape, S-shape, circular ones and oval annular ones. The series of curved measuring tube instrument are more than the series of straight measuring tube instrument. The design of curved shape is to reduce the rigidity. Compared with the straight ones, thicker wall can be adopted and the straight instrument performance is less affected by the abrasive corrosion. But the additional error can be caused by easy accumulation of gas and debris. In addition, the overall weight and size of CMF with the curved tube are bigger than the straight ones.

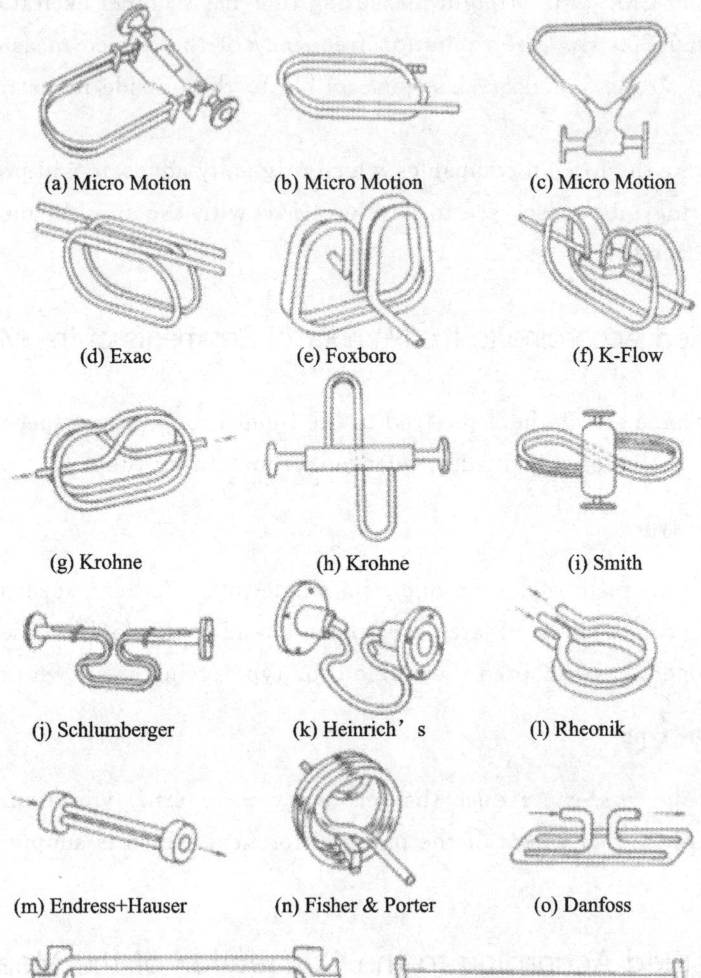

(a) Micro Motion (b) Micro Motion (c) Micro Motion
(d) Exac (e) Foxboro (f) K-Flow
(g) Krohne (h) Krohne (i) Smith
(j) Schlumberger (k) Heinrich's (l) Rheonik
(m) Endress+Hauser (n) Fisher & Porter (o) Danfoss
(p) Bailey (q) Schlumberger

Fig. 4.17 The measuring tubes with different kinds of shape and structure

2. The straight mode

CMF with straight measuring tube can hardly accumulate gas. In the vertical mounting measurement slurry, solid particles is not easily deposited within the measuring tube in the suspend operation. The flow sensor has small size and light weight. Bur the rigid is large and the tube wall is relatively thin, which means that the measured value by the abrasive corrosion have a great impact on the sensor.

Some models of CMF with straight measuring tube has a higher excitation frequency between 600 Hz and 1 200 Hz (the excitation frequency of the curved measuring tube is between 40 Hz and 150 Hz), and are less susceptible to the outside industrial vibration frequency interference.

In recent years, the foreign companies which originally advocate and product CMF with the curved measuring tube, also race to develop CMF with the straight measuring tube. It has an increasing trend.

4.6.2 Classified According to the Number of Segments of the Measuring Tube

Measurement pipe section here referred to the fluid through the respective vibration and detection of the Coriolis force is divided independent measuring tube.

1. Single-tube type

The initial development of the product is a single-tube. As it is susceptible to external vibration interference, the most of late developments tend to the double-tube type. But in recent years, developers have adopted the single-tube type again, as shown in Fig. 4. 17 (q).

2. Double-tube type

The double-tube type can reduce the sensitivity to external vibration disturbance and easily implement the measurement of the phase difference, which is adopted by the most of the instruments.

4.6.3 Classified According to the Connection of the Measuring Tube with the Double Tube Mode

1. Parallel mode

In Fig. 4. 17, according to (a), (d), (f), (h), (i), (j), (k), (l), (m), (o) and (p) shown below, the fluid that flows across the upstream pipeline of the sensor shunts into the two parallel measuring tube segment, and then flows across the same shaped shunt current collector into the downstream pipeline. Parallel form is usually for the use of more model instrument. Shunt requires the same amount allocation as much as possible. But, due to the deposition of adhesion body or abrasion, the original flow condition would be changed, which causes zero drift and produces additional errors.

2. Series mode

In Fig. 4.17, according to (b), (e), (g) and (n) shown below, the fluid flowing through the first measuring tube segment can flow across the deflector block into the second measuring tube segment. The fluid that flows through the two measurement pipe section has the same amount and does not produce the disadvantages caused by the variation of the shunt values. It is applicable to the two-shear sensitive fluid.

4.6.4 Classified According to the Layout of the Flowing Direction of the Measuring Tube

1. Parallel way

The arrangement is that the measuring tube should parallel to the flow direction of the fluid and the flow direction of the process pipeline. Models are more used in this manner, as shown in (b), (d), (f), (g), (j), (k), (l), (m), (o), (p) and (q) of Fig. 4.17.

2. Vertical way

The measuring tube is arranged perpendicularly to the process pipe. The whole flow sensor is not in the process of piping vibration interference, and has the anti-interfere ability for the pipe vibration, as shown in (a), (e), (h), (i) and (n) of Fig. 4.17.

4.7 Main Application Fields and its Characteristics

4.7.1 Application Overview

CMFs are mainly used to measure the mass flow of fluid, secondly the density. Meanwhile, it can also measure the temperature of the fluid. By the measured mass flow and the fluid density, the ratio of the components can be derived, including mass flow and volume flow. What CMF applies most is total measurement or measuring/controlling flow with the purpose of test the quality of fluid, specifically the trade settlement handover metering or corporate internal accounting measurement, partial measurement of supplement of batch process (in place of the previous weighing measurement consuming much time and labor), the configuration of blending ratio.

However, problems of CMF such as zero drift limit the application in foreign trading measurement. For example, in the 1990s API claimed that CMF applications in petroleum industrial realm was still immature. On the annual meeting of ISO/TC 28/SC2, as "the CMF technology of transportation in sealed tunnel in petroleum tube industry remained immature", WG6 was cancelled particularly who was in charge of drafting international standard of CMF. Considering the performance of CMF improves, application spreads in trading measurement in other realms and the current situation continues changing.

Density is the second parameter in CMF measuring. It is used to control some quality indicators during manufacturing process, such as preventing intentional dilution when trading, solving the solute concentration in the solution to measure the solute flow or amount in the solution like the amount of oil of oil and water mixed liquid pouring out of the well, telling the components of flowing liquid and transporting separately like differentiate the finished liquid and cleaning liquid, flowing alternately and sent into different tunnel downstream.

In the mid-1990s, CMF spreads its application in measuring the viscosity of the liquid. Through the function relationship between the CMF pressure descending and viscosity it can be used to realize online measurement with the help of differential pressure transmitter.

CMF can be applied in a large range of viscosity of measured liquid, from low viscosity liquefied petroleum gas to high viscosity Crude oil and bitumen liquid. According to one foreign instrument factory's statistics analysed in the 1990s, CMF occupied over 50% sales volume in medium and high viscosity liquid, and 10% with a viscosity of over 400 Pa • s. Besides, CMF can also be used in flow measuring of non-Newton fluid and Liquid solid two phase fluid, like Liquid latex, suspension of kaolinite, chocolate, meat paste.

Early CMF only used in liquid and later in high pressure gas. Not until the early 1990s, instrument for low pressure gas appeared. As Micro Motion Corporation claims, until 1997, there are 7 500 CMFs applied in gas measurement, among which there are around 6 000 serving CNG gas station.

In the recent years, with the development of MEMS technology, silicon micromechanical CMF's success spread into other application realm, especially in realms in need of micro flow measurements such as semiconductor industry, pharmacy and biology.

The user distribution is as follows: according to a foreign instrument manufacturer statistic analysis in the 1990s, CMF occupied 40% in chemistry industry, 20% in petroleum industry (including refining, storage and transport), 23% in food industry and 17% in others, of which food industry has a high percentage. Currently, compared with the less use in food industry, CMF is widely used in oil and petrochemical field.

4.7.2 Advantages

CMF is a typical resonant sensor, therefore possessing the general advantages resonant sensor has. Meanwhile, based on the working principle of CMF, as well as the sensing structure and the characteristic of the whole structure, this flow meter has its unique advantages for its own:

① CMF is able to measure mass flow and is under little influence of factors of the fluid, such as viscosity, density and pressure. It is at present the most precise sensor to directly access the fluid mass flow.

② Versatility. It has the capability to synchronously measure the density of the fluid (thus solve the volume flow), and solve the ratio (including volume flow and mass flow as well as their cumulative amount) between two component fluid (such as oil and water, two immiscible liquid). While on same level, this function can be spread to the measurement of two component liquid when the components are partially immiscible.

③ Signal processing. The solving of mass flow and density is directly about complete digital and periodic signal, with compatibility with computer to form a computer measuring and controlling system. The signal itself is easy for long distance transportation and solution of transient mass flow (in kg/s) and accumulative mass (in kg). At the same time, it can solve the volume flow (in m^3/s) and its cumulative amount (in m^3).

④ A wide range of measuring objects including high viscosity liquid, slurry containing solid, liquid containing small amount of gas, or medium or high pressure gas with sufficient density.

⑤ There are no obstructing and moving parts in measuring tunnel, the vibrating amplitude is small to be regarded as an inactive part.

⑥ Insensitivity towards the velocity distribution of upcoming current, therefore there is no particular requirement for upstream and downstream straight tube.

⑦ Stable performance, high precision and being more real-time.

4.7.3 Disadvantages

① The instability of CMF zero point may cause zero drift, affecting the further improvement of its precision. Therefore many patterns of instrument have to divide the total error into two parts, the basic error and instability of zero point.

② CMF cannot be used to measure low density media such as low pressure gas. The measurement will be obviously affected if the gas content in liquid exceeds certain limitation (it differs from pattern to pattern).

③ CMF is sensitive to external vibrating interference. In order to prevent the tunnel influence, most patterns of CMF have high requirement of installation.

④ Incompatibility of large tunnel with a current limitation of under 250 mm.

⑤ The erosion and deposit on the inner side of the tunnel will affect the measurement, which is more severe for CMF with thin wall measuring tunnel.

⑥ High pressure loss. It is nearly the same compared to volumetric instruments while some particular patterns of CMF are 100% bigger in pressure loss.

⑦ Most types of CMF have large volume and mass.

⑧ High price. Foreign instrument's price is 5 000 to 10 000 RMB a set, about 2 to 5 times of electromagnetic flow meter with same diameter. In the domestic, it is about 2 to 8 times compared to electromagnetic flow meter with same diameter.

4.8 Interference Factors and Their Restraining

4.8.1 Pressure Loss

The upper flow limitation of CMF, compared to other kinds of instrument such as volume type and turbine type, is up to 8～12 m/s using the water density to calculate the nominal diameter flow rate. Some patterns even have a rate of 15 m/s to 16 m/s. While the volume and turbine type has a velocity of only 3 m/s to 5 m/s. As for CMF, the velocity is even higher inside the measuring tube. Therefore the pressure loss in most pattern of CMF is enormous, with a loss of about 0.1 MPa to 0.2 MPa using in low viscosity fluids like water, which is in need of additional attention.

One of the main factors to choose the measurement of CMF according to the practical conditions is to value the instrument pressure loss (or called pressure descending) Δp, whether it is within the tube allowed values. Under the circumstances of allowed pressure descending, in order to obtain the best measuring precision, the available full-scale flow should be chosen as high as possible within the range of CMF. Usually the nominal diameter of CMF is smaller (or equal to) than the diameter of the tube, rarely bigger.

However, there are a few patterns of instrument having low pressure descending. For example, the upper flow limitation of nominal diameter of RHM series is only 2 m/s to 3 m/s, whose pressure descending is around 0.05 MPa.

The pressure of CMF increases alongside the viscosity. Fig. 4.18 shows the relation curve between the fluid flow rates with different viscosity versus pressure descending of D150 (with a diameter of 40 mm/50 mm). $\mu = 1$ mPa • s as under normal temperature and $\mu = 0.01$ mPa • s as the viscosity of most gas. From the diagram, it can be told that the pressure descending of the fluids with the viscosity of $\mu = 500$ mPa • s is ten times as of water. The high viscosity fluid flows in a laminar state within the instrument. The pressure descending Δp and flow q_m has a linear relation (namely $\Delta p = Kq_m^n$, where k is coefficient and the exponent n is equal to 1). When the viscosity is low, the fluids flow in a turbulent state, with a basically square relation (where the exponent n is equal to 2). As for fluids with medium viscosity, the relation curve is broken line. The small flow segment appears laminar flow while the medium and high flow segment locates in the transition zone from laminar flow to turbulent flow. The exponent is between 1 and 2.

The viscosity of the fluid is between the curves as shown in the diagram. According to related literature, proportion interpolation can be adopted to approximately calculated, which, however, can only be used in the high viscosity fluid layer, the laminar flow layer. As for medium and low viscosity, the viscosity-pressure loss has an exponential relationship, as are in turbulent area and transition area. Proportion interpolation is not applicable, and the value can only be given a rough estimate.

To replace other types of flow meter (such as turbine flowmeter) with CMF on the original pipeline, it is required in technology transformation, especially to account whether the lift of the power pump meets the need to overcome the pressure increased by CMF. Pump with bigger lift in need if necessary.

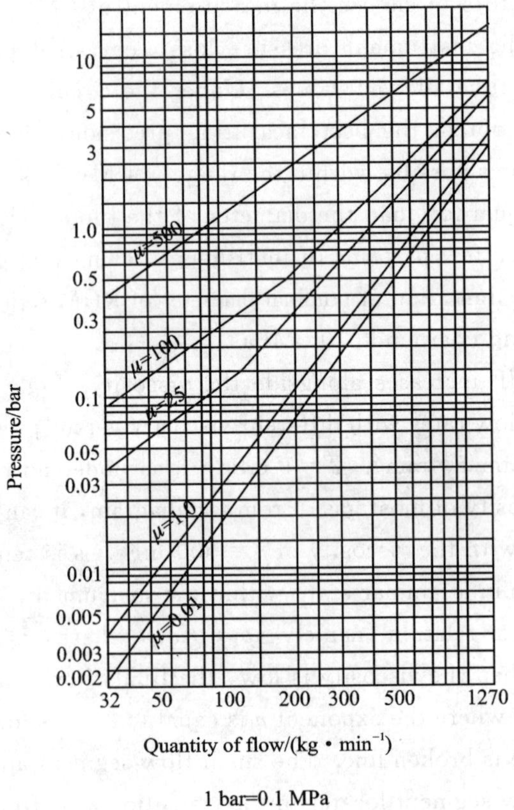

Fig. 4.18 Relationship between the mass flow and press for D150 type sensor (diameter: 40 mm/50 mm) under different viscosity

4.8.2 Measuring the Flow of Gas

Whether or not can the gas flow be measured depend on if the prescript mass flow rate value is achieved. Considering the low density of gas, value can only be achieved with a very high pressure and velocity. For example, when the flow of DS—100 (diameter 25 mm) of micro motion reaches the maximum of rated flow range, 455 kg/min, with an assumption that the gas density is 100 kg/m^2, the absolute pressure reaches 7.6 MPa and the velocity is up to 154.5 m/s. Even if the flow is only the rated minimum 68 kg/min, the velocity still reaches 23 m/s.

Some patterns of instrument have a regulation about the lower limitation of the gas density. For instance, for the TH series of Heinrich corporation, it is 2 kg/m^3; for 1.5E and 10E type of the MFS-3000 series of Krohne, it is 50 kg/m^3; it is 200 kg/m^3 for the CFS10=1/AS series of Foxboro. If to measure the air flow, the absolute pressure within the latter two types of CMF needs respectively 4.2 MPa and 16.8 MPa.

The same instrument has low performance when measuring gas compared to doing fluid. For example, the manufacturer claims that the Elite series has an error of ±0.5%R when measuring gas while the error when measuring fluid is (±0.1%+zero instability). However, in another essay of this factory, the experimental conclusion claims that all data has an error measurement of better that ±2%R. When reading the attached diagram of the essay, it can be told that when measuring low pressure gas, sometimes the error approximates or even exceeds 1%. It is because when the flow is only a few percent of the rated flow, namely low flow, the error is mainly affected by the zero instability. The repeatability is bad as well when the pressure in low, reaching 0.6%.

When measuring low pressure gas, what needs to pay attention is that the measurable flow deceases enormously. As for the 100 type CMF of the Elite series (diameter 25 mm/40 mm), when measuring the compressed air of a 0.17 MPa pressure, the maximum flow is only 4% rate flow.

Usually, CMFs designed to measure gas are not tested through gas but water to exam the instrument constant. The two are considered generally close to each other, while have their differences. Research indicates that when the fluid density varies in a large range from 1 000 kg/m^3 (water) to 2 kg/m^3 (0.17 MPa air), the instrument constant calibrated by the factory reaches high performance, like accuracy surpassing 2% and general error less that ±0.5%. The British Engineer Laboratory once conducted liquid-gas comparison experiment over 6 CMFs with diameter of 25 mm. Three of them worked among the low 1.5 MPa pressure air, two of which has nonlinearity 0.6% bigger than that in liquid, one has bad repeatability of 15% and a 10% departure in the instrument constant; while the other 3 worked under higher 6 MPa pressure, two of which has repeatability 0.3% bigger than that in liquid and nonlinearity respectively 1.5% and 3% bigger than in liquid, one is unable to work.

4.8.3 Liquid Containing Gas

Manufacturers usually claim that liquid containing a few percent of free gas has little influence on the measuring. It may be true when the gas bubble is small and distributes evenly, such as in ice cream or similar emulsion. However, the experiment the Italian Metrology

Institute conducted on 7 types of CMF about the influence of gas content indicates that 1% gas bubble (volume ratio) has no obvious influence over some patterns, some has error of about 1% to 2%, while one type of double-barreled straight tube has over 10% to 15% error. When there is 10% gas bubble, the error generally increases over to 15% to 20%, some pattern has an error even over 80%. The test from Danfoss corporation indicates as well that when the liquid contains 0.3% gas bubble, the instrument still remains its original accuracy; when the gas content reaches 5% (under normal pressure), the instrument error is over 10%. Additionally, other factors like the pressure of the fluid, velocity, viscosity and the way gas and liquid mix has different influence. Therefore, degassing is necessary before measuring fluid that may contains gas, especially when the manufacturer has no clear indication whether the instrument is capable to measure liquids containing gas.

4.8.4 Liquid Containing Solid

When measuring the flow of liquid containing little solid, different types of CMFs all share a high reliability. When the solid's content increases or the solid itself is strongly abrasive or soft (i. e. s the vegetable blocks in soup), the CMF should be considered according to the characteristics of fluid to choose the appropriate tubes.

When the liquid contains much solid or soft solid, it should be avoided to use the instrument with a much smaller measuring tube diameter than its nominal tube diameter, in case the tube is blocked. The better choice is single-tube type or the series connection of dual-tube type, because the parallel connection of dual-tube type has its problem about its current divider. If sundries are attached to it, it may change the amount of the sub-flow, causing error. What's worse is that it may not be discovered when one channel is blocked.

When measuring highly abrasive liquids, blocking problem continues. More over, current divider being corroded unevenly may also change the original dividing ratio. Therefore, parallel connection of dual-tube type is inappropriate. The better solution is the CMF with thick tube walls and single straight tube, since complex measuring tube may cause abrasion.

4.8.5 The Influence of the Parameters of Fluid Conditions or Physical Properties to the Fluid Measurement

Usually, instrument factory samples and manual file claim that the performance of Coriolis mass flowmeter may not be influenced by fluid temperature, static pressure, density or viscosity change. While, with the increasing application of experience, the fact is detected not entirely true. Fund raising commissioned third party are working on the study of amount of these factors. The factory also carried out a study of the application of technology. Some of the impact has reached the extent of the amount of correction.

If the accuracy of flow measurement of Coriolis mass flowmeter still remains the initial 0.5%R~1%R, the commonly used range of fluid conditions and the change of the physical properties can be neglected. However, the currently accuracy indicators are very high, it is up to 0.1%R~0.15%R. Then the amount of influence is even more prominent.

1. Temperature effect

Medium temperature or ambient temperature change will change the measurement of the elastic modulus of the vibrating tube and will affect the structure of the zero drift and other factors.

The instrument of most patterns has specific electric circuit to compensate the temperature coefficient of the elastic modulus to restrain its influence. The influence of zero drift is caused by the asymmetry of the vibrating tube's geometric shape and structure, is unrepeatable, therefore hard to be reduced or removed. The temperature coefficient of the elastic modulus, however, is a statistical quantity. Due to the inconsistencies of the measuring tube material lot and heat treatment process, the elastic modulus has a certain amplitude range.

In the 1990s, the British Engineering Laboratory has conducted practical flow experiments on the value of temperature influence of CMF manufactured in several factories in the foreign market. The water temperature ranged from 5℃ to 40℃. The instrument zero needs to be checked before every change on the water temperature. You can not adjust it within the temperature. In the 8 instruments, there are 3 instruments without impact, 1 instrument changes of 0.5%, 2 instruments change of 1% to 1.5%, another 2 change of 1.5% to 2%. Temperature affect on 5 instruments is ± (0.014~0.057)%/℃, it's still quite big.

2. Pressure effect

The increasing of hydrostatic pressure would cause the stiffening phenomenon of the measurement vibration tube. The bending tube also has Bourdon effect, which can produce a negative deviation. For the two kinds of pressure effect, although the impact is very small, the static pressure is quite different when it is applied or adjusted. So the value can't be ignored on high-accuracy meter. Pressure effect depends on the diameter, wall thickness and shape of the measuring tube. As for the small-caliber instrument, the ratio of wall thickness and diameter is large, the impact is small. Conversely, for the large caliber instrument, the ratio of wall thickness and diameter is small, so the impact is large. Micro motion provides the pressure data of the meters on the pressure basis 0. 2 MPa. The amount of pressure impact of Coriolis mass flowmeter of type 100 (diameter 25 mm) can reach −0. 03% R/MPa, and type 200 (diameter 40 mm/50 mm) is −0. 12% R/MPa. The pressure impact of D series is larger. D300 (caliber 80 mm/100 mm) can reach −1. 35%/MPa, and D600 (diameter 150 mm/200 mm) is −0. 75%/MPa. If the pressure has a great change in the process, instrument constants can be modified based on the actual hydrostatic correction instrument.

Eight Coriolis mass flowmeters on the market in the early 1990s had been tested by British Engineering Laboratory. The test results are shown in Tab. 4. 2. The greatest impact of the static pressure is −1. 75%/MPa, the minimum is −1%/MPa, and the average is −1. 4%/MPa.

Tab. 4. 2 Amount of pressure effect

Static pressure /MPa		2	2. 4	2. 8
Measurement error /%	Average	−2. 21	−3. 25	−3. 75
	Minimum	−1. 57	−2. 55	−2. 6
	Maximum	−3. 15	−4. 00	−4. 56

Note: The calibration pressure is as a reference.

3. Density effect

The measurement performance Coriolis mass flowmeter was previously thought that there were no medium density effects, but researches had shown that density still has some influence. Reports said that the measurement error of the meters using the water calibration on air compression test was less than ± 0. 5% R, on which density had some effect.

British Engineering Laboratory had tested eight Coriolis mass flowmeters on the market in the early 1990s. Four kinds of liquids of different densities were used for density impact test, and the density ranged from the kerosene 0. 78 kg/L to ethylene glycol 1. 11 kg/L. One of them changed +0. 5% (kerosene as the basis).

Danfoss Coriolis mass flowmeter test had also proved the existence of a density effect. Compared with the diameter instrument medium density of the flow value with 1 kg/L, the flow value with 2 kg/L had a difference of −0. 1%. The medium of 0. 5 kg/L also had a difference of +0. 06% with the medium of 1 kg/L.

4. Viscosity impact

Higher viscosity of the liquid will be more likely to absorb the energy of the the Coriolis incentive system, especially in the start of the flow. This phenomenon may lead to the temporarily stop of vibration in measurement till the normal flowing for the design of some structures of Coriolis mass flowmeter.

Eight Coriolis mass flowmeters on the market in the early 1990s had been tested by British engineering laboratory with four-viscosity liquids which were water, kerosene, gas oil, and ethylene glycol. On the test of weight flowing standard the viscosity ranged from 1 mm^2/s to 29.5 mm^2/s. One apparently had viscosity influence: the constant of large flow (2 kg/s to 5 kg/s) changed 0.25, while 20% of Q_{max} (0.5 kg/s) changed 0.5% in small flow, and 10% had the change of 2.2%.

4.8.6 The Interference of Nonlinear Vibration and Working Frequency

In the output signal of the Coriolis mass flowmeter in practical application, despite of the operating frequency f_0 of the sensor, there are $2f_0$, $3f_0$ and the operating frequency of 50 Hz. Wherein the signals in $2f_0$ and $3f_0$ are caused by the nonlinearity of the sensor itself, which is related to the structure parameters of the sensor and working state. The frequency of 50 Hz is the surround frequency of the sensor. The analysis shows that these interference signals have a larger impact on the accuracy of phase difference calculation. Therefore, before the phase difference calculation, the noise signal should be filtered out to improve the SNR. The problem can be solved by using digital band-pass filtering in DSP.

Questions and Exercises

4.1 What is the Coriolis effect? What role does Coriolis effect plays in the resonant Coriolis direct mass flow sensor?

4.2 Please describe the working principle and characteristics of the resonant Coriolis direct mass flow sensor.

4.3 Why resonant Coriolis direct mass flow sensor is a kind of multi-functional flow sensor?

4.4 Please give the detailed derivation of the Eq. (4.11).

4.5 There are two kinds of output detection implementations in the resonator Coriolis direct mass flow sensor. What are their characteristics?

4.6 Using the latest microprocessor and related technologies, design a method for the signal detection scheme and principle circuit of the resonant Coriolis direct mass flow sensor.

4.7 What is the principle of the two-components measurements by using the resonant Coriolis direct mass flow sensor? What are the conditions?

4.8 Summarizing the function of the resonant direct mass flow sensor, please describe it from the angle of test system realizing and the principle of sensor.

4.9 Please provide a kind of working scheme of a MEMS mass flow sensor, and a brief description.

Chapter 5
Surface Acoustic Wave Resonant Sensor

Basic content:

- Surface acoustic wave (SAW) IDT(interdigital transducer)
- Basic features of IDT
- IDT analysis model
- SAW resonator and its characteristics
- Typical SAW resonator sensor

5.1 Introduction

SAW resonant sensor is a special kind of resonant sensor. Its working principle is based on the regulation that the working frequency of the surface acoustic wave resonator (SAWR) is influenced by the measurand.

Surface acoustic wave is a kind of wave whose concentration of energy is spread on the surface of the ground. It was found by a British physicist Rilay in the study of the processing of seismic wave. In 1965, the American R. M. White and F. M. Voltmov invented the interdigital transducer which could incentive surface acoustic wave on the surface of piezoelectric material. This invention greatly accelerated the development of surface acoustic wave technology. A lot of distinctive surface acoustic wave components appeared successively. The surface acoustic wave technique has been applied to many subject areas, such as, telecommunications, broadcasting and television, aerospace, oil exploration and nondestructive detection, etc.

For more than twenty years, based on the fact that the frequency characteristic of SAW devices is sensitive to temperature, pressure, magnetic field, the electric field and some gas composition, the design and development of SAW is gradually attracting the attention of the sensor technology field. The history of SAW sensor is not long and the SAW sensor is not yet practical in many fields. But due to the reason that it's in line with the development direction of miniaturization, digitization, intelligence and precision of the signal system, so it arouses great attention. The SAW sensor has the following unique advantages:

① High precision and high sensitivity. For example, the sensitivity of the SAW pressure sensor is up to 0.3×10^{-6}/Pa. If the center frequency of the sensor is 300 MHz, the detector can detect the frequency change of 1 Hz, and then the sensor can reflect the changes in the pressure of 1 Pa, equivalent to the pressure change of approximately 0.1 mm H_2O. Another example is SAW temperature sensor, its theoretical resolution is up to 10 μ℃. Therefore, the SAW sensor is quite suitable for small scale measurement.

② The structure of technology is good and it's easy for us to do batch production. The structure of SAW sensor is plane, and the design of it is flexible. It has a flaky appearance, easy to combine and achieve single-chip multifunctional (the multifunction of the single-chip); it's also easy to install and can obtain good thermal and mechanical properties. A key component of the SAW sensors—SAW devices, including resonator of delay line is extremely easy to integrate. At the same time, the various functions of circuit are easy to combine and simplify. It also has robust structure, stable quality and good repeatability and reliability. The SAW sensor can reduce costs due to the fact that it's easy to achieve mass production.

③ It has the character of small size, small quality and low power consumption. Because of the plane structure, it's easy to integrate. So SAW sensor has small volume. More than 90 percent of the SAW energy is concentrated in the depth of about one wavelength away from the surface. This is the reason why it has low loss. Moreover, the SAW sensor circuit is relatively simple, so the entire sensor consumption is very small. This is particularly important for the coal mine, oil well or other explosion-proof requiring occasions.

④ SAW sensor is connected with microprocessor, and the interface is simple. SAW sensor can transfer the measured changes into frequency departure. This is quasi-digital signal, and it's convenient for transmission and processing. It's also easy to match with microprocessor directly. They made up the adaptive real-time processing system together.

5.2 SAW Interdigital Transducer

SAW IDT is a very important surface acoustic wave device. Since the emergence of the IDT, surface acoustic wave technology as well as surface acoustic wave sensor has been the rapid development of practical value. To date, the IDT is the only practical SAW transducer.

5.2.1 Basic Characteristics of IDT

1. Basic structure of IDT

The basic structure of IDT is shown in Fig. 5. 1. It consists of a number of deposited metal film electrodes on the piezoelectric substrate material. These electrodes cross with each other; both of their ends are placed together by bus bars. It is shaped as the cross of two rows of flat fingers. So it is called uniform (or non-dispersive) IDT. The interdigital cycle is $T = 2a + 2b$. Two adjacent electrodes constitute an electrode pair. The overlapping length is called effective length, that is the aperture of the transducer, denoted by W. If the overlapping length of each pair of transducer electrode is equal, it is called equal aperture transducer.

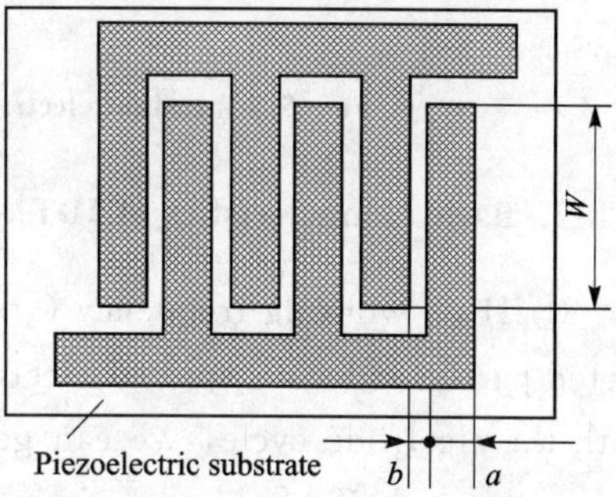

Fig. 5. 1 Basic structure of IDT

2. Physical process of SAW motivated by IDT

Using the inverse piezoelectric and direct piezoelectric effect of the piezoelectric material, the IDT can be used as not only launch transducer which can motivate the SAW, but also as receiving transducer which can receive the SAW. Thus this type of transducer is reversible.

When appropriate frequency AC signal is imposed on the launch IDT, the electric field distribution in the piezoelectric substrate is shown in Fig. 5. 2. The electric field can be decomposed into two components—E_V and E_H. Due to the inverse piezoelectric effect of the substrate, the material between the finger electric will be deformed, the particle will be displaced by the electric field. E_H can make compression (inflation) displacement parallel to the surface to the particle. E_V produce shear displacement which is perpendicular to the surface. This cyclical strain may produce SAW which is spread along the IDT by both sides of the surface. Its frequency is equal to the frequency of applied electrical signal. One side of useless wave can be absorbed by a kind of high-loss medium, the other side of wave will spread to the receiving IDT and convert into electrical signal output by the direct piezoelectric effect.

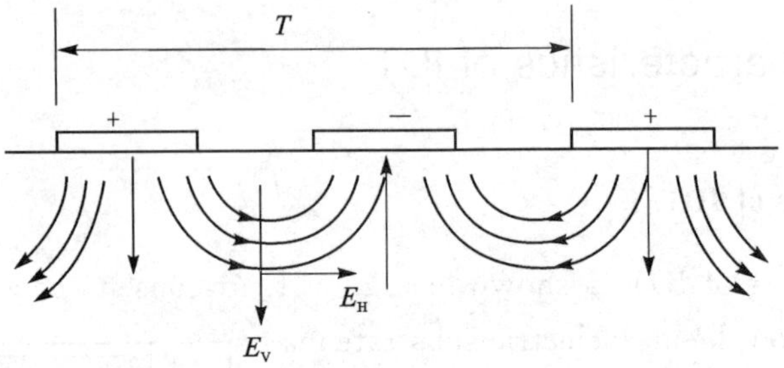

Fig. 5.2 Instant electric field distribution under the interdigital electrode

3. Basic characteristics of IDT

① High working frequency (f_0). From the Fig. 5.2, we can see that substrate is distorted partly under an applied electric field. When the acoustic wavelength is in consistent with the electrode cycle, we can get the maximum incentive (synchronous). At this moment, the cycle T of electrode is actually the acoustic wavelength, and can be expressed as

$$\lambda = T = \frac{v}{f_0} \tag{5.1}$$

where

v is SAW velocity of the material(in m/s);

f_0 is SAW frequency, i.e., the synchronous frequency of the applied electric field(in Hz).

When the finger width a is equal to the gap b, $T = 4a$, then the working frequency f_0 is

$$f_0 = \frac{v}{4a} \tag{5.2}$$

Obviously, for certain acoustic velocity, the highest working frequency of the digital transducer is only restricted by the minimum technology available electrode width(electrode width available now). Interdigital electrode is made by the planar process. With the development of the integrated circuit technology, we are now able to get the line width of about 0.3 μm. As to quartz substrates, the working frequency of the transducer can be up to 3 GHz. In fact, a lot of SAW devices whose working frequency are higher than 1.5 GHz are made. High working frequency is a major feature of this kind of devices.

② Correspondence of the time domain (impulse) response and the spatial geometry. The space location of each pair of interdigital electrode of IDT is directly corresponded with the sampling time waveform. Fig. 5.3 shows the multi-transmitter and receiver case. When we add a δ pulse to the launch transducer, the signal the receiving end gets is the superposition of the amplitude and the phase of the sound wave which reach the receiving transducer.

The energy of it is proportional to the length of finger. The pulse of the single transducer in the figure is a rectangular modulation pulse. Just as the geometric figure, the convolution output is a triangular modulation pulse.

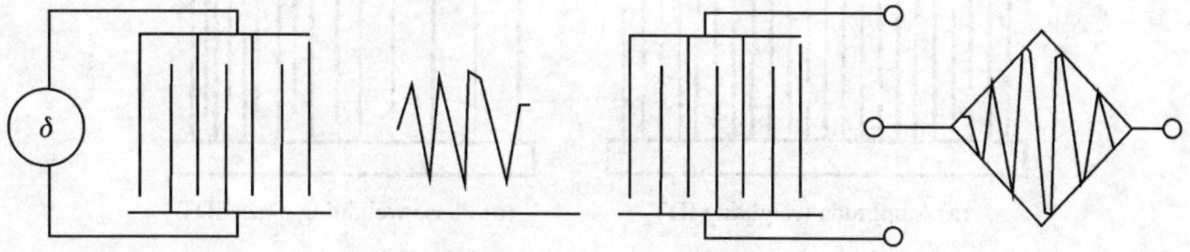

Fig. 5.3 Geometrical sketch of the pulse response for IDT

The transducer transmission (transfer) function is Fourier's transform of the impulse response. This relationship provides a very simple method for the design of the transducer.

③ The bandwidth directly depends on the finger pairs. For the IDT with equal finger width and equal spaced, bandwidth can be simply expressed as

$$\Delta f = \frac{f_0}{N} \tag{5.3}$$

where

f_0 is central frequency (or working frequency) (in Hz);

N is finger pairs.

As the Eq. (5.3) shows, when the central frequency is fixed, the bandwidth is only depended on the finger pairs. The more the finger pairs, the more narrow the bandwidth of the transducer. The bandwidth of the SAW devices has very big flexibility. Relative bandwidth can be narrow to 0.1% and can be stretched to an octave (i. e. 100%).

④ Interchangeability. IDT not only can be used to motivate SAW, but also can be used as a receive end. This is very convenient in the analysis and design, but it may also bring troubles, for example, the secondary effect such as the acoustic-electric regeneration may bring bad effect to the devices' performance.

⑤ Capable to be internally weighted. From characteristics ② we can infer that the radiation energy of each pair of interdigital is related to the effective length of the overlapped fingers, namely the aperture. This can be used to weigh the pulse signal amplitude by changing the length of overlapped fingers. Meanwhile, the location of the interdigital is the sampling of the signal phase, so we can get the phase weighting of the signal by changing the interdigital cycle deliberately, such as dispersion transducer; or we use both of them in order to obtain a specific signal spectrum, such as pulse compression filter. Fig. 5.4 simply shows this situation. (a) is an amplitude weighting transducer, (b) is a phase weighting transducer.

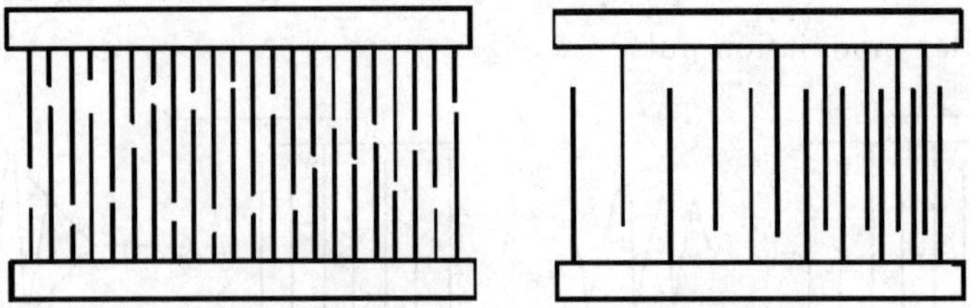

(a) Amplitude weighting IDT (b) Phase weighting phase IDT

Fig. 5.4 Inner weighting character of IDT

⑥ Easy to manufacture, good repeatability and consistency. The manufacturing process of SAW devices is similar to the semiconductor integrated circuit technology. Once the design is completed and the mask master is made, we can obtain the same device from copying. Therefore, this device has good feature of consistency, interchangeability and repeatability.

5.2.2 Basic Analyzing Model of IDT

The analyzing model of IDT needs to be studied to analyze the working mechanism and to design SAW devices which can meet the requirements. Here are two most widely used models.

1. δ function model

The electric field distribution of the cross-section of an IDT is shown in Fig. 5.5. If we approximately think that only the electric field vertical to the surface can incent SAW, we can simplify the distribution of electric field as Fig. 5.5(b) shows. At this time, we think the electric field only exists under the electrode and there is no electric field component between the electrodes. Positive and negative electric field is alternating in the electric field of each electrode. The distribution of electric field along the direction of x propagation path is shown in Fig. 5.5(c). The largest electric field gradient is at the edge of electrodes, as is shown in Fig. 5.5(d). This is a series of pulses, and those sharing same polarity alternate each other. This means that each edge of each interdigital can be considered to be the superposition of independent δ function sound output.

Supposing a IDT has N edges of finger(i. e. $N/2$ fingers): $x_1, x_2, \cdots, x_n, \cdots, x_N$; δ impulse sound source work in the middle of the finger, and the gap between two sound source

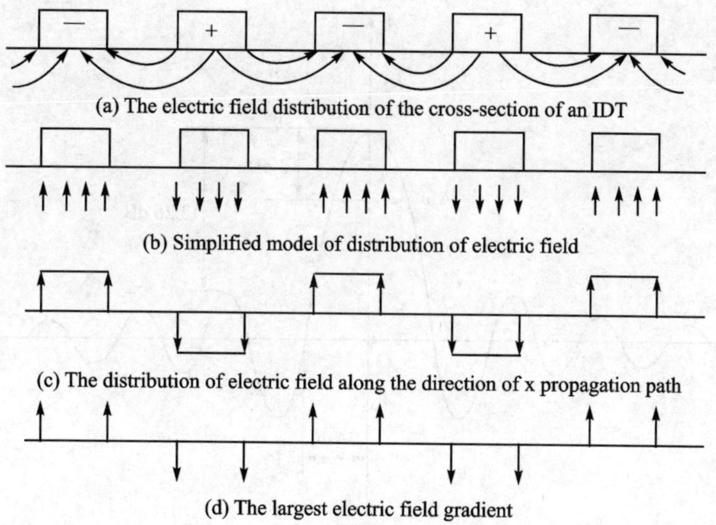

Fig. 5.5 Electric field distribution and simplified model for the δ function

pulse is $\lambda_0/2$, just as Fig. 5.6(b) shows. The length of each overlapped interdigital is the same, that is $W_1 = W_2 = \cdots = W_n = W$. Thus, for the transducer with N pairs of fingers ($2N+1$ fingers and $2N$ interval), considering $\Delta\omega/\omega_0 \ll 1$, the transfer function of it is

$$H(\omega) = 2NW \frac{\sin X}{X} e^{j\left[\omega_0 t - \left(\frac{2N-1}{2}\right)\pi\right]} \tag{5.4}$$

$$X = N\pi \frac{\Delta\omega}{\omega}$$

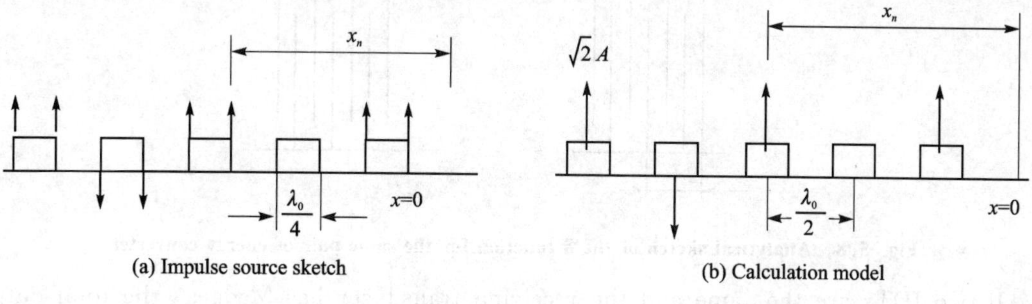

Fig. 5.6 Discrete plane waves source for the model of δ function

From the Eq. (5.4) we can see that the transfer function for the IDT with even and same diameter is sinc function. Its graphic is shown in Fig. 5.7.

The spectral width is defined as the bandwidth when main peak decreased 3 dB. From this definition we can conclude that:

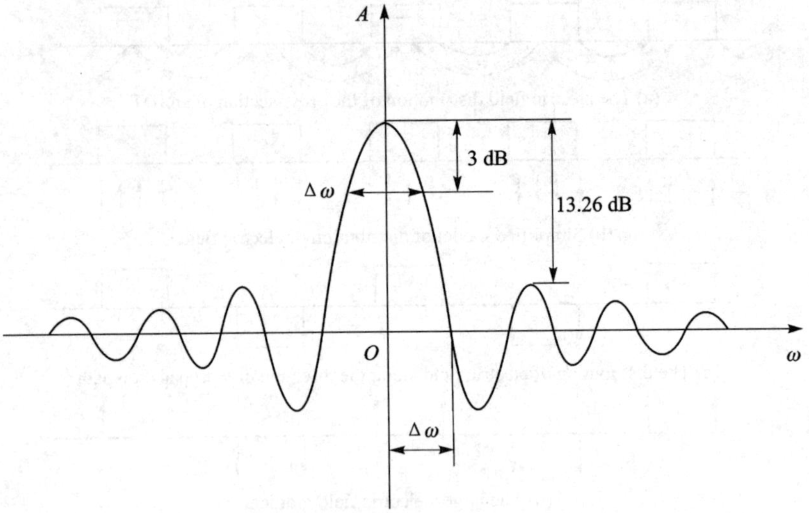

Fig. 5.7 Waveform of transfer function for the IDT with even and same diameter

$$\left(\frac{\Delta\omega}{\omega}\right)_{3\,\mathrm{dB}} \approx \frac{1}{N} \tag{5.5}$$

One SAW device has two IDTs, respectively used to transmit and receive. The location of them is shown in Fig. 5.8.

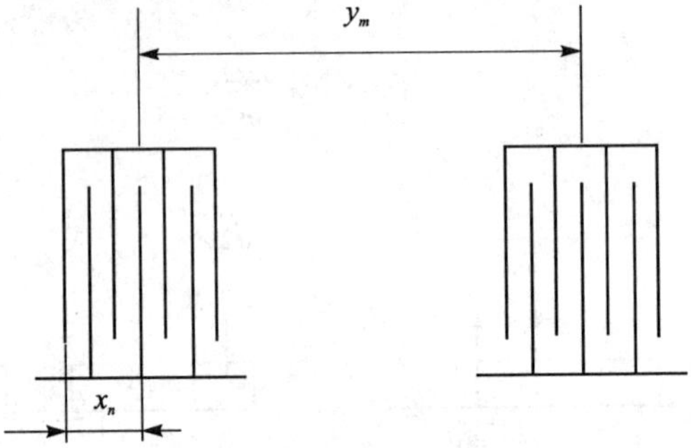

Fig. 5.8 Analytical sketch of the δ function for the same pair of energy converter

If two IDTs are the same and the receiving transducer has Medges, the total output should be the superposition of M edges' output. Its frequency response can be simplified as

$$H(\omega) = \left(\sum_{n=1}^{N} W_n e^{j\frac{2\pi f}{v}x_n}\right)^2 \cdot e^{-j\frac{2\pi f}{v}B} \tag{5.6}$$

$$B = y_m - x_n$$

where

y_m is the distance between the center of two IDTs (in m), as is shown in Fig. 5.8;

$e^{-j\frac{2\pi f}{v}B}$ is fixed delay.

The main advantages of δ function model are simple and intuitive. The δ function model can also give a better description of the shape of the pass band near the synchronous frequency.

2. Impulse Response Mode

A feature of IDT is that its impulse response has a one-to-one relationship with the geometry of the interdigital transducer. This relationship provides an intuitive and simple model—impulse response model.

As for a pair of transducers shows in Fig. 5.9, if the receiving transducer is a broadband transducer with sufficient aperture to collect all emission sound beams, the overall frequency response is

$$H(\omega) = H_1(\omega) \cdot H_2(\omega) \quad (5.7)$$

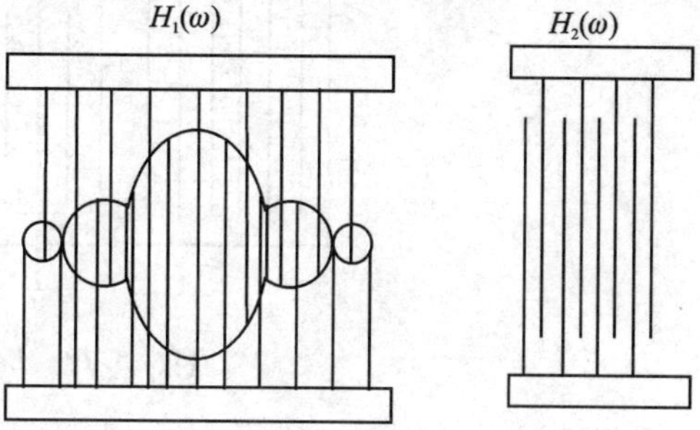

Fig. 5.9 Frequency response of a pair of IDTs

According to the convolution theorem, total impulse response $h(t)$ is the convolution of the impulse responses of two transducers. That is

$$h(t) = \int_{-\infty}^{\infty} h_1(\tau)h_2(t-\tau)\mathrm{d}\tau \quad (5.8)$$

Via the impulse response model the transfer function of the transducer can be obtained.

5.3 SAW Resonator

5.3.1 Structure and Working Principle

One key part of SAW sensors is SAW resonator. The sensor consists of SAW resonators or surface wave delay line, amplifiers and matching networks. It is the only device to achieve high quality factor in the VHF and UHF bands at present. SAWR consists of IDT and metal reflecting grizzly resonator, as is shown in Fig. 5.10. One of the two IDTs is used to launch surface acoustic wave, the other is used to receive surface acoustic wave. The IDT and the reflector is constructed by using the semiconductor integrated technology to deposite the metallic aluminum on the piezoelectric substrate material and applying the lithography technology to carve the metal film into a special structure with certain size and shape. The finger width, interdigital interval of the IDT and the trash rack width, interval of the reflector should be designed and manufactured according to the center frequency, the extent of the noise suppression and the loss size.

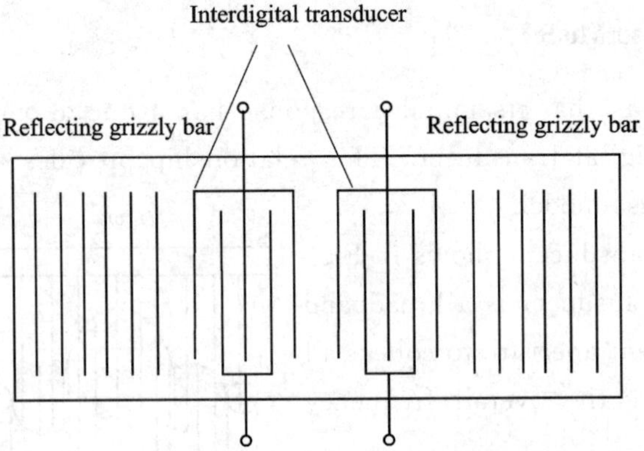

Fig. 5.10 Basic structure of SAWR

SAW resonator is a flat electrode structure. It is made on a kind of proper piezoelectric material by the photolithography technology. Its working principle is the same as that of normal piezoelectric quartz resonator. SAW is generated by IDT and IDT can change the mechanical signals into electrical signals or the other way around. SAW is limited in the resonant cavity. The value of quality factor of the resonator cavity is determined by the material's insertion loss and cavity's leak loss. SAW resonators can also have excellent frequency characteristics like body-wave quartz resonators as long as the manufacturing process is controlled strictly. Fig. 5.11 is a sketch of the three commonly used resonators. Fig. 5.11(a) is a resonator with single comb finger energy converter, the simplest one, with a single-ended pair and single-channel resonator structure.

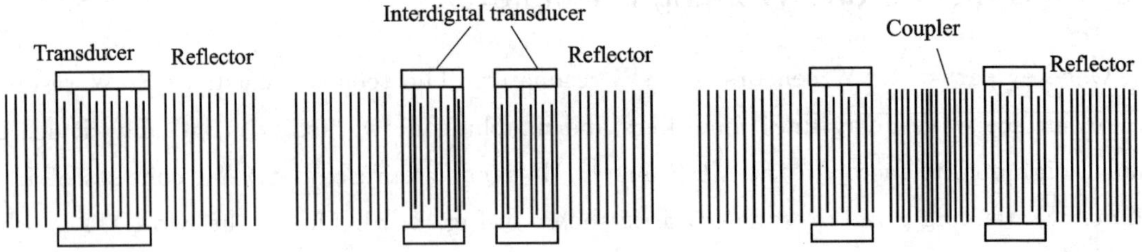

(a) Resonator with single IDT (b) Resonator with double IDTs (c) Double IDTs type resonator with coupled function

Fig. 5.11 Different kinds of resonant cavity for SAWR

The resonator shown in Fig. 5.11(a) has a low mutual interference and low insertion loss. Fig. 5.11(b) and (c) respectively shows the resonator with double comb finger energy converter and double comb finger energy converter type resonator with coupled function. The entire resonator has a high insertion loss because the acoustic signal propagation loss is high in the center of resonant cavity. But they both have a 180° phase shift necessary for oscillatory structure, which is controlled by the positive feedback resonator.

There are a lot of reflection gratings arranged periodically along the both sides of input and output transducers. When SAW wavelength is approximately twice of grating cycle, the function of reflection grating is just like a mirror. In this frequency range, all of the surface wave energy is limited in the resonant cavity consisting of two gratings. Each grating can create reflection like a transmission line with a mismatched impedance. With enough gratings, we can make the total reflection from all reflection gratings almost equal to the incident wave from the IDT. There will be a narrow-band signal of high quality factor if all reflection is superimposed together on the resonant frequency.

One-port resonator has many characteristics that meet the requirements, but its design is less flexible than that of double-port resonator. when it is used to form a resonator circuit, resonant amplifier's input signal which is the feedback of one-port resonator must be designed to have a 180° phase shift. Actually, the 180° phase shift required by one-port resonator is easy to acquire, but the phase noise of one-port resonator exceeds that of double-port resonator.

When choosing the substrate material of SAW resonator, we should consider the following factors: relative bandwidth, insertion loss, operating temperature and frequency stability which has function relationship with temperature, etc. Different kinds of piezoelectric material have different application characteristics. When the requirement are broadband and temperature coefficient is lower than 10^{-4}/℃, we can use high coupling material—lithium niobate. Quartz material is not suitable for broadband because it has big insertion loss; quartz materials are usually adopted when the requirement is narrow-band because it has good character of stability. In the temperature range of 0 ℃ to 80 ℃, we will use compensation oscillating circuit. Then the nominal frequency drift of quartz crystal SAW resonator will be less than 10^{-4}.

Fig. 5.12 shows the composed structure of the resonator for SAWR.

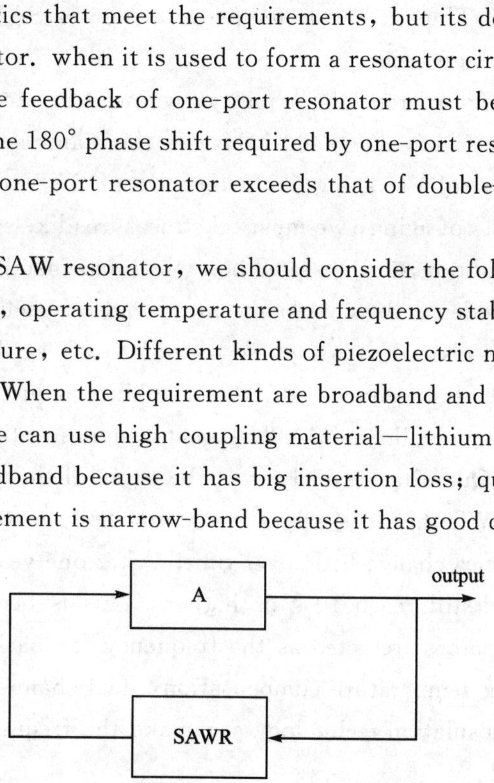

Fig. 5.12 Composed structure of the resonator for SAWR

After the output signal of SAWR is amplified, it is connected to the input end as a positive feedback. As long as the gain of amplifier can compensate loss of the resonator and connecting wire, and satisfy certain phase condition, the resonator can start oscillation and sustain self excitation. After the oscillation starts, the resonant frequency of SAW will be influenced by temperature, deformation of piezoelectric substrate material and other factors. Therefore, the SAW resonator can be used to make different kinds of sensors which are used to measure various physical quantities.

If we use the SAW delay line to make the resonator and spread a layer of material which is sensitive to certain kind of gas or humidity between the two interdigital electrodes, then it can be make as SAW gas or humidity sensor.

5.3.2 Stability of Frequency and Temperature

To improve the frequency stability of SAW resonators, a compensation circuit is in need. Thus in a wide temperature range, SAW resonators can vibrate in a given frequency with high accuracy.

To improve the stability, we need to execute the aging test within operating frequency range (i. e., 300～400 MHz) in the manufacture of SAW devices, and find out the impact factors of the aging characteristics of SAW devices. For example, in order to reduce the effects of aging, we must take measures like sealing, vacuum drying and vacuum packaging. In addition, in the sealed box which has installed SAW device, gaseous substances should be forbidden . Also, we can't spray single-molecule organic compounds or other materials in the SAW resonator cavity, so as not to affect the long-term performance of resonator or to induce the frequency drift and stability reduction. All these measures will increase the frequency stability of SAW resonator significantly.

Quantitative analysis of the aging of the resonator is a major task of the stability analysis. Whether it is quartz resonator, body wave resonators or SAW resonators, their characteristics change little over time. After one-year working, the accuracy of frequency stability could still reach 10^{-7} or higher. This is because resonators are passive devices. Usually resonators are acted as the frequency feedback to form the resonator circuit. Additionally, using temperature compensation, dual-channel SAW resonators and advanced high-vacuum encapsulation technology can make the frequency and temperature stability to reach a high level.

5.4 Typical SAW Sensor

SAW sensors are all based on the frequency characteristics of the surface acoustic resonator. That is the frequency of resonator is(The frequency of resonator is the one that) varying with the change of the measured parameters to achieve the detection of measurand.

With SAW resonator as the core, accompanied by necessary circuits and structure, sensors measuring the strain, stress, pressure, small displacement, force, temperature, etc. can be manufactured. We can also make SAW accelerometer by using the same mechanism and appropriate structure designing. The area between the two electrodes of interdigital transducer is coated with a kind of material which is sensitive to certain gases (adsorption and desorption), then it can be made into a variety of SAW gas sensor, humidity sensor and so on. The trend of development in recent years shows that variation of sensors measuring chemical and physical measurands using SAW resonators has a promising application prospects. This section briefly describes some typical kinds of SAW sensors.

5.4.1 SAW Strain Sensor

SAWR substrale will initiate strain when they are under the action of force or torque of the measurand, as is shown in Fig. 5.13. Therefore, we can detect the measurand by measuring the strain caused by it.

Usually, the resonant frequency of SAWR can be expressed as

$$f = \frac{v}{\lambda} \tag{5.9}$$

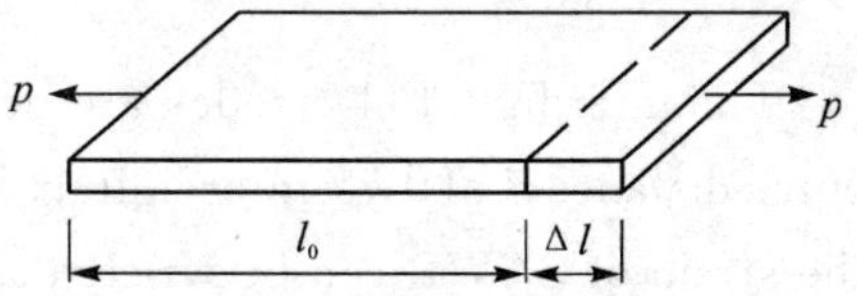

Fig. 5.13 Deformed sketch of the SAWR when force is applied on it

where

v is the propagation velocity of acoustic waves on the surface of piezoelectric substrate material(in m/s), $v \approx \sqrt{E/\rho}$;

λ is wave length of acoustic wave (in m);

ρ and E are respectively density (in kg/m^3) and elastic modulus of material(in Pa).

As to uniform distribution interdigital transducer, the wavelength of acoustic wave λ has a relationship with the center distance d between two adjacent electrodes of the interdigital transducer. That is

$$\lambda = 2d \tag{5.10}$$

We can easily get the relationship of Eq. (5.10) through Fig. 5.14. If the finger width a is equal to finger space b, then

$$a = b = \frac{\lambda}{4} \tag{5.11}$$

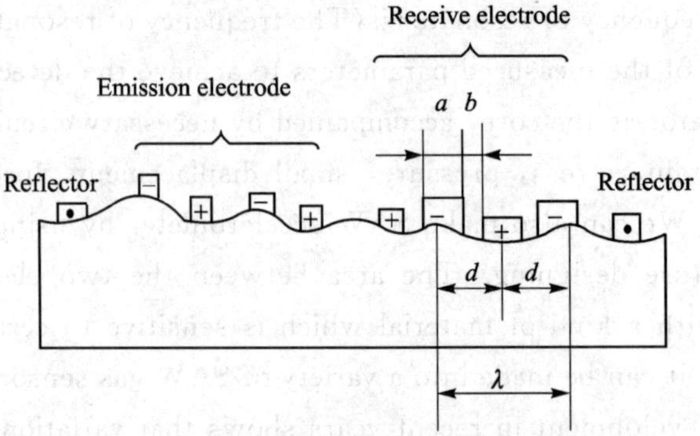

Fig. 5.14 Excited sketch of the surface status for the SAWR

When SAWR is not loaded, the propagation velocity of surface acoustic wave is v_0, the wavelength is λ_0, the resonant frequency is

$$f_0 = \frac{v_0}{\lambda_0} \tag{5.12}$$

When the force is applied on the SAWR substrate alone the propagation direction, strain ε is generated consequently(see Fig. 5.13)

$$\varepsilon = \frac{\Delta l}{l} \tag{5.13}$$

The interdigital electrodes are deposited on the piezoelectric substrate material, so the center distance d of two interdigitals is changing due to the stain changes of the substrate. The strain of SAWR can be written as

$$\varepsilon = \frac{\Delta d}{d_0} \tag{5.14}$$

SAW devices will produce strain after they are affected by forces. At that time, the center distance of interdigital d has a relationship with the strain ε:

$$d(\varepsilon) = d_0 + \Delta d = d_0 + \varepsilon d_0 = d_0(1+\varepsilon) \tag{5.15}$$

Because $\lambda_0 = 2d_0$, then

$$\lambda(\varepsilon) = 2d(\varepsilon) = 2d_0(1+\varepsilon) = \lambda_0(1+\varepsilon) \tag{5.16}$$

From Eq. (5.16), the SAW length on the surface of piezoelectric materials increases alongside the strain ε.

At the same time, strain of the piezoelectric material can cause changes in the density (ρ) of material. And it will affect the propagation velocity of acoustic wave. Strain ε may affect the propagation velocity (v), it can be described as follows:

$$v(\varepsilon) = v_0(1 + k_\varepsilon \varepsilon) \tag{5.17}$$

where k_ε is material constant.

Therefore, the resonant frequency of SAW resonator is related to its strain, the relationship is

$$f(\varepsilon) = \frac{v(\varepsilon)}{\lambda(\varepsilon)} = \frac{v_0(1 + k_\varepsilon \varepsilon)}{\lambda_0(1 + \varepsilon)} \tag{5.18}$$

The absolute changes of resonant frequency caused by its strain is

$$\Delta f = f(\varepsilon) - f_0 = f_0\left(\frac{1 + k_\varepsilon \varepsilon}{1 + \varepsilon} - 1\right) = f_0 \frac{\varepsilon(k_\varepsilon - 1)}{1 + \varepsilon} \tag{5.19}$$

Usually, owing to $\varepsilon < 10^{-3}$, the ε in denominator in Eq. (5.19) is negligible. The approximate linear relationship is as follows:

$$\Delta f = f(\varepsilon) - f_0 \approx f_0 \varepsilon(k_\varepsilon - 1) \tag{5.20}$$

$$f(\varepsilon) = f_0 + \Delta f \approx f_0(1 - k\varepsilon) \tag{5.21}$$

where

$$k = 1 - k_\varepsilon$$

If the substrate material is quartz crystal, then

$$k_\varepsilon = -0.4$$

$$k = 1 - k_\varepsilon = 1 - (-0.4) = 1.4$$

so

$$f(\varepsilon) = f_0 + \Delta f = f_0(1 - 1.4\varepsilon) \tag{5.22}$$

Theoretical analysis shows that the electrode of interdigital transducer may affect the propagation velocity (v) of acoustic wave over the surface of substrate material. The extent of its impact is related to thickness of the mass covered layer, as is shown in Fig. 5.15. Actually, the interdigital electrode is made of very thin metal coating ($d/\lambda = 0.01 \sim 0.1$). Therefore, interdigital electrode has little effect on the propagation velocity. The effect can be ignored in calculation.

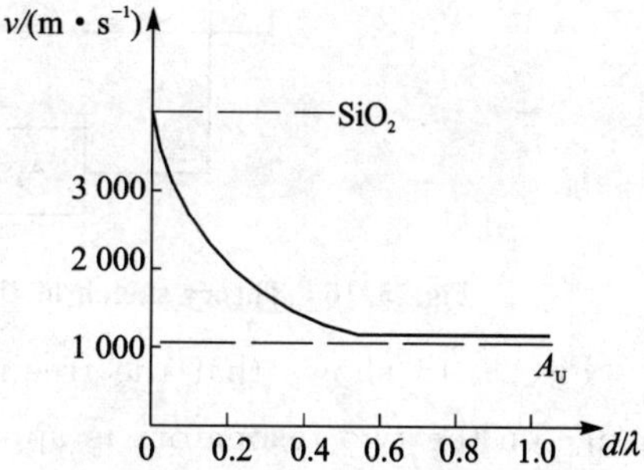

d = thickness of the mass covered layer;

λ = wavelength of the sonic wave

Fig. 5.15 The influence of the covering mass on the quartz crystal having on the sound wave propagating velocity

5.4.2 SAW Pressure Sensor

What is shown in Fig. 5.16 is a theory sketch for SAW sensor with a differential structure for temperature compensation. The essential component of the SAW sensor is the pressure sensing chip manufactured on the quartz diaphragm, on which two identical SAW resonators(SAWR1 and SAWR2), are symmetrical placed, near the centre.

Two SAW resonators are connected separately to the feedback loop of the amplifier to form the resonator with a frequency output. The two channels of output are processed via frequency mixing, low-pass filter and amplifying to get a difference frequency output corresponded to external pressure.

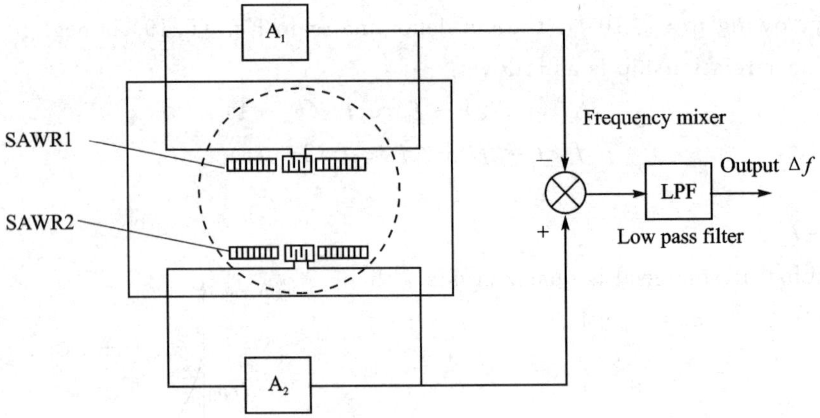

Fig. 5.16 Theory sketch of the differential SAW press sensor with double resonator

Fig. 5.16 shows that the frequency departure caused by the interfere of environment change on the two resonators is approximately equal considering that they are placed quite close to each other on the diaphragm. Therefore by picking up the difference frequency signal via mixing, the interfere can be compensated or neutralized. It means that a SAW pressure sensor with a differential structure is able to realize the temperature compensation.

In the two oscillation circuits, the designed amplifier gain can compensate the insertion loss of the resonator; meanwhile certain phase condition is satisfied. Then the system will be able to oscillate, working in a closed loop. With the help of Eqs. (1.12) and (1.13), starting conditions can be described as

$$G_A > L_S(f) \tag{5.23}$$

$$\phi_R + \phi_A = 2n\pi \quad (n \text{ is integer}) \tag{5.24}$$

where

G_A is the amplifier gain;

$L_S(f)$ is the insertion loss of the resonator;

ϕ_R and ϕ_A are respectively phase displace of the resonator and amplifier (in(°)).

It can be told that the resonant frequency of SAW is

$$f = \frac{v}{\lambda}$$

As for uniform interdigital transducer, the wavelength λ of SAW is twice the distance of Interdigitated electrodes center distance, namely $\lambda = 2d$. Therefore, when external pressure is exerted on the sensitive diaphragm, the substrate will generate strain resulting from stress, causing displacement of the Interdigitated electrodes center distance, namely the wavelength λ is changed. At the same time, the modulus of elasticity and the density alter, resulting in the change of SAW wave velocity. The differences of wavelength λ and velocity v cause the departure of the frequency output of the resonator. There is a relationship between the changes and the exerted pressure.

Considering that the velocity (v) propagating in the substrate and the interdigitated electrodes center distance (d) are both functions of pressure and temperature, therefore the resonant frequency as well, which can be described as

$$f(p,T) = \frac{v(p,T)}{\lambda(p,T)} \tag{5.25}$$

According to Eq. (5.25), it is told that:

$$\mathrm{d}f = \frac{\mathrm{d}v}{\lambda} - \frac{v\mathrm{d}\lambda}{\lambda^2} = \frac{v}{\lambda}\left(\frac{\mathrm{d}v}{v} - \frac{\mathrm{d}\lambda}{\lambda}\right) \tag{5.26}$$

Additionally,

$$\mathrm{d}v = \frac{\partial v}{\partial p}\mathrm{d}p + \frac{\partial v}{\partial T}\mathrm{d}T \tag{5.27}$$

$$\mathrm{d}\lambda = \frac{\partial \lambda}{\partial p}\mathrm{d}p + \frac{\partial v}{\partial T}\mathrm{d}T \tag{5.28}$$

Thus referring to Eqs. (5.26)~(5.28), when pressure and temperature both changes, the relative variation of resonant frequency caused by the two factors is

$$\frac{\mathrm{d}f}{f} = \frac{\mathrm{d}v}{v} - \frac{\mathrm{d}\lambda}{\lambda} = -(\alpha_P\mathrm{d}p + \alpha_T\mathrm{d}T) \tag{5.29}$$

where

$$\alpha_p = \frac{1}{\lambda}\frac{\partial \lambda}{\partial p} - \frac{1}{v}\frac{\partial v}{\partial p} \tag{5.30}$$

$$\alpha_T = \frac{1}{\lambda}\frac{\partial \lambda}{\partial T} - \frac{1}{v}\frac{\partial v}{\partial T} \tag{5.31}$$

α_p and α_T are respectively the first-order coefficients of pressure (in 1/Pa) and temperature (in 1/℃).

Each electrodes center distance (d) of interdigital transducer is related to the strain (ε_l) along the SAW propagation direction, and the relationship is

$$d + \Delta d = d(1 + \varepsilon_l) \tag{5.32}$$

According to Eq. (5.32) and $\lambda = 2d$, it can be concluded

$$\frac{1}{\lambda} \cdot \frac{\partial \lambda}{\partial p} = \frac{1}{d} \cdot \frac{\partial d}{\partial p} = \frac{\partial \varepsilon_l}{\partial p} \tag{5.33}$$

And the relationship between SAW velocity and strain can be described as

$$v + dv = v(1 + \delta_1 \varepsilon_1 + \delta_2 \varepsilon_2) \tag{5.34}$$

where

ε_1 and ε_2 are respectively separately the strains parallel and vertical to the SAW propagation direction;

δ_1 and δ_2 are respectively the strain coefficient measured by experiment. As for ST-cut quartz, $\delta_1 = -0.044 \pm 0.002$, $\delta_2 = -0.164 \pm 0.01$.

Insert Eqs. (5.33) and (5.34) into Eq. (5.32), the first-order pressure coefficient is

$$\alpha_p = \frac{\alpha \varepsilon_1}{\partial p} - \delta_1 \frac{\alpha \varepsilon_1}{\partial p} - \delta_2 \frac{\alpha \varepsilon_2}{\partial p} \tag{5.35}$$

Usually, the influence of temperature and pressure all over the diaphragm is not uniform, therefore the pressure coefficient is nonlinear, and the mathematical expression between the two is complicated, which can be only described approximately. If it is practical to average the pressure coefficient over every point within effective area on the resonator substrate and the aperture of interdigital transducer is shorter than one fourth of the radius of the diaphragm (namely smaller than $R/4$), then the average pressure coefficient of the SAW resonator on the clamped diaphragm is

$$\bar{\alpha}_p(x_0, y_0) = \bar{\alpha}_{p0}\left[(\delta_1 - 1 + \delta_2) - \left(\frac{x_0}{R}\right)^2 (\delta_1 - 1 + 3\delta_2) - \left(\frac{y_0}{R}\right)^2 \left(\delta_1 - 1 + \frac{\delta_3}{3}\right)\right] \tag{5.36}$$

$$\bar{\alpha}_{p0} = \frac{3(1-\mu^2)}{8E}\left(\frac{R}{H}\right)^2 \tag{5.37}$$

where

R and H are respectively the radius and the thickness of the diaphragm (in m);

E and μ are respectively modulus of elasticity and Poisson's ratio of the material of the substrate. As for ST-cut quartz, $E = 8.3 \times 10^{10}$ Pa, $\mu = 0.26$.

(x_0, y_0) is the location of SAW interdigital transducer's center point.

Considering that the two resonators are quite close to each other on one diaphragm, thus the environment temperature interfere is regarded approximately equal, the variation of temperature (ΔT) is the same during working.

By simple derivation of Eq. (5.29), the two channels' frequency outputs are

$$\left.\begin{aligned} f_1 &= f_{10}[1-\bar{\alpha}_{p1}p-\bar{\alpha}_T\Delta T] \\ f_2 &= f_{20}[1-\bar{\alpha}_{p2}p-\bar{\alpha}_T\Delta T] \end{aligned}\right\} \tag{5.38}$$

where

f_{10} and f_{20} are respectively the output frequency of resonator 1 and resonator 2 located in the middle and on the edge of the diaphragm on a condition without pressure (in Hz);

$\bar{\alpha}_{p1}$ and $\bar{\alpha}_{p2}$ are respectively the average pressure coefficient in the middle and on the edge of the diaphragm(in 1/Pa);

$\bar{\alpha}_T$ is the average temperature coefficient of the substrate (in 1/℃);

ΔT is the variation of the temperature (in ℃).

According to Eq. (5.38), the different frequency output of the sensor is

$$f_D = f_1 - f_2 = f_{D0} - (\bar{\alpha}_{p1}\cdot f_{10} - \bar{\alpha}_{p2}\cdot f_{20})p - \bar{\alpha}_T(f_{10} - f_{20})\Delta T \tag{5.39}$$

where

$f_{D0} = f_{10} - f_{20}$ is the different frequency output of the two resonator without pressure.

Therefore the departure of frequency caused by external pressure is

$$\Delta f_{Dp} = f_D - f_{D0} = -(\bar{\alpha}_{p1}\cdot f_{10} - \bar{\alpha}_{p2}\cdot f_{20})p \tag{5.40}$$

The drift caused by temperature departure is

$$\begin{aligned} \Delta f_{DT} &= -\bar{\alpha}_T f_{D0}\Delta T = -\bar{\alpha}_T(f_{10} - f_{20})\Delta T = \\ &-(\bar{\alpha}_T f_{10}\Delta T - \bar{\alpha}_T f_{20}\Delta T) = \Delta f_{1T} - \Delta f_{2T} \end{aligned} \tag{5.41}$$

where Δf_{1T} and Δf_{2T} are the frequency departure of the two resonators caused by temperature changes (in Hz).

By analyzing Eq. (5.40), as long as the parameter is chosen acceptably, its sensitivity will be largely raised than one channel structure. Via Eq. (5.41), if $f_{D0} = f_{10} - f_{20} \ll f_{10}$ (or f_{20}), the difference frequency departure caused by temperature is much smaller than the one channel frequency departure caused by temperature. Thus, we got a high sensitivity SAW resonant pressure sensor with temperature compensation.

5.4.3 SAW Acceleration Sensor

SAW acceleration sensor uses cantilever elastic sensing structure. The SAW resonator structure is placed on the surface of the cantilever made up from piezoelectric material (like piezoelectric crystals). The sensitive mass attached to the free end of the cantilever is used to sense the measured acceleration, generating inertia force on the mass, causing surface displacement on the resonator area, changing the SAW wave velocity and resulting in the departure of the center frequency of the resonator. Therefore, the SAW acceleration sensor is actually an acceleration-frequency transducer. After a few processing of the frequency output, measured acceleration will be gained.

Fig. 5.17 shows an one end clamped cantilever with a length L, width b and thickness h. The free end has a mass with a radius of d to sense the acceleration.

The lowest-order natural resonant frequency of the cantilever sensing structure is

$$f_{B,m} = \frac{1}{4\pi}\sqrt{\frac{Ebh^3}{L^3 m}}m = \frac{\rho\pi d^2 b}{4} \quad (5.42)$$

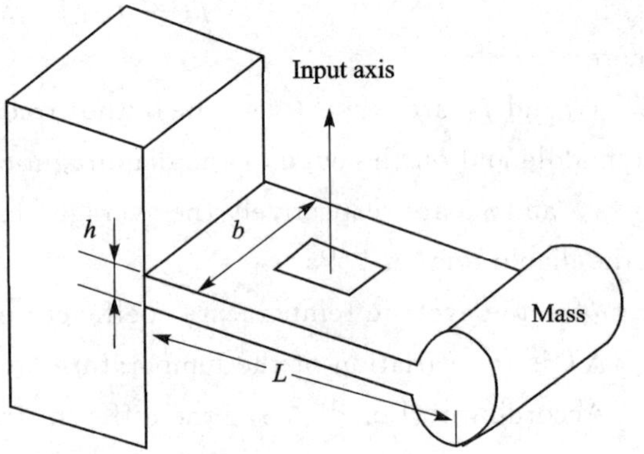

Fig. 5.17 Structure sketch of the SAW accelerometer

where

$f_{B,m}$ is the lowest-order natural resonant frequency of the cantilever with a sensitive mass m (in Hz);

m is the weighing of the sensitive mass (in kg);

E and μ are respectively modulus of elasticity (in Pa) and Poisson's ratio of the substrate material.

Through Eq. (5.42), we can design structure parameters related to the cantilever and the sensitive mass aiming at the lowest-order natural resonant frequency of the acceleration sensor.

As a matter of fact, it can be told that the positive strain along the x direction on the upper surface of the cantilever caused by the acceleration a is

$$\varepsilon_x(x) = \frac{-6ma(L-x)}{Ebh^2} \quad (5.43)$$

When the SAW resonator is placed at the point of (x_1, x_2), the average strain sensed by the SAW resonator is

$$\bar{\varepsilon}_x(x_1, x_2) = \frac{-6ma[L-0.5(x_1+x_2)]}{Ebh^2} \quad (5.44)$$

By Eq. (5.44), we can design structure parameters related to the cantilever and the sensitive mass aiming at the measuring sensitivity of the acceleration sensor.

5.4.4 SAW Gas Sensor

In recent years, there's a growing number of people care more about the gas which may pollute the environment. Home use gas, liquefied petroleum gas, gas which contains sulfur compounds or even poisonous chemical gas from battlefield, etc, can cause great pollution to the environment. Therefore, corresponding gas sensors are necessary for the detection of specific gases. Currently, scientific workers are stepping up on the development of high sensitivity, good selectivity, small volume and inexpensive gas sensors. The development of SAW gas sensor is particularly prominent.

The most popular method of detecting small concentration of specific gas in ambient air is employing semiconductor gas sensor. Semiconductor is different from metal, it is the material that can control the number of carriers. The semiconductor will show an apparent surface effect when there are only a small number of carriers. The basic working principle of semiconductor gas sensor is changing the number of carriers by using the adsorption reaction on its surface, namely using the resistance change as a means of gas concentration detection. Now there are some relatively successful kinds of semiconductor gas sensor, such as sintered, contact combustion and conduction type. But they all have a common drawback that they are required to work in a heated state. The general heating temperature is 300℃ to 500℃. This may lead to continuous growth of internal grains of semiconductor material. In addition, it may bring bad effects on the performance of sensor. For example, it will have low sensitivity, poor stability, short life span and slow response speed. At the same time, semiconductor gas sensors usually are resistive or capacitive. Its output signal is analog signal. In order to interface with the microprocessor, the analog signal should be converted into digital signal by A/D conversion. This process may also decline the accuracy.

In 1979, American scientists raise a proposal of using SAW devices to sensitive different kinds of gas composition. They used the SAW devices which are made of ST-cut quartz or LiNbO to detect gas chromatography. It achieved great success. Since then, the research rapid (development) in SAW gas sensors is fast. The types' number of gases which can be detected is lager and larger. Presently, main gas which can be detected are SO_2, H_2O, acetone, H_2, H_2S, CO, CO_2, NO_2 and so on.

SAW gas sensor can output quasi-digital signal, which can easily interface with microprocessor and compose of real-time processing system.

1. The mechanism of SAW gas sensor

Early SAW gas sensor is based on one channel delayline oscillator. On the propagation path of the delayline of SAW is covered with a layer of selective absorption film, with an absorbing effect on the gas that is needed to be sensed. Film absorbing gas will cause the change of the frequency of the SAW oscillator and by precisely measure the difference, the targeted gas viscosity will then be told. At present, most of the SAW gas sensor use dual-channel delayline structure to compensate the influences of common-mode interference like environment temperature. Fig. 5.18 shows the structure sketch of the double channels SAW gas sensor.

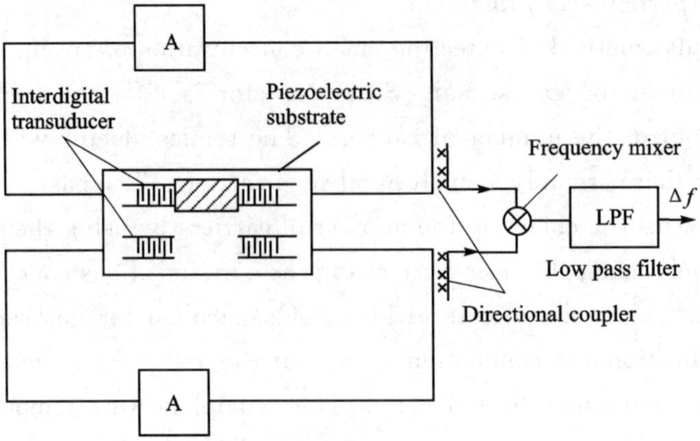

Fig. 5.18 Structure sketch of the double channels SAW gas sensor

In the structure of duel-channel SAW delayline oscillator, the propagation path of one of the two channels is covered with gas-sensitive film to sense the component of measured gas, while the other channel is raw as a reference. The frequencies of two oscillators are processed by mixed and the output is in the form of difference frequency, to compensate the common-mode interference (mainly the changes of environment temperature).

In a SAW gas sensor, the most important component besides SAW delayline is the selective gas-sensitive film. The sensing mechanism of the SAW gas sensor alters alongside the adapted type of film. When the film is developed from isotropic insulating materials, its absorbing effect towards gas will transferred into the change of the density of the layer. The mass loading effect along the propagation path of the SAW delayline causes the change of the velocity of wave and further cause the departure of the frequency of the SAW oscillator.

Concerning such circumstance, the signal provided by the SAW gas sensor can be approximately describe as:

$$\Delta f = f_0^2 h\rho(k_1 + k_2 + k_3) \tag{5.45}$$

where

Δf is the frequency departure of the SAW oscillator caused by layer absorbing gas (in Hz);

k_1, k_2 and k_3 are the material constant of piezoelectric substrate (in $m^2 \cdot s \cdot kg^{-1}$).

f_0 is the initial resonant frequency of the SAW oscillator, namely the working frequency with a zero measured gas viscosity (in Hz);

h is thickness of the film (in m);

ρ is density of the material of the film (in kg/m^3).

Table 5.1 lists the sound velocity v_R and material constant k_1, k_2, k_3 of some of the common piezoelectric substrate material.

When conducting material or metal oxide semiconducting material is used in the film, the electrical conductivity changes along with the change of the viscosity of the absorbing gas, which also causes the change in velocity and attenuation of the SAW wave. The frequency alters as well. Here, the output responses of the SAW gas sensor can the described using Eq. (5.46), namely

$$\Delta f = -f_0 \cdot \frac{k^2}{2} \cdot \frac{\sigma_0^2 h^2}{\sigma_0^2 h^2 + v_R^2 c_f^2} \tag{5.46}$$

Tab. 5.1 the material constant of some common piezoelectric substrate

Substrate	Trimming type	Direction of propagation	$v_R/(m \cdot s^{-1})$	k_1	k_2	k_3
				$\times 10^{-9}\ m^2 \cdot s \cdot kg^{-1}$		
Quartz	y	x	3 159.3	−41.65	−10.23	−93.34
$LiNbO_3$	y	z	3 487.7	−17.30	0	−37.75
$LiTaO_3$	y	z	3 229.9	−21.22	0	−42.87
ZnO	z	$x+45°$	2 639.4	−20.65	−55.40	−54.69
Si	z	x	4 921.2	−63.32	0	−95.35

where

k is electromechanical coupling coefficient;

c_f is material constant of the film (in A/V);

σ_0 is electrical conductivity of the film (in $A \cdot (V \cdot s^{-1})$);

v_R is sound velocity of SAW (in $m \cdot s^{-1}$).

It can be told from Eq. (5.46) that when conducting film or metal oxide semiconducting film is adopted, the change of film electrical conductivity is the main mechanism to measure the responses of the SAW gas sensor.

2. Relation between the film and characteristics of the sensor

The covered film is the direct sensing component of the SAW gas sensor, and its characteristics are closely related to the performance of the sensor itself.

(1) Film and selectivity of sensor

This is one essential performance indicator of SAW gas sensor. The selectivity the film having over gas determines the selectivity of SAW gas sensor itself. Different types of chemical film are in need concerning the specific type of gas. At present, sensing film used in SAW gas sensor mostly are Triethanolamine film (sensing SO_2), Pd film (sensing H_2), WO_3 film (sensing H_2S), Phthalocyanine film (sensing NO_2). Truly, as long as particular film sensing certain gas is successfully developed, it is within the capability to manufacture the corresponding SAW sensor to detect the gas. Therefore, the preparation for the absorbing film with better selectivity is one important task.

(2) Film and reliability of sensor

As a sensor, its output responses should be repeatable and reliable at the same time. The reliability of the output of SAW gas sensor is determined by the stability of the sensing film, especially the reversibility and high stability of the characteristic of the sensing film, the basic requirement for the sensing film.

Reversibility means that the sensing film is able to absorb and realize the desorption of certain gas. When the gas concentration rises, the amount of gas absorbed by the film increases. The amount decreases along with it when the concentration goes down. Absorption and desorption ought to be exactly mutually inverse, as well as rapid, which lays the foundation of the gas sensor's normal and reliable working status.

The stability of the film relies on its mechanical quality. If the internal stress and the adhesion between the film and the substrate are inappropriate, the film will creep, crack or shed. The mechanical quality of the film is largely determined by its structure, namely the deposition method. Usually the film developed by sputtering has a smaller internal stress. Meanwhile, because of the high energy injecting particles during the developing process, flaws appear on the substrate and the binding energy increases as well. Therefore, the adhesion condition of sputtering film is better than those developed through other methods.

(3) Film and response time of sensor

SAW gas sensor is alike other sensor, with a perception of a shorter response time, which is closely related to the thickness of the sensing layer and the working frequency of the delayline resonator. With a higher working frequency, the speed of gas spreading and counterbalance is higher, the responding speed increases consequently. What needs to take into consideration is that high frequency brings bigger substrate noise, which impedes the detection of gas with a low concentration. Considering that the time of gas spreading is proportional to the film's thickness square, when the thickness of sensing film decreases, it will surly shorten the response time of the sensor. Generally, the response time of SAW gas sensor will largely decline with the raising of SAW resonator's working frequency and the utility of much thinner film.

(4) Film and resolution of sensor

The resolution of SAW gas sensor mainly relies on the stability of the sensing film. Research indicates that the resolution and stability stays at the same order of magnitude. Currently, gas sensor with high resolution has already developed. For instance, SAW H_2S sensor has a performance of repeatable response with the lowest concentration of H_2S below 10^{-8}.

When the film was covered on the SAW's delay path, not only the resonant frequency of covered delayline resonator shifts, but the SAW signal attenuates. When the concentration of the gas to be measured is high enough and the film has already absorbed enough gas, the SAW signal might attenuate so fast while propagating along the delayline covered by the film that the resonator can hardly work properly. This generates the problem of the detection upper limit of the sensor. One effective way to lift the upper limit is to shorten the delayline length covered by the film in order to weaken the attenuation of SAW. This method also causes a decline in the sensitivity of the sensor. Therefore it takes a thorough consideration to design each performance indicator.

In the end, what need to point out is that the manufacture of the film is an essential process of developing SAW gas sensor. Readers can refer to other related materials, and this book will not further introduce.

5.4.5 SAW Flow Sensor

1. structure and mechanism of the sensor

SAW flow sensor mainly consists of SAW delayline, heater, amplifier and the channel for flowing fluids, as is shown in Fig. 5. 19. Its basic working mechanism is below: the heater warms up the SAW substrate and the temperature of the substrate stabilizes when in the status of heat balance; when fluids flow through, the heat of the substrate loses, causing the SAW wave speed altering as well as the shifting of resonator. By measuring the changes of frequency the flow will be told. It actually is the process of modulation on the working frequency of the SAW delayline resonator by the flow. The departure of frequency corresponds to the flow change.

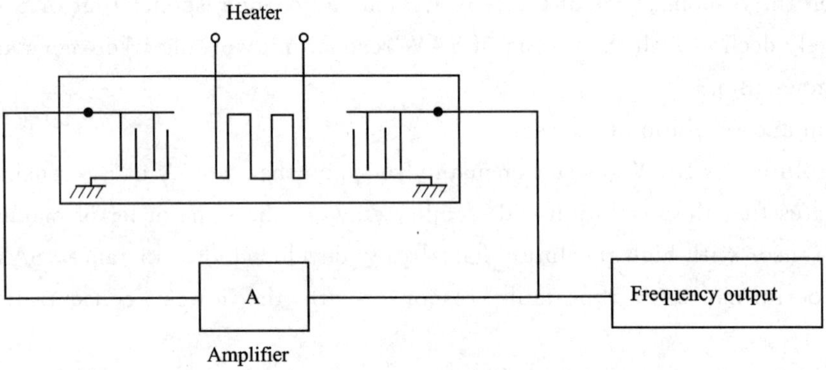

Fig. 5.19 Structure sketch of the SAW flow sensor

2. Fundamental functions

When gas flows, the heat attenuates through three methods: thermal conductivity, natural convection and thermal radiation, below:

$$q_{cond} = G_{th}(T_{SAW} - T_0) \tag{5.47}$$

$$q_{nc} = h_n A(T_{SAW} - T_0) \tag{5.48}$$

$$q_{rad} = k\varepsilon A(T_{SAW}^4 - T_0^4) \tag{5.49}$$

where

q_{cond}, q_{nc} and q_{rad} are respectively loss through thermal conductivity, natural convection and thermal radiation separately (in W);

T_0 and T_{SAW} are respectively ambient environment temperature and SAW substrate temperature (in K);

G_{th} is coefficient of thermal conductivity between the substrate and the environment (in W/K);

h_n is coefficient of natural convection (in $W \cdot (m^2 \cdot K)^{-1}$);

A is surface area of the substrate (in m^2);

k is Boltzmann constant, $k = 1.381 \times 10^{-23}$ J/K;

ε is radiation coefficient of the substrate (in $m^{-2}\cdot s^{-1}\cdot K^{-3}$).

Usually, the radiation attenuation is relatively small to be ignored, so under the thermal balance status when the heat input power P_{th}

$$P_{th}=q_{cond}+q_{nc}=(G_{th}+h_nA)(T_{SAW}-T_0) \tag{5.50}$$

In a SAW flow sensor, the SAW device is thermally separated from surrounding objects. If an insulator with a thickness of d is adopted in the device to separate it from the cage, then

$$G_{th}=\frac{KA}{d} \tag{5.51}$$

where K is the coefficient of thermal conductivity of the insulator.

Furthermore,

$$\left.\begin{aligned} G_0 &\overset{\text{def}}{=\!=} G_{th}+h_nA \overset{\text{def}}{=\!=} Ag_0 \\ g_0 &= \frac{K}{d}+h_n \end{aligned}\right\} \tag{5.52}$$

where G_0 is the effective thermal conductivity between the substrate and the environment without gas flow.

Based on Eqs. (5.50) and (5.52), it can be concluded that:

$$T_{SAW}-T_0=\frac{P_{th}}{G_0}=\frac{P_{th}}{Ag_0} \tag{5.53}$$

When additional heat loss resources are introduced in the air flow to form the forced convection, its loss q_{fc} can be described as

$$q_{fc}=h_fA(T_{SAW}-T_0) \tag{5.54}$$

where h_f is the forced convection coefficient ($W/(m^2\cdot K)$). It is a function of v_f, therefore can be described as $h_f(v_f)$.

When forced convection cooling takes place, Eq. (5.53) converts to

$$T_{SAW}-T_0=\frac{P_H}{A[g_0+h_f(v_f)]} \tag{5.55}$$

According to Eq. (5.55), the flow changing caused by SAW substrate temperature variation

$$\Delta T_{SAW}=\frac{-P_H\Delta h_f(v_f)}{A[g_0+h_f(v_f)]^2}=\frac{-(T_{SAW}-T_0)\Delta h_f(v_f)}{g_0+h_f(v_f)} \tag{5.56}$$

The relationship between the changes of SAW resonator frequency Δf and ΔT_{SAW} is

$$\frac{\Delta f}{f_0}=\alpha\Delta T_{SAW} \tag{5.57}$$

where

f_0 is the resonator frequency when the flow speed is zero (in Hz);

α is the frequency/temperature coefficient (in 1/K or 1/℃).

$$\alpha = \frac{\Delta f}{\Delta T_{SAW} f_0} = \frac{1}{\Delta T_{SAW}}\left(\frac{\Delta v}{v} - \frac{\Delta l}{l}\right) \tag{5.58}$$

where $\frac{\Delta v}{v}$ and $\frac{\Delta l}{l}$ are respectively the relative change of SAW velocity and route caused by the substrate temperature change.

Through Eqs. (5.57) and (5.58), it can be concluded that:

$$\Delta f = \frac{-\alpha f_0 (T_{SAW} - T_0) \Delta h_f(v_f)}{g_0 + h_f(v_f)} \tag{5.59}$$

The function relationship between volume flow Q_V and average velocity is

$$Q_V = A_c v_f \tag{5.60}$$

where A_c is the cross-sectional area of floating gas passing above the substrate (m^2).

According to Eqs. (5.59) and (5.60), the fluid's volume flow can be solved by using the frequency departure Δf of SAW resonator (sensor).

According to Eqs. (5.60), in order to gain a high sensitivity, it is required that the substrate has a bigger frequency/temperature coefficient α, and make sure that a better thermal insulator (namely g_0 is quite small) exists between the substrate and the environment. When the sensor is working in a relatively higher temperature environment, a higher sensitivity will be achieved. In order to lower the heating power, it is participated that the SAW device's surface area should be small with a given substrate environment.

Questions and Exercises

5.1 What is surface acoustic wave (SAW)? What are its characteristics?

5.2 Briefly introduce the characteristics of the SAW resonator sensor.

5.3 Draw the schematic diagram of the interdigital transducer and briefly introduce its working mechanism.

5.4 What are interdigital transducer's main performance indicators? What are the relationships between the transducer and the indicators?

5.5 How many types of basic analyzing model of SAW interdigital transducer are there? What are their application characteristics?

5.6 How to realize the SAW resonator and compare the two typical SAW resonators.

5.7 Concerning the delayline SAW resonator, based on its mechanism, briefly analyze the structure parameter designing problem of emitting interdigital transducer and receiving interdigital transducer.

5. 8 Briefly introduce the mechanism and application characteristic of SAW resonator strain sensor.

5. 9 Briefly introduce the mechanism and application characteristic of the SAW resonant flow sensor shown in Fig. 5. 19.

5. 11 Briefly compare the mechanism of the SAW resonant flow sensor shown in Fig. 5. 19 with that of regular thermal gas mass flow sensor.

5. 12 Briefly introduce the working mechanism and application characteristic of SAW resonant gas sensor.

5. 13 Briefly introduce the necessity of SAW resonant sensor working at differential mode.

5. 14 The width between the teeth *a* and the gap *b* of the interdigital transducer manufactured by using piezoelectric quartz substrate, try solving out its working frequency.

Appendix A
Dynamic Characteristics Analysis of Vibrating Cylinder Pressure Transducers (VCPT)①

Zhuang Haihan　Fan Shangchun　Guo Zhanshe　Zheng Dezhi

Abstract: The dynamic character of the vibrating cylinder pressure transducers (VCPT), which have frequency output ability, is studied in this paper. The oscillator's parametric excitement characteristic has been analyzed first. An approximate solution for the parametric excitement function is established through the theoretical analysis on the oscillator's parametric excitement characteristic and is validated by the numerical simulation. It was proved that the dynamic pressure demodulation is correlated with the average pressure, and the parametric excitement phenomenon will not lead to the variance of the phase difference between the input pressure and the frequency output. Then, the following theoretical and computational fluid dynamics (CFD) simulation analysis on the cavity effect and the Pipe-Damping Effect shows that the two effects generate a first order low pass filter like RC lowpass filter which makes the VCPT critical frequency as low as about 4～6 Hz, which is also verified by experiments data. So the main source that influences the dynamic characteristic drastically has been found out. And the corresponding design process flow for structural parameters optimization is proposed in the end to fulfill the specific dynamic index demands of VCPT, which is helpful to expand the VCPT application field.

Keywords: Vibrating cylinder shell, Dynamic characteristic, Parametric excitement, Cavity effect, Pipe-Damping Effect

1　Introduction

The VCPT measures the pressure applied on the thin vibrating cylinder shell (TVCS) inner surface by measuring the vibration signal's frequency of the TVCS. Because frequency or periodicity could achieve extremely high measurement accuracy, the VCPT has very high

① Sensors and Actuators A: Physical, 2010, 157(2): 219-227.

static measurement accuracy index which could reach the magnitude as low as 10^{-5}. Additionally, the frequency is a kind of quasi-digital signal, which could be connected with digital signal processing system directly without any A/D converter. Therefore, the VCPT are widely applied in the air data computing system in aviation industry to indicate the speed and the altitude of the aircrafts for the pilots and the guiding units on the ground[1,2,4,5]. The VCPT have been widely applied in the worldwide due to the precision static characteristics [5,7,8].

The VCPT have very high static measurement accuracy, while quit a few analysis on dynamic characteristic could be found. Only reference [9] reported the passband of SONIX VCPT products to be no less than 10 Hz. It is a pity that the parametrical excitement characteristics and that how to influence the transfer character of VCPT are not analyzed. The analysis on the dynamic characteristics is the essential method to improve the performance and expand the apply domain, such as the pressure monitors for motors, the engines and the boilers where a higher critical frequency index more than 10 Hz is demanded.

As early as 1957, H. F. Olson researched the fluid transfer behavior in the capillary tube with a small end cavity in reference [10]. The theory was applied in solving the fluid transmission problems in the capillary tube in the books[11,12] written in 2003 and 2005. The theory shows that time delay do exist between the outlet and the inlet, but it does not fit the situation under which the tube is a capillary connected with a "big" hole at the end, which would make more time delay. In addition, the actual pipe is not so straight and the inner surface is not so smooth as that applied in the theoretical model, which makes the experiment result is far different from the theory result. The critical frequency calculated from the formula in [12,13] is about 100 Hz, while practical data indicates it is only about 4~6 Hz.

This paper analyzed the parametrical excitement characteristics based on the basic VCPT working principle, and introduced the computational fluid dynamics (CFD) simulation method to help analyze the VCPT fluid transmission character, which influence the VCPT lower 50 Hz frequency response drastically. Also a pipe structure optimization method was established for a demanded dynamic index at the end.

2 Working Principle of the VCPT

A vibrating cylinder pressure transducer (VCPT) changes the nature frequency of its core part, a thin vibrating cylinder shell (TVCS), to present the absolute pressure applied on the inner surface of the shell as what is shown in the schematic view in Fig. 1. To maintain the resonant status for VCPT, the drive coil is forcing the TVCS. The pickup coil is detecting the vibrating signal. During this detection procedure the pressure applied on the inner surface of TVCS will change the nature frequency of the TVCS. And the frequency of the vibration signal from the pick up coil presents this pressure load. [1,6,12]

VCPT's characters depend on the vibration character of the TVCS. The TVCS can vibrate at a set of different frequencies with different shape, which is named "mode" presenting the pair of frequency and shape. Each mode has different sensitivity to the pressure load. Because of the highest sensitivity of pressure load and lowest vibrating energy, (1,4) mode of TVCS is chosen to be the working mode of the TVCS. Fig. 2 shows the (1,4) mode shape simulated by ANSYS. "(1,4)" means that the number of the wave in the axial direction is 1 and 4 in the circumferential direction. The vibration theory shows that the different vibrating mode is orthogonal to each other, and each of them could be considered as a single 2nd order vibration system which could be described as

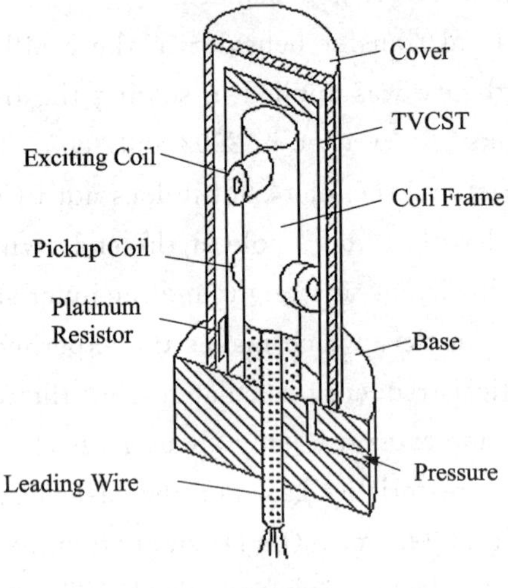

Fig. 1 Schematic view of VCPT[2]

Fig. 2 VCPT working mode, the (1,4) mode

$$\left.\begin{aligned} m_{1.4}\frac{\mathrm{d}^2 T}{\mathrm{d}t^2} + K_{1.4}T = 0 \\ \omega_{1.4} = \sqrt{\frac{K_{1.4}}{m_{1.4}}} \end{aligned}\right\} \tag{1}$$

where

T is TVCS vibration displacement;

$K_{1.4}$ is TVCS (1,4) mode equivalent stiffness;

$m_{1.4}$ is TVCS (1,4) mode equivalent mass;

$\omega_{1.4}$ is (1,4) mode resonant angular frequency.

The TVCS vibration stiffness $K_{1.4}$ changes with the change of the pressure applied on the inner surface of the TVCS, while the equivalent mass maintains the same, which makes the (1,4) mode resonant frequency changes. The monotonic function relationship between the input pressure and the frequency of vibration signal(sinusoidal wave) is as below

$$f = f_0 \sqrt{1+Sp} \tag{2}$$

where

$f=\omega_{1.4}/2\pi$ is TVCS frequency output;

f_0 is no-load resonant frequency of TVCS;

S is transducer's coefficient;

p is the pressure load of the TVCS[1-6,12].

3 Parametric Excitement Theoretical Modeling of TVCS under Alternating Pressure Load

No TVCS vibration model with alternation load exits in the current references. This article attempts to establish a parametrical exciting vibration model of TVCS based on the basic 2nd order vibration model introduced in section 2. If the pressure is a constant value with an alternating part, which is

$$p = p_0 + p_c \cos \omega_p t \tag{3}$$

where

p_0 is average pressure load;

p_c is amplitude of the alternating pressure;

ω_p is frequency of the alternating pressure.

The VCPT working with a periodical load can be described as a 2nd order vibration system with an alternating coefficient

$$\frac{d^2 T}{dt^2} + (\widetilde{K} - 2\widetilde{\alpha}\cos \omega_p t) T = 0 \tag{4}$$

where

$\widetilde{\alpha}=\omega_0^2 S p_c$ is intensity of parametric excitement;

$\widetilde{k}=\omega_0^2(1+sp_0)$is stiffness constant of vibration system;

$\omega_0=2\pi f_0$ is no-load resonant angular frequency of TVCS.

Then rescale argument $z=\pi_p t/2$ in Eq. (4),

$$\left(\frac{\omega_p}{2}\right)^2 \frac{d^2 T}{dz^2} + (\widetilde{K} - 2\widetilde{\alpha} \cos 2z)P = 0 \tag{5}$$

The normalized form of Mathieu function[14] could be described as

$$\frac{d^2 T}{dz^2} + (K - 2\widetilde{\alpha} \cos 2z)T = 0 \tag{6}$$

In which $K = 4\widetilde{K}/\omega_p^2$, $\alpha = 4\widetilde{\alpha}/\omega_p^2$.

Till now, the vibration characteristics of the TVCS with a sinusoidal alternation load for a specific mode could be expressed as Mathieu parametric excitement system. The solution of Eq. (4) could be achieved by simply rescaling the solution for Eq. (6), $T = T(\omega_p t/2)$.

The alternating frequency of the pressure is always much slower than the constant part of TVCS working frequency, which means $\omega_p \ll \omega_0 = \sqrt{\widetilde{K}}$, sometimes 2 or 3 orders of magnitude gap between them. So the coefficient K in Eq. (6) is bigger than 4×10^4, and α is comparatively small. With this kind of limit parameters, the first order approximate solution could be achieved by small parameter perturbation method[14].

$$T(z) = C\cos\left(\sqrt{K}z - \frac{\alpha}{2\sqrt{K}}\sin 2z + \varphi\right) \tag{7}$$

In which C and φ. are associated with the initial condition. The frequency response of the signal is just the differentiation of the part in parenthesis by z, which is

$$\omega(z) = \frac{d}{dz}\left(\sqrt{K}z - \frac{\alpha}{2\sqrt{K}}\sin 2z + \varphi\right) = \sqrt{K} - \frac{\alpha}{\sqrt{K}}\cos 2z \tag{8}$$

After the time scale transformation, the approximate formula of output signal from TVCS with the limit parameters could be written as

$$T(t) = C\cos\left[\sqrt{\widetilde{K}} \cdot t - \frac{\widetilde{\alpha}}{\omega_p \sqrt{\widetilde{K}}}\sin \omega_p t + \varphi\right] \tag{9}$$

And the frequency output of the TVCS is the differentiation of the part in parenthesis.

$$\omega(t) = \frac{d}{dt}\left[\sqrt{\widetilde{K}} \cdot t - \frac{\widetilde{\alpha}}{\omega_p \sqrt{\widetilde{K}}}\sin \omega_p t + \varphi\right] = \widetilde{K} - \frac{\alpha}{\sqrt{\widetilde{K}}}\cos \omega_p t =$$

$$2\pi f_0\left(\sqrt{1 + Sp_0} - \frac{S}{\sqrt{1 + Sp_0}} p_c \cos\omega_p t\right) \tag{10}$$

From Eq. (10) the following two primary conclusions can be drawn.

① With the limit parameters, that is, modulation amplitude of the alternating pressure is much smaller than the constant pressure and the frequency of the alternating pressure is much slower than the TVCS working frequency, the amplitude of output frequency signal $(f_0 S/\sqrt{1+Sp_0})p_c$ is in direct proportion with the alternating pressure amplitude p_c, and in inverse proportion with the average frequency. $\overline{f}=f_0\sqrt{1+Sp_0}$.

② There is no phase difference between the frequency output and the pressure load with above limit parameters.

The conclusions illustrated that the amplitude of frequency output is not only associated with the amplitude of pressure, but also related with the average pressure p_0. Therefore, the average TVCS frequency $\overline{f}$ should be calculated first to demodulate the average pressure p_0, and then the alternating part of pressure load can be demodulated by Eq. (10).

In section 7 of this article, the ideal frequency output will be calculated by Eq. (10) based on the known pressure p_0 and p_c to calibrate the frequency output, which will be discussed in detail thereafter.

4 The Numerical Simulation for the TVCS Parametric Exciting Normalized Function

The numerical solution of Eq. (6) obtained by the Runge-Kutta method could validate the precision of the approximate solution of Eqs. (7) and (8) in the VCPT parameters range. Fig. 3 shows the numerical responses with three sets of parameters, in which (a) is the stable solution and (b) the partial enlarged detail whose parameters are chosen from the stable area, (c) and (d) is the neutrality solution and the dismissed solution respectively. It is illustrated in (a) and (c) that the amplitude is modulated with the change of the frequency in a parametric excitement system. When the system stiffness (the center line) reaches its peak value, the frequency of the response signal also rise to its max while the amplitude decades to its minimum. It reverses when the stiffness reaches its floor. "PES" in Fig. 3 presents "Parametrical Exciting System".

The error between approximate solution and the numerical results will be evaluated following below three steps.

① Observing the amplitude of frequency response by fixing α and changing K;

② Observing the amplitude of frequency response by fixing K and changing α;

③ Observing the phase difference between the parameter modulation signal $-2\alpha\cos(2z)$ and the frequency output within the $\alpha-\sqrt{K}$ plane to verify the conclusion drawn in section 3 that no delay happens between the two signals.

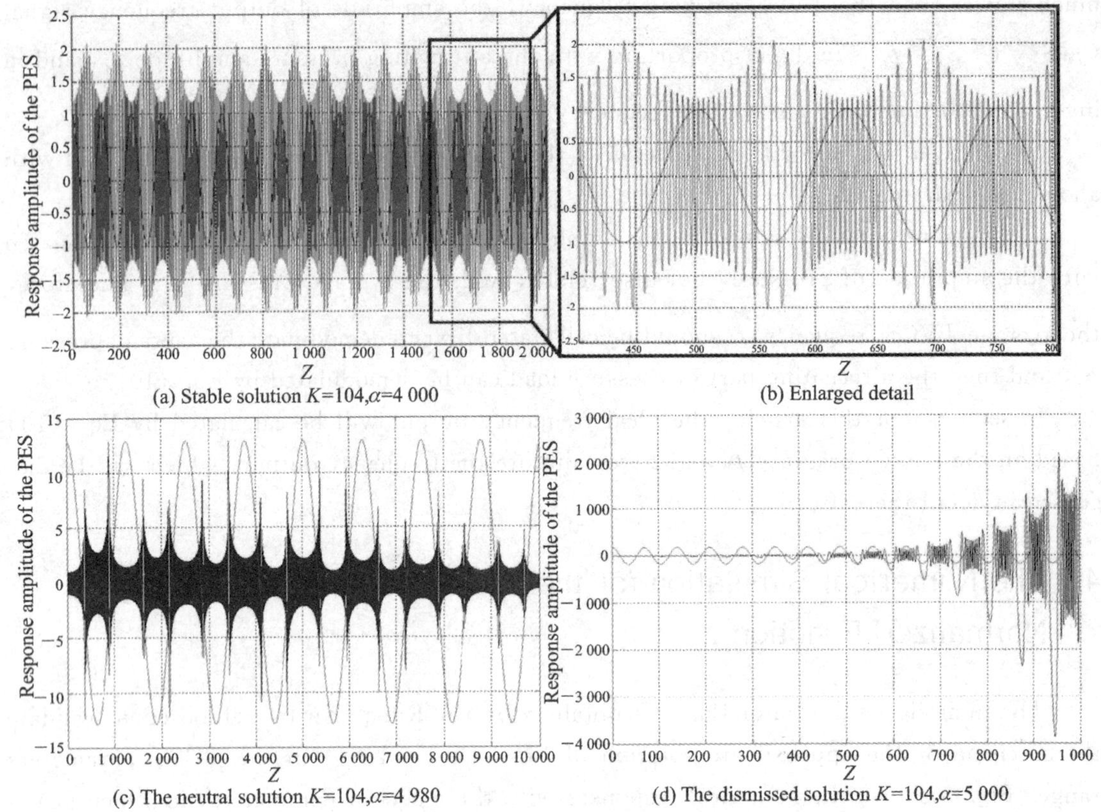

Fig. 3 The numerical solution of Mathieu function (the center line presents the alternating part of the stiffness)

The first step: The frequency output will be in inverse ratio to the average stiffness, following Eq. (8). The compare result of the two methods is illustrated in Fig. 4, where "AFA" presents "Alternating Frequency output's Amplitude", and "AFO" presents "Amplitude of Frequency Output". The ratio (y axis) between frequency output amplitude and 2α is illustrated in the upper subgraph with the theoretical solution drawn by line and the numerical method drawn by circle at the special points, while the stiffness $\sqrt{K}$ (x axis) changes and α is fixed to 500. If the TVCS center working frequency is 5 kHz, and the pass band of the VCPT is lower than 20 Hz, the span of $\sqrt{K}$ is within (500, 1 600). The lower subgraph is the relative error between theoretical solution and the numerical solution that indicates the error is always lower than 0.002%.

The second step: The frequency output amplitude will increase with the changing modulation intensity α in direct ratio. The upper subgraph in Fig. 5 shows the comparison of the

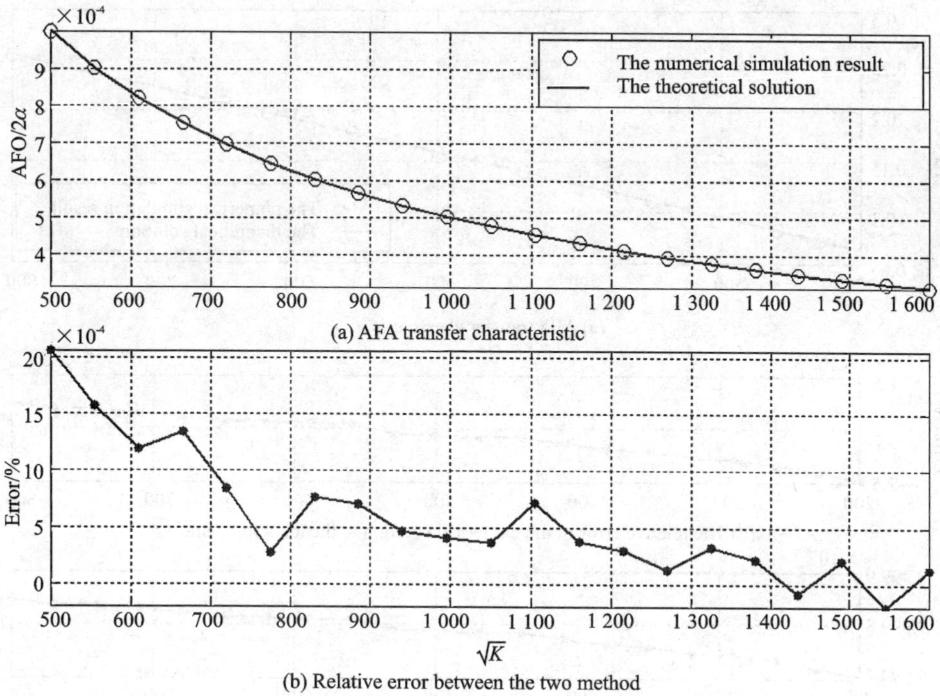

Fig. 4 Frequency output amplitude of parametrical excited TVCS achieved form the numerical and the theoretical method, and their comparison, when $\alpha=500$

theory result (in line) and the simulation result (in circle), where $\sqrt{K}=500$ and α within (200, 800). The middle subgraph shows the relative error between the two results that indicates the error is always lower than 0.002 5%. The lower subgraph illustrates the system transmission gain of $A(\omega)/2\alpha$, which is the ratio between the frequency output amplitude and the parametric modulation intensity. The value is around $9.999\,8\times10^{-4}$, while the theoretical value is $1/2\sqrt{K}=10^{-3}$ that is quit close to the simulated one.

The third step: Compare the phase difference among all the numerical solutions and the parametrical modulation signal ($-90°$), within the parameter range involved in the previous two steps, to verify if the frequency output could follow the change of the parameter output. In Fig. 6, the phase difference between frequency output of the numerical solution and the parametrical modulation signal $-2\cos(2z)$ is illustrated, which means the phase difference is kept in $\pm1.22\times10^{-4}$ rad $=\pm7\times10^{-3}$° within the α-$\sqrt{K}$ plane.

From the comparison of the upper 3 steps, the numerical solution's frequency output is close enough with theoretical Eq. (7) or Eq. (9). The approximate solution could be regar-

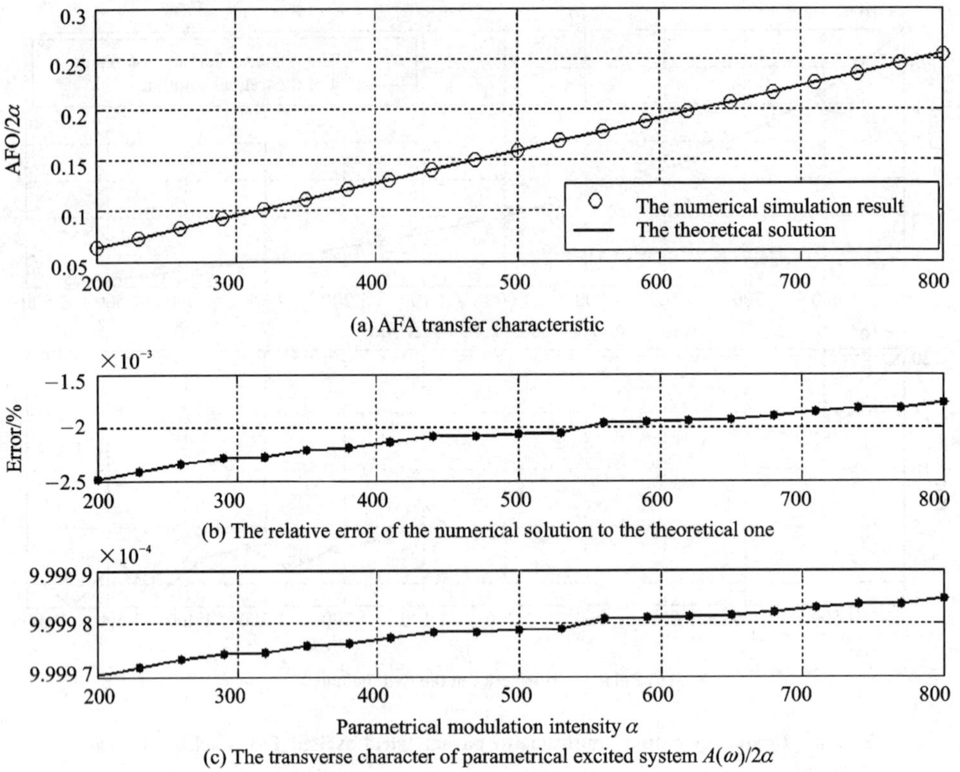

Fig. 5 The frequency output amplitude of the parametric excited TVCS achieved form the numerical and the theoretical method, and their comparison, where $\sqrt{K}=500$

ded as the enough accurate result to illustrate the real condition of the parametric excitement input, and be helpful for the parameter design in engineering. Therefore, from above analysis work on the parametric excited system, the conclusion could be drawn directly that the TVCS frequency output can follow the change of the pressure load applied on its inner surface without any time delay. The parametrical excitement effect would only influence the output amplitude through alternating the constant part of the stiffness. The bigger the constant K the smaller the frequency output amplitude, which means the passband of the parametrical excited system keeps infinite until the frequency of the alternating parameter is close to the working frequency of the TVCS. Nevertheless, the passband of the VCPT is not as wide as discussed above, and instead it is very narrow compared with the working frequency, 5 Hz for example. Therefore, in my opinion, the failure of the dynamic response is mainly due to two key factors that the cylinder cavity and the pipe through which the fluid transfer-

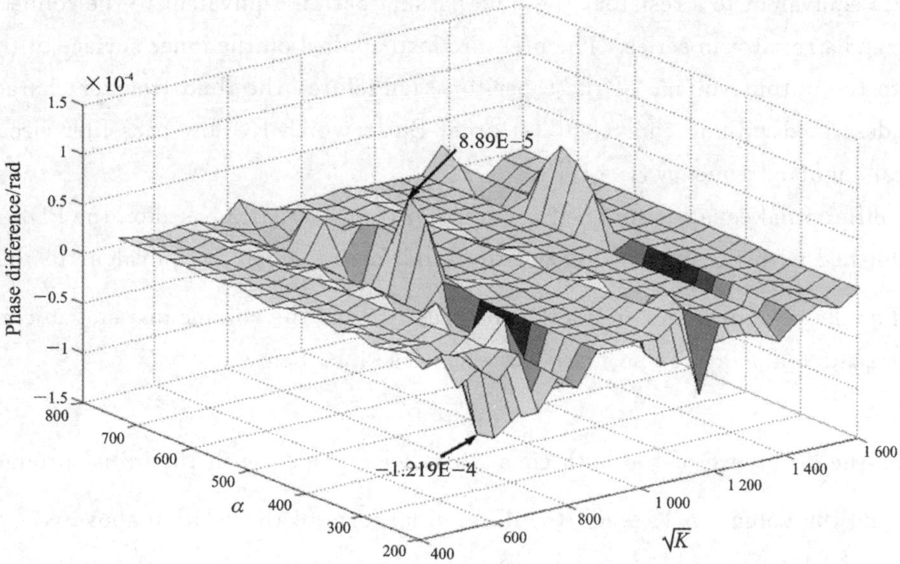

Fig. 6 The phase delay from the numerical solution to the parametric modulation signal in the α-$\sqrt{K}$ parameter plane

ring. In the following section, VCPT dynamic characteristics will be analyzed further with the consideration of the Pipe-Damping Effect and the Cylinder's Cavity Effect.

5 Theoretical Modeling of the Cylinder's Cavity Effect (CCE) and Pipe-Damping Effect (PDE) of VCPT

The air mass in the cylinder will change with the changing of the pressure load applied in the VCPT cavity because of the air's compressibility, the higher the pressure the heavier the air mass. The effect could be considered as the cavity absorbability of the air just as the capacitance accumulating the electrical charge in a circuit, which could be called the Cylinder's Cavity Effect.

Because of the fluid's viscosity and contacting with the wall, there exists a pressure drop between the pipe outlet and the inlet. That is to say, a pressure loss will take place in the viscous fluid. If take the pressure drop as the voltage and the mass flow as the electric current, the effect can be interpreted that an electric current causes a voltage drop when passing through a resistor, which could be called the Pipe-Damping Effect.

Considering the two effects above together that the CCE is equivalent to a capacitor and the PDE is equivalent to a resistor, the fluid passing path is equivalent to the connection of a capacitor and a resistor in series. The pressure load applied on the inner surface of the TVCS is equal to the output voltage of the capacitor. Therefore, the fluid transfer characteristics could be described with an equivalent 1st order Butterworth RC low pass filer circuit, with the critical angular frequency $\omega_c = 1/RC$.

The differential equation for the fluid transfer characteristics based on the PDE and CCE was established with the intermedium of the fluid mass, which is equivalent to the electric charge. For the ideal gas, the product of the pressure and the volume just state a constant $\widetilde{R}$ under the same temperature and mass, as what described below.

$$pV = \widetilde{R} \tag{11}$$

When the difference of the both equal sides was divided with the initial product of the pressure and the volume $p_0 V_0 = \widetilde{R}$, the differential form of the relation above is

$$\frac{\mathrm{d}p}{p_0} + \frac{\mathrm{d}V}{V_0} = 0 \tag{12}$$

where V_0 is the volume of the VCPT cavity. Because the air is compressed, the mass in the cavity of the cylinder increase with time

$$\frac{\mathrm{d}m}{\mathrm{d}t} = -\rho_0 \frac{\mathrm{d}V}{\mathrm{d}t} = -\rho_0 \frac{\mathrm{d}V}{\mathrm{d}p}\frac{\mathrm{d}p}{\mathrm{d}t} = \rho_0 \frac{V_0}{p_0}\frac{\mathrm{d}p}{\mathrm{d}t} \tag{13}$$

where

ρ_0 is gas density under the static pressure p_0;

m is the whole mass of the gas in cavity;

V is the fixed mass gas volume.

While the formula above describes the CCE, the PDE could be considered approximately that the air mass flow causes a pressure drop linearly multiplied by coefficient ξ.

$$p_{\mathrm{in}} - p = \xi Q \tag{14}$$

where Q is the mass flow in the pipe. The mass flow Q equal to the mass variation in the cylinder cavity per unit time, $Q = \mathrm{d}m/\mathrm{d}t$. The pressure transmission equation can be derived combining Eqs. (12) and (13).

$$\frac{\xi \rho_0 V_0}{p_0}\frac{\mathrm{d}p}{\mathrm{d}t} + p = p_{\mathrm{in}} \tag{15}$$

It can be concluded from the model above that the CCE and the PDE work as a low pass filter that filter the pressure signal during fluid transmission with the critical angular frequency of

$$\omega_c = \frac{p_0}{\xi \rho_0 V_0} \tag{16}$$

6 The CFD Simulation Model of CCE and PDE of VCPT

The coefficient ξ mentioned in Eq. (14) is very hard to achieve from the theoretical analysis because the actual situation of the pipe is very complex. Therefore, CFD simulation method is introduced here based on the Ansys Fluent 6.3 to calculate the coefficient ξ. The process of the simulation could be described as follow. First, set up the whole flow field structure including the pipe and the cylinder cavity to achieve a flow field preliminary solution by transient analysis and to check if there is a pressure transfer phase difference between the tube inlet and the cylinder cavity and thereafter give out the range of the quantities of the air mass flow. Second, set up the space mesh structure of the pipe and apply a series of mass flow to obtain the pressure drop of the inlet and the outlet of the tube. Then the coefficient ξ could be derived by the ratio between the pressure drop and the mass flow as Eq. (14) illustrated.

In the first step, set the surroundings pressure 1 atm, and the input alternating pressure amplitude set to 5 kPa. An accurate solution of the fluid field could not achieved only by the transient analysis, because the algorism itself would accumulate the errors with the increase of time step, which will make the result unreliable when the time goes too long. Therefore, this step could only certify whether the phase difference exists and the amount of the mass flow to determine the load applied in the next step.

Fig. 7 is the CFD mesh model for fluid field including the pipe and the Cylinder's Cavity. The fluid in the field is the ideal air with the density $\rho = 1.235\ \text{kg/m}^3$, the viscosity $\eta = 1.7894 \times 10^{-5}\ \text{kg/ms}$, pressure circumstance 1 atm $= 101\,325$ Pa, the inlet pressure (alternating pressure) $p_{\text{in}} = 5\cos(62.831853 \cdot t)$ kPa.

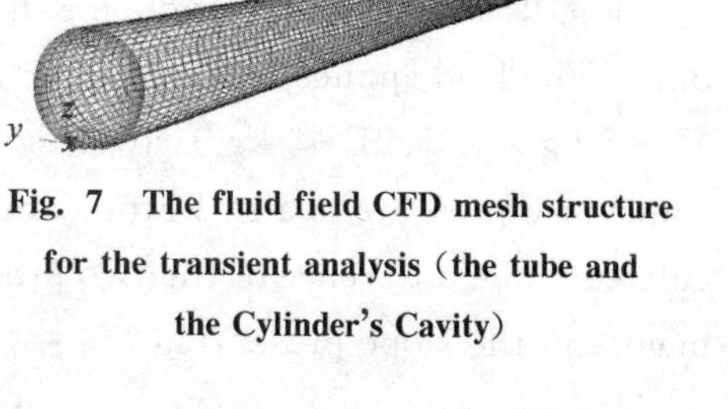

Fig. 7 The fluid field CFD mesh structure for the transient analysis (the tube and the Cylinder's Cavity)

96 h calculation result is shown in Fig. 8, with the black line representing the average pressure in the inlet cross-area and the gray line representing the outlet cross-area average pressure. To indicate the phase difference between the two lines obviously, the outlet pressure weights to the magnitude

level as much as the inlet pressure.

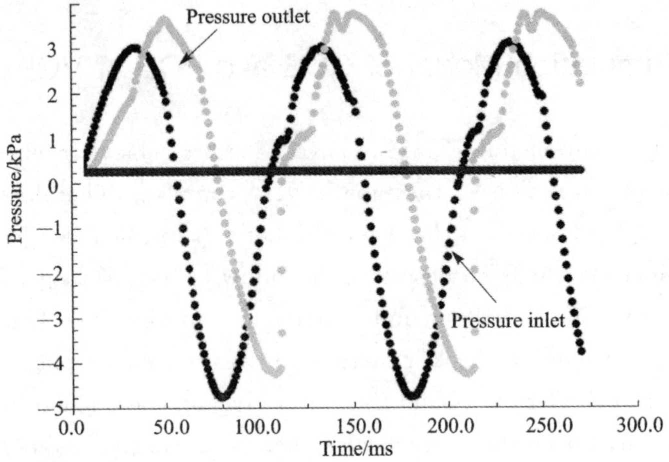

Fig. 8 The simulation result of the transient CFD analysis for the whole fluid field structure

The forward and the backward air flow in the curved structure cannot stay the same and the flow in the pipe is intense turbulent flow, which can be regarded as the nonlinearity of the fluid flow and the numerical calculation errors especially for the transient analysis will interfuse in the results as well. Considering the factors mentioned above, the lines in Fig. 8 do not look as sinusoidal function and are not smooth enough. Nevertheless, the phase difference between the outlet pressure and the inlet pressure do exist and almost up to about 60° when the frequency is 10 Hz. The assumption that the VCPT dynamic characteristics in the range 0～50 Hz are determined by the PDE and the CCE is proved partially by the CFD transient analysis, though the accuracy of which is poor.

From the transient simulation process the peak value of the mass flow is about 10^{-5} kg/s, when applying a 1 atm static pressure plus a 5 kPa alternating pressure to the inlet of the tube. Then more accurate ξ could achieve by steady flow analysis process after achieving the amount of mass flow.

Fig. 9 is the specific model for the calculation of ξ, in which tube structure is involved in only. The load applied on the inlet of the tube is set to a series of the mass flow ranging from 4E－6 kg/s to 3.2E－5 kg/s covering the result calculated by the transient analysis. The circumstance pressure is also 1 atm. The pressure of the inlet and the outlet of the tube are calculated. The difference of the two pressures is the function value with the mass flow as argument and the slope presenting the ξ.

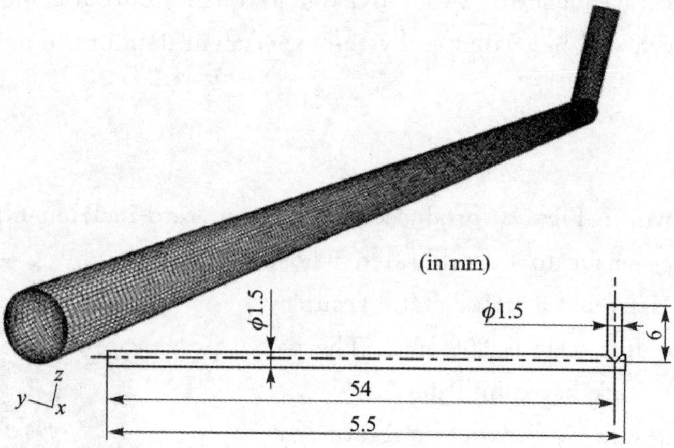

Fig. 9 The Pipe-Damping Effect (PDE) flow field's CFD simulation mesh model(upper) and the dimensions (lower)

The result lists in Tab. 1. The ξ is given by the ratio between the difference of adjacent pressure drop and the difference of adjacent mass flow, which means the curve slope of the pressure drop vs. the mass flow.

Because the actual condition of the pipe is not as ideal as simulated one, which has a much bigger roughness and a more complex of the pipe shape, the damping coefficient of the actual pipe is higher than the simulated one. Thus the superior limit of the value listed in Tab. 1, $\xi = 2.013\,4\text{E}8$ Pa · s/kg, is adopted for the approach of the actual situation. The critical frequency of the VCPT dynamic characteristic is derived following Eq. (16) as

$$\omega_c = \frac{p_0}{\xi \rho_0 V_0} = \frac{101.325}{2.013\,4\text{E}8 \times 1.2 \times 13.995\ \text{E}-6} = 4.769\,3 \times 2\pi\ \text{rad/s} \tag{17}$$

Tab. 1 The pressure drop and the corresponding slope (damping coefficient) to the mass flow

Mass flow $Q/(\text{kg} \cdot \text{s}^{-1})$	Pressure drop $\Delta p/\text{Pa}$	Damping coefficient $\xi/(\text{Pa} \cdot \text{s} \cdot \text{kg}^{-1})$
4.111 371 3E−6	1 194.550 1	—
8.222 742 5E−6	1 652.847 1	1.114 9E8
1.644 548 5E−5	2 576.343 8	1.244 6E8
1.843 092 7E−5	2 813.148 5	1.192 7E8
2.212 423 0E−5	3 542.803 8	1.975 6E8
2.466 817 4E−5	4 048.941 2	1.989 6E8
3.289 097 0E−5	5 704.519 7	2.013 4E8

Thus, the critical frequency of VCPT dynamic frequency output could be estimated to be around 4.7 Hz, which will be evaluated by the experiment data in the next section.

7 Experiment

The VCPT shown in Fig. 10 produced by Taiyuan Aero-Instruments CO., LTD. was adopted here as the sensor to be calibrated whose working frequency is around 5 kHz, static accuracy is 0.01%FS and the full scale is 300 kPa. The parameters of the VCPT are listed in Tab. 2.

The experiment was carried out in 1 atm circumstance while the inlet of the VCPT was exposed to an alternating standard pressure plus an around 200 kPa constant absolute pressure measured and controlled by another standard dynamic pressure sensor having a very excellent dynamic characteristic compared with VCPT, as what is photographed in Fig. 11. The calibrating system is composed of 4 parts which are pore plate sinusoidal pressure generator, pressure divider, standard dynamic pressure sensor and the signal sampling device.

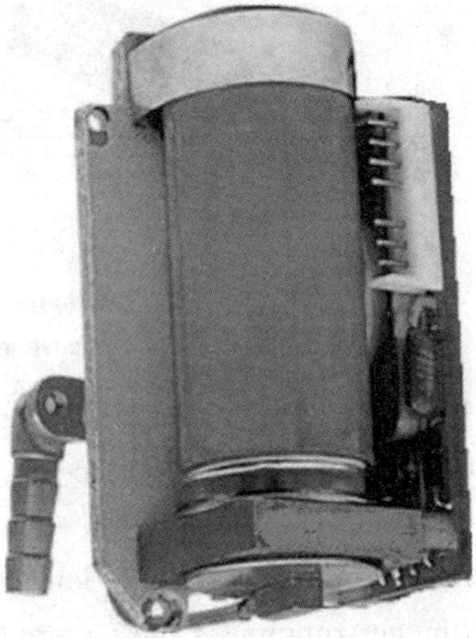

Fig. 10 The VCPT produced by Taiyuan Aero Instruments CO., LTD. [15]

The pore plate pressure generator sends out a sinusoidal pressure signal with a controllable average value and amplitude. The pressure is applied on the standard sensor and the calibrated sensor simultaneously through the pressure divider. The standard sensor is a silicon membrane piezo-resistor pressure sensor, whose nominal accuracy is 0.05%FS and the dynamic passband is larger than 1 kHz. The direct output signal of the VCPT, the signal from the pickup unit, and the pressure signal from the standard sensor are sampled by the signal sampling device with the sample frequency $f_s = 50$ kHz meanwhile. The sampled crude signal is drawn in the lower subgraph in Fig. 12. The frequency output is extracted by the zero passing method and is plotted by dot in the upper subgraph.

Tab. 2 The parameters of the VCPT for the experiment

f_0/Hz	S/MPa^{-1}	Nominal accuracy	Full scale/kPa
4 228.095 76	1.131 479 942	0.05%FS	0～346.6

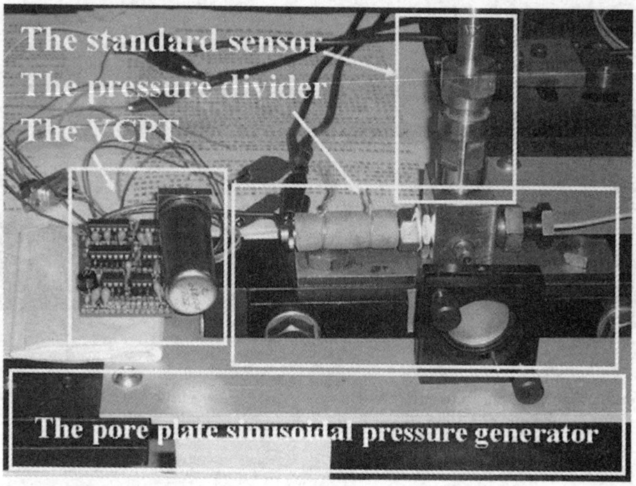

Fig. 11 The dynamic characteristics calibration assembly for the air pressure sensors

The calibrating pressure measured by the standard dynamic pressure sensor is converted to the ideal frequency output according to Eq. (10) deduced in section 3, and plotted by a line in the upper subgraph in Fig. 12. The transform process could be general-ized as follow: first extract the average pressure p_0 and then obtain the alternating frequency output by subtracting p_0 from the pressure signal and multiplying the coefficient $f_0 S\sqrt{1+Sp_0}$. Finally the ideal frequency output, which is the frequency output without consideration of the PDE and CCE, is the plus of the alternating frequency and the average frequency $f_0\sqrt{1+Sp_0}$.

As what shown in Fig. 12, the frequency output of the VCPT can follow the change of the pressure when the pressure alternating frequency is 1 Hz, see Fig. 12 (a). But if the frequency increases as high as 10 Hz, see Fig. 12(b), the amplitude of frequency output decreases suddenly accompanied with a time delay, which means the VCPT is entering the stop band. Till 20 Hz, see Fig. 12 (c) the amplitude of the frequency output decreases dramatically while the phase difference is almost 90° which means 20 Hz is in VCPT deep stop band.

The amplitude ratio between alternating part of the frequency output and the changing pressure is plotted as the amplitude frequency characteristics in the upper subgraph of Fig. 13, while the phase difference of the two signals is drawn in the lower subgraph as the phase frequency characteristics. The frequency of the pressure load is the x axis. Evidently, the VCPT represents a 1st order low pass filter characteristics within 50 Hz band wide with a horizontal line in the passband and attenuating by a −20 dB slope in the stopband. Further more, the nonlinear regression method for curve fitting is introduced to calculate the lowpass filter pa-

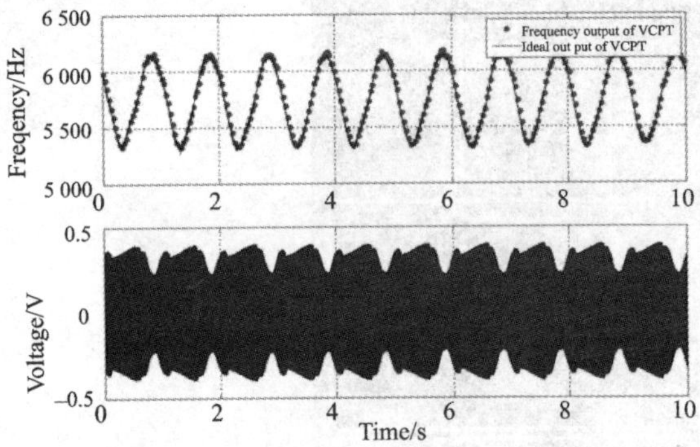

(a) Condition when the pressure alternating frequency is 1 Hz

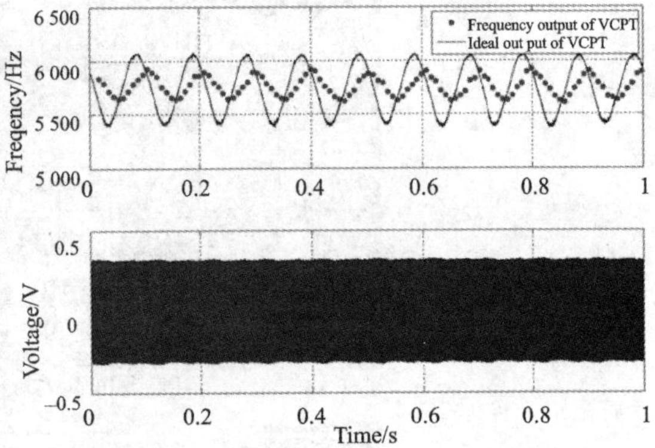

(b) Condition when the pressure alternating frequency is 10 Hz

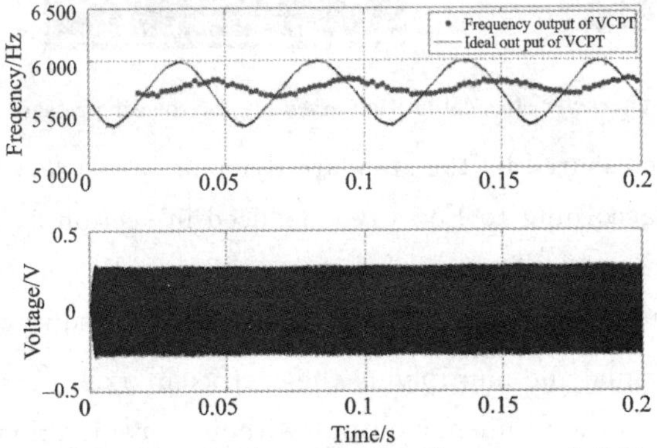

(c) Condition when the pressure alternating frequency is 20 Hz

Fig. 12 Comparison of the desire and actual frequency output of the VCPT at 1 Hz, 10 Hz and 20 Hz

rameters shown as the line in Fig. 14. The R-Square is 99.86% meaning the correlation of the curve and the experimental data coincide perfectly. The curve fitted shows the critical frequency of the lowpass filter to be

$$\widetilde{f}_c = 4.717\,6\ \text{Hz}$$

The VCPT critical frequency is $f_c = \omega_c / 2\pi = 4.769\,3$ Hz achieved form theoretical analysis with the CFD simulation result from the foregoing section, which is so close to the experiment data processing result $\widetilde{f}_c$ that the dynamic characteristic within 50 Hz could be illustrated by the theory model set up in this paper. The relative error of the two results is

$$e = \frac{4.769\,3 - 4.717\,6}{4.717\,6} \times 100\% = 1.096\%$$

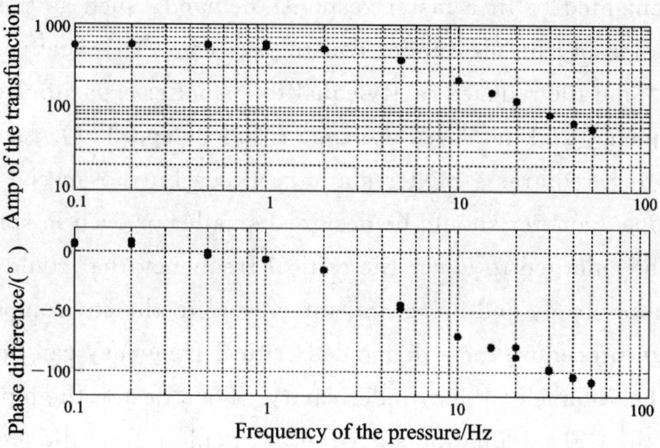

Note: The upper subgraph is the amplitude ratio between the frequency output and the pressure load. The lower subgraph is the phase difference of the two signals.

Fig. 13 The experimental data under 196 kPa absolute average pressure

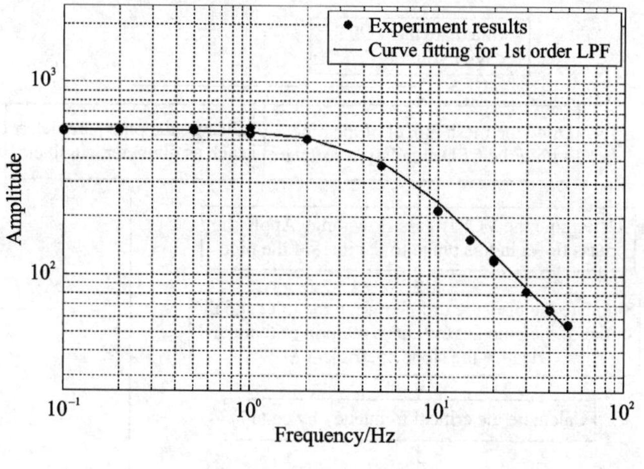

Fig. 14 The curve fitting result of the amplitude frequency response data for the amplitude frequency transmission function of 1st order lowpass filter

According to the theoretical analysis, the Cylinder Cavity Effect and the Pipe-Damping Effect could be determined to be the primary critical factor for the VCPT dynamic characteristics within 50 Hz.

The theory derived from all the analysis above presents the VCPT dynamic character within 50 Hz, which could be used for the structure parameter optimization for a demanded dynamic performance. The pipe diameter expansion or the volume of the cylinder cavity reduction can be implemented to fit a faster response demand, such as the pressure measurement in engines, boiler or the air tunnel control system. An expanded pipe has a lower damping coefficient ξ and the reduced cavity equal to a smaller capacitor, which would lead to a higher critical frequency and improved dynamic response speed. Or for the application of a pressure standard which requires a higher static accuracy but no special demand for the response speed, the pipe diameter should be designed smaller or even in spiral shape. The cylinder cavity should be enlarged to lower the critical frequency that could make the measurement result much more steady and being anti-interfering to the environmental disturbance.

Finally, a design procedure for a demanded critical frequency can be generalized as follow. First, obtain the volume of the cylinder cavity. Then design the pipe structure parameters with the aid of the CFD simulation software to calculate the ratio ξ of the pressure drop between inlet and outlet of the tube and the mass flow. The last, calculate the critical frequency following Eq. (17) to check if the structure can fulfill the dynamic demand and decide whether to repeat the process, just as what is shown in Fig. 15.

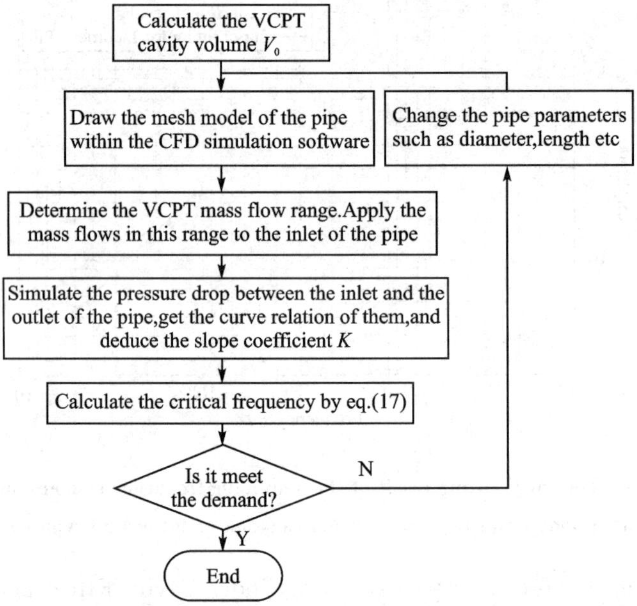

Fig. 15 Optimization flow process of the VCPT pipe structure parameters for a demanded dynamic character

8 Conclusions

① The vibration characteristic of the TVCS with an alternating load is equivalent to a parametrical excitement system which could be described by the Mathieu function. The approximate relationship between the TVCS transfer and the alternating load is derived in the theoretical way and validated by the numerical method. The corresponding demodulation method for the transient pressure load is achieved.

② It is proved that the phase delay from the TVCS frequency output to the pressure load is not caused by parametric exciting characteristic.

③ The critical factors influencing the dynamic character of the VCPT are the Cylinder Cavity Effect and the Pipe-Damping Effect, which is verified by both of the theoretical and experimental way. The critical angular frequency decided by the two effects is $\omega_c = p_0/\xi\rho_0 V_0$.

④ A design process for the pipe and cylinder structure for a demanded dynamic characteristic is implemented in the end of the paper following the analysis result, which is very helpful to expand the VCPT application field.

Acknowledgements

This study is supported by the Aviation Research Foundation, China with grant No. 2007ZD51050 and Ph. D. Innovation Foundation of BUAA, China.

References

[1] Weng Shanchen, Lin Youde. Foundation of the Instruments Elastic Element Design. Machinery Industrial Publisher of China, 1982: 330-347.

[2] Fan Dajun, Liu Guangyu. Design of the New Elastic Sensing Element. National Defense Industry Publisher of China, 1995:321-393.

[3] Fung Y C, Sechler E E, Kaplan A. On the vibration of thin cylindrical shells under internal pressure. J. Aeronaut. Sci., 1957(24) 650-660.

[4] Leissa A W. Vibration of Shells, NASA Report SP—288. Government Printing Office, Washington, DC, 1973.

[5] Copley D W. The application of vibrating cylinder pressure transducers to nonmilitary usages// Proceedings of the IEEE. National Aerospace and Electronics Conference, 23-27 May, 1994,2: 1252-1258.

[6] Fan Shangchun. Master Dissertation, Double-Mode Piezoelectric Excited Vibrating Cylinder Pressure

Transducer. Beijing University of Aeronautics and Astronautics, 1986.

[7] Qin Jie, Fan Shangchun. A Piezoelectric Excited Vibration Cylinder Pressure Transducer: China, 200310109681.7[P].

[8] Zhang Guang F, Liu Guangyu. Vibration Cylinder Pressure Transducers: China, 200310109682.1[P].

[9] Love J Scott, Juanarena Douglas B. Design and Application of Vibrating Cylinder Based Pressure Standards, ISA, Paper #88—0739, 1988(88):337-345.

[10] Olson H F. Elements of Acoustical Engineering. 2nd ed. D. Van Nostrand Co., New York, 1947: 87-90.

[11] Du Gonghuan, Zhu Zhemin, Gong Xiufen. Foundation of Acoustics, Njupress, China, 2003: 277-279.

[12] Fan Shang Chun. Aviation Testing System. Beijing: Beihang University Press, 2005: 91-95, 101-109.

[13] Doebelin Ernest O. Measurement Systems Application and Design. 5th ed. Mc Graw Hill, 2004: 515-520.

[14] McLachlan N W. Theory and Application of Mathieu Function. Dover Pub., Inc., 1964:90-95.

[15] http://www.taihangybc.com.

Appendix B
Mathieu Equation with Application to Analysis of Dynamic Characteristics of Resonant Inertial Sensors①

Li Yan   Fan Shangchun   Guo Zhanshe   Li Jing   Cao Le   Zhuang Haihan

Abstract: In this paper, Mathieu equation is applied to analyze the dynamic characteristics of resonant inertial sensors. Unlike previous work, Mathieu equation is not just a differential equation and analyzes the stability of the transition curves, but become an important method in analyzing parametric resonant characteristics and approximate output of resonant inertial sensors. It is demonstrated that the mathematical model of resonant inertial sensors is described by Mathieu equation. The relevant Mathieu equation theory and dynamic characteristics analysis methods were proposed, which include both stability and dynamic linear output. Finally, theoretical and experimental analysis show that the correlation of the theoretical curve and the experimental result coincide so perfectly, which means proposed analysis methods for Mathieu equation could be used to analyze the dynamic output characteristic of resonant inertial sensors. The theoretical analyzing approach of Mathieu equation and experimental results of resonant inertial sensors are obtained, which provide an application area for Mathieu equation and a reference for the robust design for resonant inertial sensors.

Keywords: Mathieu equation, Application, Dynamic characteristics, Resonant inertial sensors

1 Introduction

The majority of equations used in technical and applied mathematics have originated as the result of investigating practical problems. Mathieu equation was introduced by their originator in 1868[1], when he determined the vibrational modes of a stretched membrane having an elliptical boundary.

---

① Commun Ications in Nonlinear Science and Numerical Simulation, 2013, 18(2): 401-410.

The Mathieu equation

$$q''(z) + (\delta + \varepsilon\cos 2z)q(z) = 0 \tag{1}$$

occurs in a wide variety of practical problems, which has been studied in great detail since its discovery. Mathieu equation is an excellent paradigm to study parametric resonance, henceforth referred to as PR, and it exemplifies the response of a system to the kind of periodic force. In fluid dynamics, many examples of waves may be excited by PR. Other examples can be found in acoustic instabilities in flames[2] and Rayleigh-Benard convection[3]. In geophysical fluid dynamics, Poulin et al. [4] and Pedlosky and Thomson[5] derived ordinary differential equations of the Hill type to describe the stability of barotropic and baroclinic shear flows, respectively. A rediscovery of damped Mathieu equation appears in Ref. [6], where it is shown that the elastic motion of a Large Space Structure. But the reports about Mathieu equation applied in kinetics sensors have not been detailed described.

The inertial sensors are targeted for accelerometers and gyroscopes. Although the inertial sensors have been the subject of intensive research in the past few years[7,14], they are just researched about novel structure, control strategies, and the working principles of the sensors were not resonant exciting and resonant sensing. Resonant inertial sensors are not only advantageous for micromechanical actuators but also for resonant sensing. Resonant sensing often involves the detection of input measurand by means of resonant frequency shift in the sensing device. Therefore resonant inertial sensors technique benefit from being highly sensitive, have the potential for large dynamic range, good linearity, low noise and potentially low power.

However, there are not too many units taking the research about resonant inertial sensors and their dynamic characteristics. Seshia[15] reported integrated microelectromechanical resonant inertia sensors, but the dynamic characteristics of resonant inertial sensors were not introduced in detail. Moussa and Bourquin[16] reported the theory of direct frequency output vibratory gyroscope, but the gyroscope they researched that the vibratory mode is out of plane, and the detection principle is not resonant sensing to measure the Coriolis force directly. There are some other literatures about resonant inertial sensors, but they are all focusing on driving control and fabricated process[17-22].

The main reason for resonant inertial sensors develops so slowly is that the dynamic characteristics of resonant inertial sensors just are parametrically excited systems, which could be described by differential equations in which the input appears as a time dependent coefficient. They are typified by the Mathieu equation. This work extends Mathieu equation dynamic characteristics analysis to resonant inertial sensors. A series of differential equations are necessary to model the resonator dynamics. The resonant frequency is difficult to model and there is considerable uncertainty in the parameters, so the dynamic characteristics analysis is needed, and Mathieu equation with application to analysis of dynamic characteristics of resonant inertial sensors is necessary. Moreover, Mathieu equation cannot be solved analytically in terms of standard functions, the reason is that one of the coefficients is not constant but time-dependent.

This work using the notion of parametric resonance to create dynamic characteristics analyzing methods for resonant inertial sensors systems. About the analyzing methods, recently several authors have studied the stability of and approximate solution for Mathieu equation. In references[23-25], the authors divided the parameter space into regions of stability/instability. Arrowsmith and Mondragon[26] analyzed how the size of parametric regions of stability for the Mathieu equation can be enlarged. In references [27-31], the authors analyzed the of quasi-periodic nonlinear Mathieu equation, the extensions to the Mathieu equation contain two periodic terms in different frequencies, which are not applicable in our research.

This work focuses on analyzing of Mathieu equation with application to dynamic characteristics of resonant inertial sensors, enabling the resonant inertial sensors to robust design. Since this is the focus, here the applications of each of the resonant inertial sensors studied are introduced using simple method.

The rest of this paper is organized as follows. section 2 derive the proposed Mathieu equation relevant theory with particular emphasis on how to analyze dynamic characteristics of resonant inertial sensors. The method is then applied to resonant acceleration sensors and resonant angular rate sensors in sections 3 and 4 summarizes the conclusions.

2 Theoretical Development

2.1 Mathematical modeling of resonant inertial sensors

Resonance is a property of a system to describe an enhanced response at a certain characteristic natural frequency that is wholly determined by parameters of the system[15]. The specific frequency is one where the system retains input energy with minimum loss. At a microscopic scale, operating systems at resonance enhances the effects of small forces and the device signal-to-noise ratio. Resonant sensing has been implemented in numerous devices for the measurement of pressure[32], humidity[33], temperature[34], acceleration[35], mass flow[36], specific gas[37], biological detection (immunosensors[38], cytometers[39]), force (AFM cantilevers[40]) and magnetic field[41]. The resonator sensor element is often built into a large device that transmits the effect of the parameter to be measured as a variation in spring constant of the resonant sensor element. While changing spring constant changes the frequency of the system. The resonant sensor element can take a number of forms such as a cantilever, a double-ended tuning fork or a singly clamped beam.

For the resonant inertial sensors we researched, the functional schematic of a resonant inertial sensor is shown in Fig. 1, the time-varying axial force (acceleration sensor: dynamic acceleration, angular rate sensor: Coriolis force) act on the proof mass, then the sense-mode system transmit this force onto two double-ended tuning fork (DETF) resonators placed on either side to provide for a differential output. The two tines of each tuning fork vibrate anti-phase to each other subjected to a drive force (F_d). The periodic compression and tension of the tuning fork tines by the time-varying axial force at the proof mass driving frequency modulates the resonant frequencies of these force sensors. Each force sensor comprises of the tuning fork mechanical structure embedded in the feedback loop of an oscillator circuit.

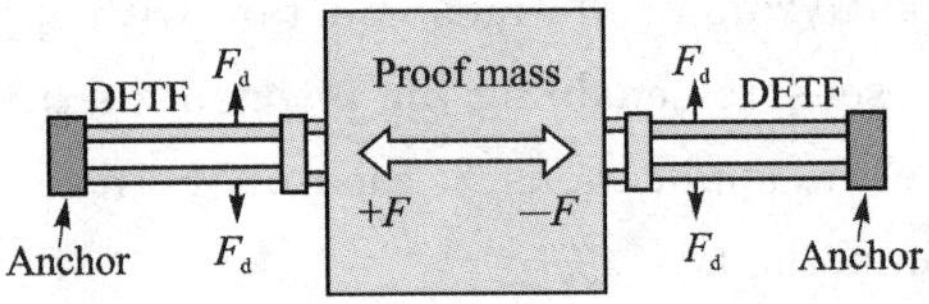

Fig. 1 Functional block diagram schematic of resonant inertial sensors

We will start with a general theory of mechanical resonance. In essence, we will analyze a resonating element where the parameter to be measured is coupled in some form so as to change the resonance characteristics of the resonant inertial sensors. The analysis proceeds by beginning with the one-dimensional Euler – Bernoulli equation and following this by proceeding with the Rayleigh – Ritz energy method. The response of a resonant system to an external axial load has been analyzed analytically by several sources [42,43].

In the condition of driving force F_d, resonant beam working in its resonant frequency, and the dynamic equation could be described as Eq. (1)

$$q''(z) + (\delta + \varepsilon\cos 2z)q(z) = 0$$

Eq. (1) is known as the standard Mathieu equation, which exactly represents the parametric excitement dynamic characteristic of resonant inertial sensors. From the Eq. (1), the resonant beam vibration stiffness $k = \delta + \varepsilon\cos 2z$ changes with the changing periodic axial force applied on the resonant beam, which makes the resonant frequency of resonant beam change. The parameter δ which represents natural resonant frequency of resonant beam. The significant perturbation term $\varepsilon\cos 2z$ which represents modulation stiffness of resonant beam at the periodic axial force. Hence, it is of interest to characterize the dynamic characteristics of resonant inertial sensors by analyzing Mathieu equation.

2.2 Review of relevant Mathieu equation theory

Periodic solutions to this equation are called Mathieu functions, known in the form of infinite series. These do not exist for all value of δ and ε. Given a value of δ, there are only a countably infinite number of δ's for which periodic solutions to the Mathieu equation exist. These δ are called characteristic numbers, and the plot of δ vs. ε for a given Mathieu function is termed its characteristic curve. The characteristic curves for Mathieu functions of integral order separate the $\delta-\varepsilon$ plane into stable and unstable domains (Fig. 2). Parameter pairs (ε, δ) belonging to the shade areas correspond to unstable behavior, for parameter pairs (ε, δ) belonging to the white areas all solutions are bounded. Thus, once the characteristic curves of Mathieu functions of integral order have been plotted, the $\delta-\varepsilon$ plane is separated once and for all into distinct domains.

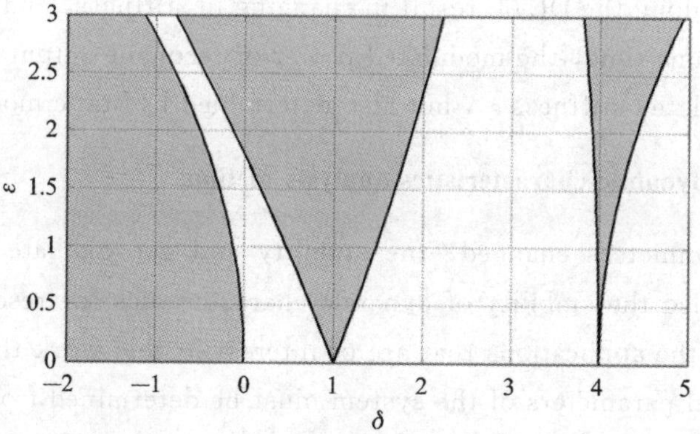

Fig. 2 Stability of the Mathieu equation

Since the input into the resonant inertial sensors is modulated by a periodic function, its output must be demodulated into stability-state, the solution of the parametric excitement Eq. (1) must in its stable reg ions. For resonant inertial sensors, with limitation of parameters δ and ε, the approximate solution could be derived in its stable regions. About the analyzing method of approximate solution for Mathieu equation. In references[44-47], the authors considered second and cubic order with two small parameters and terms in sin and cos, but the extension to the Mathieu equation is nonlinear. Guennoun et al. [48], based on the notion of kinematics similarity and an estimate of the growth rate of solutions, relaxed this assumption and derived two sufficient conditions. Based on an earlier contribution of Pipes[49], Richards[50-52] suggested the replacement of the time varying term by a trapezoidal function, thereby converting it into Hill's equation, which admits a closed form solution. It is provide an excellent approximation to the lossless Mathieu equation.

In reference[53], the authors described WKB method to obtain the power series solution for second order linear ODE, which give a significant direction for the solution of Mathieu equation. Referring to the methods in reference [53] and the physical significations in Mathieu equation parameters, the approximate solution of Mathieu equation $q(z)$ as follows (detail is provided in Appendix B):

$$q(z) = \frac{B}{(\delta + \varepsilon\cos 2z)^{\frac{1}{4}}}\cos\left[\int(\delta + \varepsilon\cos 2z)^{\frac{1}{2}}\,\mathrm{d}z + \theta\right] \tag{2}$$

$$\omega(t) = \sqrt{\delta'}\sqrt{1 + \beta\cos\omega_{\mathrm{d}}t} = \omega_0 + \frac{\varepsilon'}{\omega_0}\cos\omega_{\mathrm{d}}t \tag{3}$$

Eq. (2) is only valid for parameters $|\delta| > |\varepsilon|$. Form Eq. (3), it is easy to see that the changing in tension along the DETF result in changing in stiffness, and shifting the resonant frequency. At the same time, the modulated power of resonant output frequency is not only determined by modulated stiffness ε', but also determined by static modulated stiffness δ'.

2.3 Proposed dynamic characteristics analysis method

With model parameters changed, the stability and approximate solution of Mathieu equation changed, also the stability of dynamic characteristics for resonant inertial sensors system changed. In the applications that are of interest in this work the sensor input is not known, so the model parameters of the system must be determined from the dynamic stable characteristics only. To do this, the relation between modulated frequency output and input time-varying axial force is that

$$\hat{\omega}(t) = \frac{\varepsilon'}{\omega_0}\cos\omega_{\mathrm{d}}t = \omega_0 \cdot S \cdot F(t) \tag{4}$$

where $F(t)$ is the input time-varying axial force (acceleration sensor: dynamic acceleration inertial force, angular rate sensor: Coriolis force). In practice, one cannot obtain linear output signal only. In order to obtain a linear output of modulated frequency, some transformations in one period transform are desired.

For resonant acceleration sensors,

$$F_a(t) = \frac{m\,a\cos\omega_d t}{4} = \frac{m\,aCy_{m1}(t)}{4} \tag{5}$$

$$\hat{\omega}_1(t) = \omega_0 \cdot S \cdot \frac{Cm}{4} \cdot a \cdot y_{m1}(t) \tag{6}$$

$$y_{output1} = \int_T \hat{\omega}_1(t) \cdot y_{m1}(t)\mathrm{d}t \Big/ \int_T y_{m1}^2(t)\mathrm{d}t =$$
$$\int_T \omega_0 \frac{CSm}{4} a y_{m1}(t) \cdot y_{m1}(t)\mathrm{d}t \Big/ \int_T y_{m1}^2(t)\mathrm{d}t = \frac{CSm}{4}\omega_0 \cdot a \tag{7}$$

For resonant angular rate sensors,

$$F_c(t) = \frac{m}{4} \cdot \omega_d \cdot \Omega \cdot y_{m2}(t) \tag{8}$$

$$\hat{\omega}_2(t) = \omega_0 \cdot S \cdot \frac{m}{4} \cdot \omega_d \cdot \Omega \cdot y_{m2}(t) \tag{9}$$

$$y_{output2} = \int_T \hat{\omega}_2(T) \cdot y_{m2}(t)\mathrm{d}t \Big/ \int_T y_{m2}^2(t)\mathrm{d}t =$$
$$\int_T \frac{Sm}{4}\omega_0\omega_d\Omega y_{m2}(t) \cdot y_{m2}(t)\mathrm{d}t \Big/ \int_T y_{m2}^2(t)\mathrm{d}t = \frac{Sm}{4}\omega_0\omega_d \cdot \Omega \tag{10}$$

where $y_m(t)$ is the vibrating amplitude of proof mass, C is a arbitrary constants. The dynamic frequency output characteristics of the resonant inertial sensors based on Mathieu equation approximate solution are ultimately determined by model parameters. All above are just theoretical analyses, which will be evaluated by the experiment date in the next section.

The proposed routine of Mathieu equation with application to analysis dynamic characteristics of resonant inertial sensors can be summarized as follows:

① Give a series of resonant inertial sensors model parameters.

② Calculate the parameters (δ and ε) in Mathieu equation.

③ Simulate transition curves in Mathieu equation using the method of small parameter perturbation.

④ Estimate whether (δ, ε) lies in the stable regions for Mathieu equation.

⑤ Regulate geometrical parameters of resonant inertial sensors to make (δ, ε) lies in a stable region for Mathieu equation.

⑥ Calculate the approximate solution of Mathieu equation.

⑦ Transform some transformations in one period in order to obtain a linear output of modulated frequency.

3　Application: Resonant Inertial Sensors

The resonant inertial sensors were used to evaluate the proposed dynamic characteristics analyzing method of Mathieu equation and designed to meet the following performance specifications:

- It must make the solution of the motion equation in its stable regions.
- It must make output signals are linear modulated frequency.

As shown in Tab. 1, the parameters dimensions and material properties of resonant inertial sensors are considered for the dynamic characteristic analysis. Once the parameters have been chosen in Tab. 1, the (δ, ε) was (4 816. 3, 3. 692), as shown in Fig. 3, which is apparently in the stable regions of Mathieu equation.

Tab. 1 The parameters of resonant inertial sensors for the application

Symbol	Description	Numerical values	Unit
ρ	Density	7. 8E3	kg/m^3
E	Young's modulus	2E11	Pa
m	Proof mass	0. 261 495	kg
L	DETF beam length	55E 3	m
w	DETF beam width	3E 3	m
h	DETF beam height	0. 5E 3	m
I	Area moment of inertia of the DETF beam	3. 125 0 E14	m^4
ω_0	Natural resonant frequency of the beam	860. 32	Hz
ω_d	Excitation frequency of the periodic axial force	30	Hz

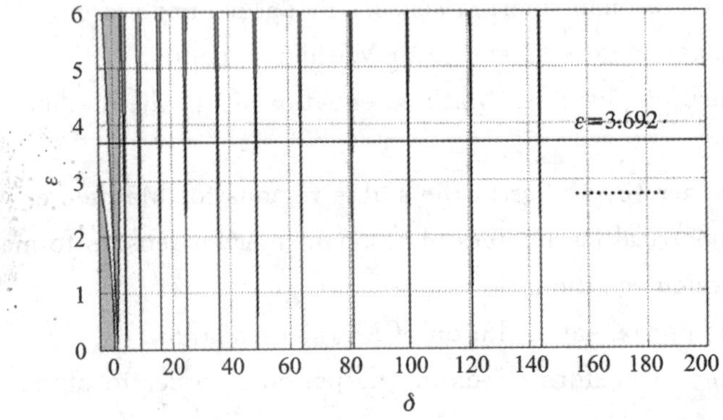

Fig. 3 Stability of the Mathieu equation

Some studies based on stability of the solution for Mathieu equation have been proved. In order to evaluate the translated approximate solution for Mathieu equation, and whether it can make output signals are linear modulated frequency in the experimentation, some related experiments are carried out. The resonant inertial sensors are produced by TAIYUAN AEROINSTRUMENTS CO. LTD, which was produced here as the sensor to be calibrated.

For resonant acceleration sensors, the experiment was carried out while the resonant acceleration sensor was exposed to standard acceleration from 0 g to 10 g. The schematic sketch of calibration assembly as shown in Fig. 4(a), and the photographed as shown in Fig. 5. The calibrating system is composed of 3 parts which are the resonant acceleration sensor, Centrifuge (LXJ—70) and signal sampling device. The output of the resonant acceleration sensor can not follow the linear frequency output. As shown in Fig. 6, on the condition of transformative output in Eq. (7), the output instant frequency of the resonant acceleration sensor is achieved linear output.

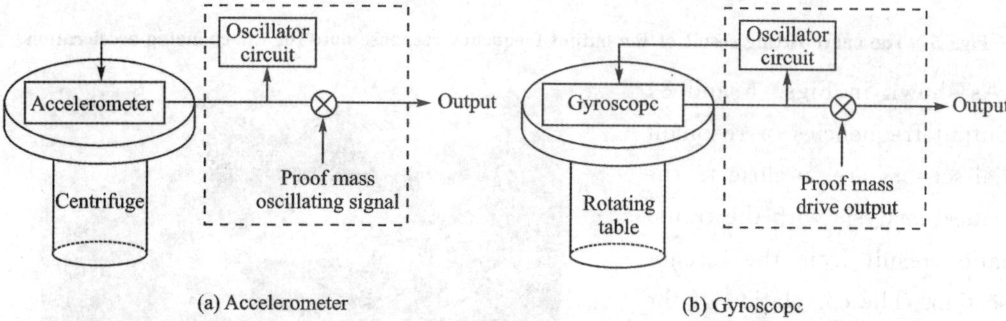

Fig. 4 The schematic sketch of calibration assembly for resonant inertial sensors

For resonant angular sensors, the experiment was carried out while the resonant angular sensor was exposed to standard angular velocity from $-200(°)/s$ to $200(°)/s$. The schematic sketch of calibration assembly as shown in Fig. 4(b), and the photographed as shown in Fig. 7. The calibrating system is composed of three parts which are the resonant angular sensor, rotating table (CIMM00003719) and signal sampling device. The output of the resonant angular sensor can not follow the linear frequency output. On the condition of transformative output in Eq. (10), as shown in Fig. 8, the output changed frequency of the resonant

Fig. 5 The calibration assembly for the resonant acceleration sensor

angular sensor is achieved linear output.

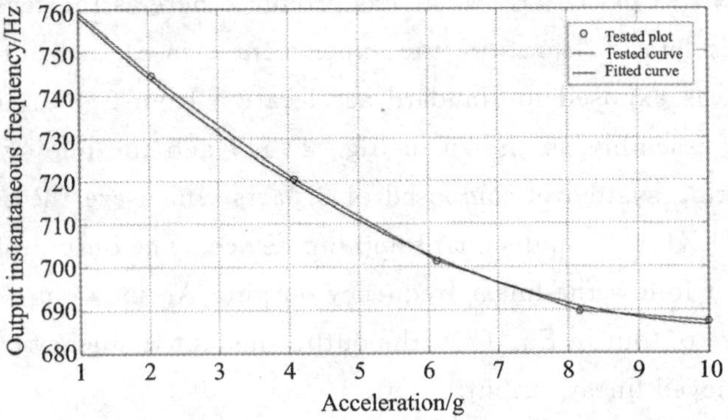

Fig. 6　The curve fitting result of the output frequency response date for the changing acceleration

As shown in Figs. 6 and 8, the output frequencies of resonant inertial sensors are so close to the theoretical analysis with the transformation result from the foregoing section. The correlation of the theoretical curve and the experimental data processing result coincide so perfectly, which means the approximate solution for Mathieu equation could be used to analyze the dynamic output characteristic of resonant inertial sensors.

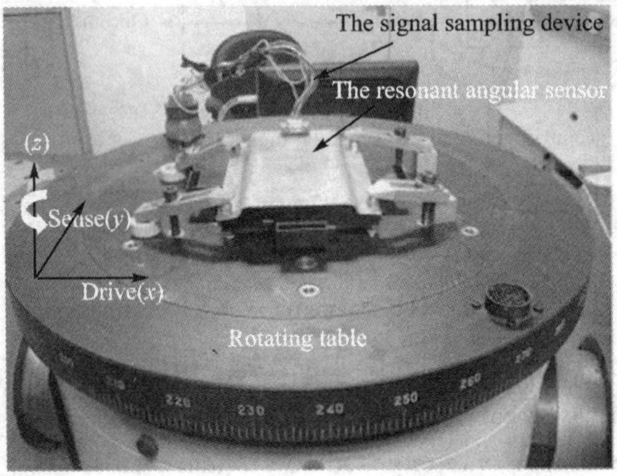

Fig. 7　The calibration assembly for the resonant angular sensor

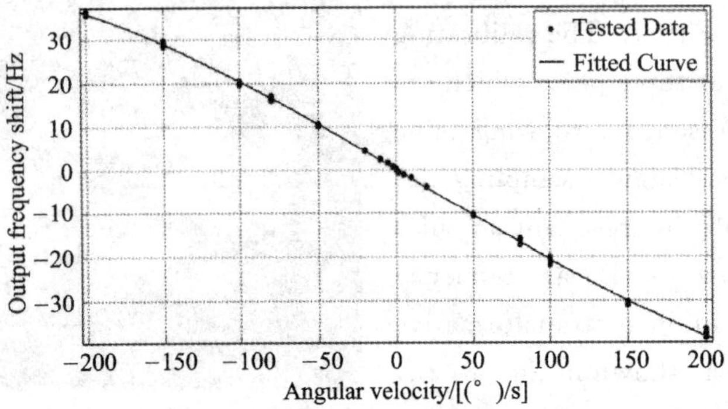

Fig. 8　The curve fitting result of the output frequency response date for the changing angular velocity

4 Conclusions

In this paper, we have found that the dynamic characteristic of resonant inertial sensors is equivalent to a parametric resonant system which could be described by Mathieu equation. Furthermore, we have studied the dynamic characteristic analysis methods using Mathieu equation theory. The proposed analytical methods for resonant inertial sensors have been implemented. The correlation of the theoretical curve and the experimental result coincide so perfectly, which means experiment results illustrated the effectiveness of the proposed Mathieu equation analytical methods.

Acknowledgments

This study was supported by the National Natural Science Foundation of China (No. 60927005), the Innovation Foundation of BUAA for Ph. D. Graduates (YWF-12-RBJ-026), and Fundamental Research Funds for the Central Universities (YWF-10-01-A17).

References

[1] Mclachlan N W. Theory and application of Mathieu functions. USA: Dover Publications Inc., 1964.

[2] Sammarco P, Tran H H, Gottlieb O, et al. Subharmonic resonance of Venice gates in waves. Part 2: Sinusoidally modulated incident waves. J. Fluid Mech, 1997,15:327-349.

[3] Vittori G. Oscillating tidal barriers and random waves. J. Hydraul Eng, 1998. 4:124-406.

[4] Poulin F J, Flierl G R, Pedlosky J. Parametric instability in oscillatory shear flows. J. Fluid Mech,2003, 481:329-353.

[5] Pedlosky J, Thomson J. Baroclinic instability of time-dependent currents. J. Fluid Mech, 2003,490:189-215.

[6] Maharana P K, Shrivastava S K. Stability of large flexible damped spacecraft modelled as elastic continua. Acta Astronaut, 1984,11:103-109.

[7] Acar C, Shkel A M. Structural design and experiential characterization of tensional micromachined gyroscopes with non-resonant drive mode. J. Micromech Microeng, 2004,14:15-25.

[8] Yang J B, Jiang L J, Chen D C H. Dynamic modeling and control of a rotating Euler. Bernoulli beam. J. Sound Vib, 2004,274:863-875.

[9] Mohite S, Patil N, Pratap R. Design, modeling and simulation of vibratory micromachined gyroscopes. J. Phys, 2006,34:757-763.

[10] Mahmoodi S N, Jalili N, Khadem S E. An experimental investigation of nonlinear vibration and frequency response analysis of cantilever viscoelastic beams. J. Sound Vib, 2008,311:1409-1419.

[11] Ansari M, Esmailzadeh E, Jalili N. Coupled vibration and parameter sensitivity analysis of rocking-mass vibrating gyroscopes. J. Sound Vib, 2009,327:564-583.

[12] Dong Peitao, Li Xinxin, Heng Yang. High-performance monolithic triaxial piezoresistive shock accelerometers. Sensors Actuator A, 2008,141:339-46.

[13] Huang Weidong,Cai Xia, Xu Bulu. Packaging effects on the performances of MEMS for high-G accelerometer with double-cantilevers. Sensors Actuator A, 2003,102:268-278.

[14] Wang Zuankai, Zong Denggang, Lu Deren. A silicon micromachined shock accelerometer with twin-mass-plate structure. Sensors Actuator A, 2003,107:50-56.

[15] Seshia A A. Integrated micromechanical resonant inertial sensors for inertial measurement systems. Ph. D. thesis. Berkeley, USA: Engineering-Electrical Engineering and Computer Science, University of California, 2002.

[16] Moussa Hassan, Bourquin Roger. Theory of direct frequency output vibrating gyroscopes. IEEE Sensors J., 2006,6:310-315.

[17] Moghimi M Z, Ahmadian M T. Vibrational analysis of electrostatically actuated microstructures considering nonlinear effects. Commun Nonlinear Sci. Numer Simul, 2009,14:1664-1678.

[18] Jeong Chihwan, Seok Seonho, Lee Byeungleul, et al. A study on resonant frequency and Q factor tunings for MEMS vibratory gyroscopes. J. Micromech Microeng, 2004,14:1530-1536.

[19] Gallacher B J, Burdess J S, Harish K M. A control scheme for a MEMS electrostatic resonant gyroscope excited using combined parametric excitation and harmonic forcing. J. Micromech Microeng, 2006,16: 320-331.

[20] Park Sungsu, Tan Chin-Woo, Kim Haedong, et al. Oscillation control algorithms for resonant inertial sensors with applications to vibratory gyroscopes. Sensors, 2009,9:5952-5967.

[21] Song M T, Cao D Q, Zhu W D. Dynamic analysis of a micro-resonator driven by electrostatic combs. Commun Nonlinear Sci. Numer Simul, 2011,16:3425-3442.

[22] Ahmadian M T, Borhan H, Esmailzadeh E. Dynamic analysis of geometrically nonlinear and electrostatically actuated micro-beams. Commun Nonlinear Sci. Numer Simul, 2009,14:1627-1645.

[23] Poulin Francis J, Flierl Glenn R. The stochastic Mathieu's equation. Proc. R. Soc. A, 2008,464:1885-1904.

[24] Coisson Roberto, Vernizzi Graziano, Yang Xiaoke. Mathieu functions and numerical solutions of the mathieu equation// International workshop on open-source software for scientific computation (OSSC), Guiyang, China, September 18-20, 2009:1-5.

[25] Rand R H, Sah S M, Suchorsky M K. Fractional Mathieu equation. Commun Nonlinear Sci. Numer Simul, 2010,15:3254-3262.

[26] Arrowsmith D K, Mondragon R J. Stability region control for a parametrically forced Mathieu equation. Meccanica, 1999,34:401-410.

[27] Guennoun K, Houssni M, Belhaq M. Quasi-periodic solutions and stability for a weakly damped nonlinear quasi-periodic Mathieu equation. Nonlinear Dyn, 2002,27:211-236.

[28] Wooden Susan M, Sinha S C. Analysis of periodic-quasiperiodic nonlinear systems via Lyapunov-Floquet transformation and normal forms. Nonlinear Dyn, 2007,47:263-273.

[29] Water Thomas J. Stability of a 2-dimensional Mathieu-type system with quasiperiodic coefficients. Nonlinear Dyn, 2010,60:341-356.

[30] Zounes R S, Rand R H. Global behaviour of a nonlinear quasiperiodic Mathieu equation. Nonlinear Dyn, 2002,27:87-105.

[31] Arrowsmith D K, Mondragon R J. Stability region control for a parametrically forced Mathieu equation. Meccanica, 1999,34:401-410.

[32] Esashi M. Resonant inertial sensors by silicon micromachining// Proceedings of the IEEE international Frequency Control Symposium, Hilton Hawaiian Village, Honolulu, Hawaii, USA, June 5-7, 1996: 609-614.

[33] Boltshauser T, Schonholzer M, Brand O, et al. Resonant humidity sensors using industrial CMOS-technology combined with postprocessing. J. Micromech Microeng, 1992,2:205-207.

[34] Burns D W, Herb W R, Zook J D, et al. Optically driven resonant microbeam temperature sensors for fiber optical networks//Technical digest solid-state sensor and actuator workshop, Hilton Head, SC, June 2-6, 1996: 294-298.

[35] Burns D W, Horning R D, Herb W R, et al. Sealed-cavity resonant microbeam accelerometer. Sensors Actuator A, 1996,53:249-255.

[36] Enoksson P, Stemme G, Stemme E. A silicon resonant sensor structure for Coriolis mass-flow measurements. IEEE/ASME J. Microelectromech Syst, 1997,6:119-125.

[37] Hagleitner C, Lange D, Kerness N, et al. A gas detection system on a single CMOS chip comprising capacitive, calorimetric, and mass-sensitive microsensors// Technical digest IEEE international solid-state circuits, San Francisco, USA, February 20-24, 2002: 430-431.

[38] Florin E L, Reif M, Lehmann H, et al. Sensing specific molecular interactions with the atomic force microscope. Biosensors Bioelectron, 1994,10:895-901.

[39] Ilic B, Czaplewski D, Zalalutdinov M, et al. Single cell detection with micromechanical oscillators. J. Vacuum Sci. Technol. B (Microelectron Nanometer Struct), 2001,19:2825-2828.

[40] Albrecht T, Grutter P, Horne D, et al. Frequency modulation detection using high-Q cantilevers for enhanced force microscope sensitivity. J. Appl. Phys., 1991,69:668-673.

[41] Kadar Z, Bossche A, Mollinger J R, et al. Magnetic-field measurements using an integrated resonant magnetic-field sensor. Sensors Actuator A, 1998,70:225-232.

[42] Roessig T A. Integrated MEMS tuning fork oscillators for sensor applications. Ph. D. thesis. Berkeley, USA: Dept. of Mechanical Engineering, University of California, 1998.

[43] Meirovitch L. Fundamentals of vibrations. New York: McGraw Hill, 2001.

[44] Sun Dexian. A note for a second order periodic linear differential equation. Commun Nonlinear Sci. Nu-

mer Simul, 2010,15:3339-3348.

[45] Butchera Eric A, Sari Ma'en, Bueler Ed, et al. Magnus' expansion for time-periodic systems: parameter-dependent approximations. Commun Nonlinear Sci. Numer Simul, 2009,14:4226-4245.

[46] Younesian D, Esmailzadeh E, Sedaghati R. Asymptotic solutions and stability analysis for generalized non-homogeneous Mathieu equation. Commun Nonlinear Sci. Numer Simul, 2007,12:58-71.

[47] Younesian D, Esmailzadeh E, Sedaghati R. Existence of periodic solutions for generalized form of Mathieu equation. Nonlinear Dyn, 2005,39:335-348.

[48] Guennoun K, Houssni M, Belhaq M. Quasi-periodic solutions and stability for a weakly damped nonlinear quasi-periodic Mathieu equation. Nonlinear Dyn, 2002,27:211-236.

[49] Pipes L A. Matrix solutions of equations of the Mathieu-Hill type. J. Appl. Phys., 1953,24:902-910.

[50] Richards J A. Stability diagram approximation for the Lossy Mathieu equation. Appl. Math., 1976,30:240-247.

[51] Richards J A. Modeling parametric processes—a tutorial review. Proc IEEE, 1977,65:1549-1557.

[52] Richards J A. Analysis of periodically time varying systems. New York: Springer Verlag, 1983.

[53] Bender Carl M, Orszag Steven A. Advanced mathematical methods for scientists and engineers// Asymptotic methods and perturbation theory. New York: Springer Verlag, 1978.

Appendix C
Vibration Membrane Liquid Density Sensor①

Zhang Yuxiao　Fan Shangchun

Abstract: A new vibration membrane-type liquid density sensor that is based on resonance theory is presented in order to measure the density of different kinds of liquids. This sensor, which consists of resonant membrane, exciter, and signal picking system, possess the advantages of easy fabrication, excellent repetition, high stability and fine resolving ability. Density of the liquid is measured by comparing the natural frequency variation in air and liquid respectively. In this paper, the calculating model of the sensor is simplified with one degree freedom system, in which equivalent mass m_e and equivalent elasticity coefficient k_e are all of importance. Finally, density of water is measured at 20℃ by this sensor. The measured natural frequency in air and water is 5 359 Hz and 3 769 Hz respectively. Corresponding Q factor (quality factor) of the sensor is in the range of 500～600. Experiment results have shown that this sensor has good repeatability and it is a kind of effective method to measure the density of liquid.

Keywords: liquid density sensor, liquid density, resonant membrane, natural frequency, Q factor

1 Introduction

In many industrial processes and in laboratories, it is important to measure the density of different kinds of liquids. There are several basic principles for density measurements, density sensor or densitometers, which can roughly be divided into the following five categories: sensors sensitivity to the inertial mass[1-2]; the gravitational mass; the buoyant force; the hydrostatic pressure; the attenuation of gamma ray[3]. Among the many applications of densitometers are quality control and metering in the food & beverage, pharmaceutical, chemical, petroleum and paper industries. The most commonly used densitometers are based on the principle of vibrating tubes. Those weigh the mass of the liquid passing through a

① Proceedings of the Society of Photo-Optical Instrumentation Engineers (SPIE), 2006, 6358: S3580-S3580. doi: 63580S.

specific tube and calculate the liquid density. Sometimes ultrasonic density sensor for liquid is utilized in the industry. The density ρ of a liquid is obtained from its acoustic impedance Z and sound velocity with the help of the relation $Z=\rho \cdot v$. The sound velocity is calculated from the transit time of an ultrasonic pulse along a known distance. The acoustic impedance can be calculated from the measurement of the reflection coefficient R of sound at the interface of a material with known acoustic impedance (reference material) and the unknown liquid. It is difficult to measure the reflection coefficient, the ratio between incident and reflected sound pressure, directly.

A resonant density sensor is achieved if one of the density measurements principles, e. g. the inertial mass change, is used to determine the mechanical resonant frequency of the sensor system. The main advantages of a resonant sensor are the directly microcomputer interface ability and the possibility of having a comparatively and stable sensor arrangement.

In this paper we present a real-time, low-cost resonant liquid density sensor based on membrane techniques. Membrane technology provides the means to make a resonant density sensor with good performance in terms of high density sensitivity, high resolution and good accuracy. The resonance sensor model is simplified a degree freedom system which consists of equivalent quality m_e and equivalent elasticity coefficient k_e. They both affect inherent frequency of the system. A new vibration membrane-type density sensor for liquid is designed on the basis of the model.

2 Design Principle and Structure

2.1 Structure of the sensor

The vibration membrane liquid density sensor consists of membrane, exciter, and pick up and basic of membrane. The heart of the sensor is circular flat membrane of uniform thickness, edge fixed and sticking on two piezoelectric components located in the edge area (they are on the same side). One piezoelectric component operates as exciter. It generates the motions which make the membrane to vibrate as an oscillator. The related movement is detected by the other which supplies a proportional piezoelectric signal. The basic of membrane and the membrane was welded. The structure is kept in resonant condition by an electronic circuit which makes use of a phase-locked loop (PLL) technique. Fig. 1 show main structure of vibration membrane liquid density sensor and part of electronic circuit.

2.2　Working principle

The membrane vibrates on the resonant frequency. The frequency will be changed while the sensor immerges in liquid. The reason of frequency variety is the mass of the whole system must be added. The value of the liquid density is calculated by the added mass that is brought on vibration of the membrane. The Eq. (1)[4] is the nature frequency of the membrane.

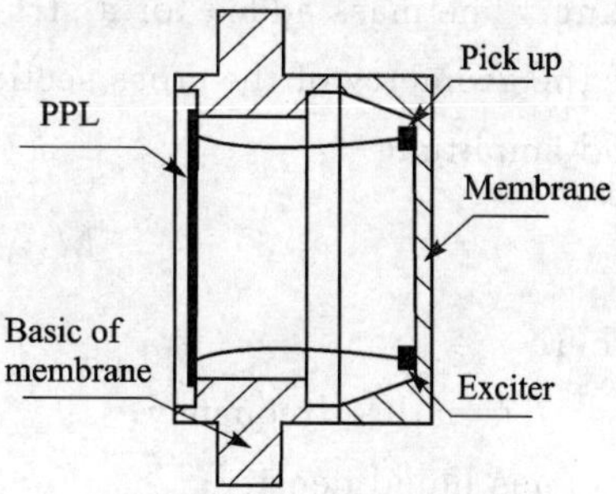

Fig. 1　structure of the vibration membrane liquid density sensor

$$f = \frac{K_n}{4\sqrt{3}\pi}\sqrt{\frac{Et^2}{\rho r^4(1-\mu^2)}} \tag{1}$$

where

f is nature frequency;

K_n is constant coefficient;

n refer to the mode of vibration;

E is modulus of elasticity;

t is the thickness of the circular membrane;

r is the radius of the circular membrane;

μ is Poisson's ratio.

If the membrane is immersed in a liquid, its natural frequencies may be considerably altered. In order to take the mass of the liquid into account for the fundamental mode of vibration, we should replace expression by the Eq. (2)[4]:

$$f = \frac{K_n}{4\sqrt{3}\pi\sqrt{1+\eta}}\sqrt{\frac{Et^2}{\rho r^4(1-\mu^2)}} \tag{2}$$

where

$$\eta = 0.6689\frac{\rho_1}{\rho}\frac{r}{h} \tag{3}$$

In which, $\frac{\rho_1}{\rho}$ is the ratio of the mass density of the liquid to the mass density of the material of the membrane. η is the mass added coefficient. The total effect is equivalent to increasing the mass added coefficient of the membrane.

2.3　The effect factor of mass added

As the structure vibrates in the fluid, it will meet the problem of the fluid-structure coupling. The mass added for a structure in a liquid with zero mean flow is generally functions of the geometry of the cross section, analogous Reynolds number, and the vibration direction and amplitude[5]:

$$M_{A,\text{still fluid}} = \rho F\left(\text{geometry}, \frac{x_0}{D}, \frac{fD^2}{\nu}\right) \tag{4}$$

where

f is nature frequency;

ρ is liquid density;

x_0 is the amplitude;

D is the geometry of the cross section;

ν is the kinematical viscosity.

Geometry includes both the overall shape of the cross section and the roughness of the surface.

If the viscosity and the compressibility of the liquid are neglected we make use of the potential flow function theory to analyse the added mass of the structure. And the added mass is just relation to the structure shape of the cross section:

$$M_{A,\text{potentail fluid}} = \rho F(\text{geometry}) \tag{5}$$

Study of the experiment and analyse of the theory show that the Mach number of liquid and the amplitude of vibration is small and that the analogous Reynolds number is big enough the Eq. (5) have a good accuracy.

3 The Signal-conditioning Electronic Circuit

To sustain the resonant condition in the mechanical oscillator an electronic circuit operating according to feedback control technique has been introduced. A voltage controlled exciter provides the excitation signal, amplified by a proper stage at a 5 V peak value, to the actuator to the generation of flexural strains in the membrane. The motion detected by picker in the same side and amplified by inducing a voltage signal, whose phase is shifted with respect to at the resonant frequency (since the membrane vibrates in condition of forced exciter). This result is essentially related to electronic devices effects which must be compensated by means of phase regulator during calibration procedures. A change of the resonant frequency due to a density change affects this phase relation, which is however promptly restored taking advantage of the Phase-Locked Loop (PLL) techniques. The PPL has been implemented with the integrated circuit which essentially consists in a phase-comparator connected with the exciter. The output of the comparator is an analogic signal, proportional to the phase difference; this signal drives the exciter which restores the required phase relation and so the mechanical system vibrates at the new value of the resonant frequency.

4 Experimental Results

To test the performance of the density structure it was mounted in a measurement set-up where a stainless steel fixture clamps the sensor frame. The sensor was then immersed in liquid which was calibrated before by using standard gravitational method. The frequency response of the sensor prototype was characterized using a gain-phase analyser with capabilities from 10～100 MHz. The gain amplitude shows a sharp resonance near to 5 400 Hz (1 mode of flexural vibration) and the phase has a rapid decrease in the vicinity of the resonance. Either of the curves can be used to determine the Q factor. Q values near 500～600 can be calculated. The fluid density sensitivity of the sensor was determined by measuring the resonance frequency of the selected flexural mode at a fixed fluid temperature of 20.

The resonant frequency measured showed a good degree of stability and repeatability. This implies that there was no special problem in filling the sensor completely without gas trapping on the vibrating membrane surface. The resonant frequency of the selected vibrating mode was 5 359 Hz in air and 3 769 Hz in water.

This temperature dependence is mainly due to the temperature dependence of the young's modulus together with the mismatching between the thermal coefficients of expansion[6] of the criterion and the stainless steel. However a complete temperature compensation can be easily improved acquiring both temperature and density information microprocessor.

5 Conclusion

We have presented the first membrane sensor for fluid density measurements. The sensor is based on a ceramic and membrane system sensitive to changes in the inertial mass which arises as a result of changes in density of the fluid. This structure is an on line, low-cost sensor well suited to be connected with single-chip microcontroller unit.

6 Acknowledgements

This paper is supported by the National Natural Science Foundation of China

(60274039).

References

[1] Enoksson P, Stemme G, Stemme E. Fluid density sensor based on resonance vibration. Sensor and Actuators A: physical, 1995, 47(1-3): 327-331.

[2] Stemme E, Ekelof J, Nordin L. Measuring liquid density with a tuning-fork transducer. IEEE Transactions on Instrumentation and Measurement, 1983, 32(3): 434-437.

[3] Gopel W, Hesse J, Zemel J N, et al. silverman, HP. Sensors, A Comprehensive Survey. Journal of the American Chemical Society, 1994, 16(15): 6993.

[4] Timoshenko S, Young D H. Vibration Problems in Engineering. Wiley-Interscience, 1990.

[5] Blevins R. Formulas for Nature Frequency and Mode Shape. Krieger Publishing Company, 2001.

[6] Warren y, Richard B. Formulas for Stress and Strain. Mcgraw, 2011.

Appendix D
Finite-element Modelling and Simulation on Frequency Characteristics of the Silicon Beam Resonator Attached to an E-type Round Diaphragm for Measuring the Concentrated Force[①]

Fan Shangchun　Liu Guangyu　Lee Man Hyung

Abstract: This paper establishes the finite-element method(FEM) model of a practical silicon beam resonator attached to an E-type round diaphragm, which is used for measuring the concentrated force in detail, based on the sensing mechanism of the microsensor. We calculate, analyse and investigate the relationship between the basic natural frequency of the beam resonator and the concentrated force by making use of the above FEM model. Some important simulation results on the vibration features of the above beam resonator and the microsensor are obtained. Finally, based on the differential output mean, a set of optimum parameters of the above sensing unit is determined.

Keywords: Beams, E-type round diaphragms, Microsensors, Force, Finite-element method

1 Introduction

During the past decade, many experts in sensor technology have noticed rapid development in silicon resonant sensors[1-4]. The silicon microstructure resonant sensors are noted for the advantages of generalized resonant sensors, such as direct digital output(without A/D conversion), long-term stability, low hysteresis and high repeatability[5], and the advantages of silicon material, namely, excellent mechanical properties, high strength, freedom from mechanical hysteresis. ability to batch process at low cost, and the compatibility of mechanical

① Sensors and Actuators A, 1997.

and electrical properties. Meanwhile. the dynamic characteristics of silicon resonant sensors are much better than those of conventional resonant sensors, due to their high working frequency. In addition, the temperature characteristics of silicon resonant sensors are much better than those of another important silicon sensor. the piezoresistive sensor. It is much easier to interface them with a micro processor in order to develop smart or "intelligent" sensors or sensor systems, due to their unique operating principle based on the relationship between the natural frequency and measured signal instead of the silicon piezoresistive effect. Now a days, it has been reported that some prototypes, with different structure and configuration. and even a few batch products of silicon resonant sensors for measuring pressure, differential pressure or force, have been developed and manufactured in Japan, the USA, Netherlands and Britain. Although the properties of these silicon resonant sensors have not reached the designed criteria, they must be used widely in many technology fields in the near fulure.

As compared with a round or square diaphragm, the E-type round diaphragm is one of the advanced elementary sensing units in silicon microstructure resonant sensors. Its major advantages are as follows: stress concentration on the effective ring sensing area, flexible stress distribution design, easier to achieve a differential measuring scheme in order to increase the sensitivity and reduce some disturbances such as temperature, etc.

The objective of this paper is to present a finite-element method (FEM) model of an advanced silicon resonant concentrated force sensor. Based on the model, some simulation calculations on the relationship between the output frequency and the applied concentrated force are carried out.

2 Nomenclature

B, A, H = Inner, outer radii and thickness of the E-type round diaphragm;

r = Radial coordinate of the E-type round diaphragm in polar coordinate of the E-type round diaphragm;

F = Concentrated force;

L, b, h = Length, width and thickness of the beam which is attached to the E-type round diaphragm;

R_0, R_0+L = Radial coordinates of the beam in polar coordinate of the E-type round diaphragm;

E, ρ, μ = Young's modulus, density and Poisson's ratio of the sensing structure;

w = The normal displacement of the E-type round diaphragm;

D = The flexural rigidity of the diaphragm;

Q = The shearing force of the circle of radius r;

$\sigma_r(r), \sigma_\theta(r)$ = Radial and circular stresses of the E-type round diaphragm;

$u(s,z,t), w(s,t)$ = Axial and normal vibrating displacements of the beam in Cartesian coordinate of the beam;

t = Time;

s, z = Axial and normal coordinates of the beam in Cartesian coordinate of the beam;

$\varepsilon_s(s,z,t), \sigma_s(s,z,t)$ = Axial strain and stress of the beam;

$\sigma_s^0(s)$ = Initial axial stress of the beam;

V, S = Integrated volume and length of the beam;

$\omega[\text{rad} \cdot \text{s}^{-1}], w(s)$ = Natural frequency and its corresponding vibrating shape along the axial direction of the beam;

S_j, S_{j+1} = The jth and the $(j+1)$th node of the beam element;

l = Half length of the beam element;

N = Total number of beam elements;

$f(F), f(0)[\text{kHz}]$ = Basic natural frequency of the beam for concentrated force F and for concentrated force $F=0$;

$\Delta f[\text{kHz}] = f(F) - f(0)$ = Variation of the basic natural frequency of the beam within $(0, F)$;

$\beta = [f(F) - f(0)]/f(0)[\%]$ = Relative variation of the basic natural frequency of the beam within $(0, F)$;

$R_B, R_B + L$ = Radial coordinates of the beam's blind spot in polar coordinate of the E-type round diaphragm.

Superscripts

j = Element number;

T = Transpose of matrix.

Subscripts

j = Element number;

T = Total.

3 Sensor Structure and Sensing Mechanism

Fig. 1 shows the structure of a silicon resonant sensor for measuring concentrated force. The preliminary sensing unit is an E-type round diaphragm. The measured concentrated force acts perpendicularly on the lower plane of the E-type round diaphragm and yields the radial and circular stresses. The final sensing unit is a beam which is attached to the upper plane of the E-type round diaphragm. On the other hand, the thickness of the beam h should be much less than the thickness of the E-type round diaphragm H. Based on the above structural feature, an appropriate initial stress is applied along the axial direction of the beam, which is identical with the radial stress of the E-type round diaphragm at the same place. Because the principal vibration of the beam is along its normal direction (direction of the beam's thickness) and the above initial stress is applied along the axial direction of the beam (direction of the beam's length), thus the natural frequency of the beam is varied with the radial stress on upper plane of the diaphragm. Therefore, the applied concentrated force will be measured via the change of the natural frequency of the beam. The advantage of the above sensing structure is that the beam resonator is of very high Q factor because it can be packaged within a vacuum housing. Moreover, the sensor's quality will be much better.

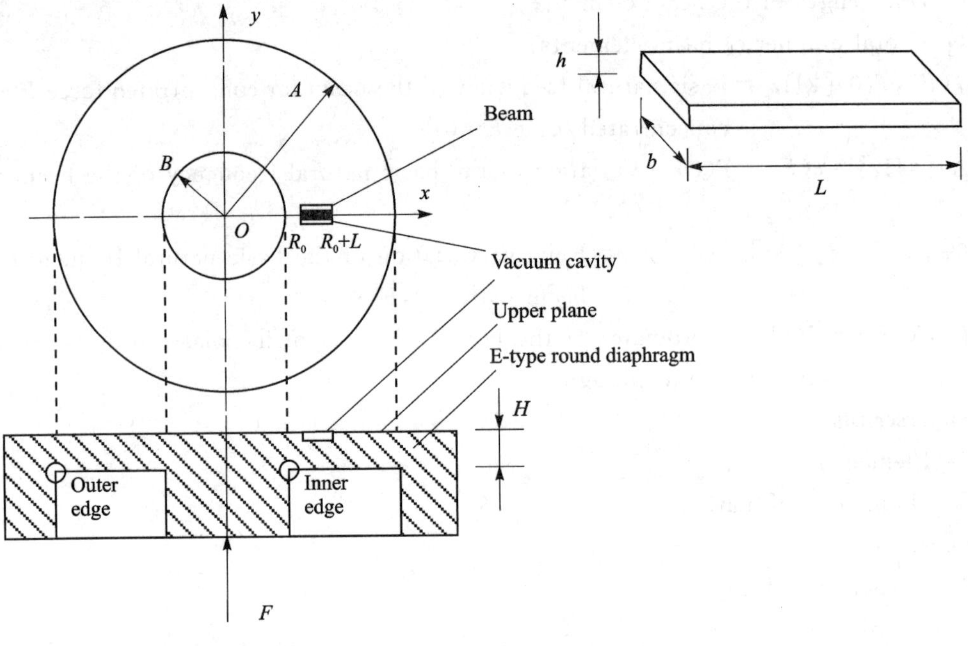

Fig. 1 Sensing structure of the sensor (Oxy: reference frame of the diaphragm)

4 Stresses on the upper Plane of the E-type Round Diaphragm

According to the structural feature and the overall design demands, the E-type round diaphragm is within the linear and small deflection range. Because a concentrated force which is a symmetric load is applied to the E-type round diaphragm, the differential equation of the round diaphragm can be written as follows[6]:

$$\frac{\mathrm{d}}{\mathrm{d}r}\left[\frac{1}{r}\frac{\mathrm{d}}{\mathrm{d}r}\left(r\frac{\mathrm{d}w}{\mathrm{d}r}\right)\right]=\frac{Q}{D} \tag{1}$$

where

D is the flexural rigidity of the diaphragm, and Q is the shearing force of the circle of radius r, given by

$$D=\frac{EH^3}{12(1-\mu^2)},\quad Q=\frac{F}{2\pi r}$$

The boundary conditions at the inner and outer edges of the E-type diaphragm are as follows:

$$\left.\begin{aligned} r=B,\quad & \frac{\mathrm{d}w}{\mathrm{d}r}=0 \\ r=A,\quad & w=\frac{\mathrm{d}w}{\mathrm{d}r}=0 \end{aligned}\right\} \tag{2}$$

From Eqs. (1) and (2), we can obtain the stresses on the upper plane of the E-type round diaphragm as follows:

$$\left.\begin{aligned} \sigma_r(r) &= \frac{-3F}{4\pi H^2}\left[\left(2\ln R+2-C-\frac{C}{R^2}\right)+\left(2\ln R-C+\frac{C}{R^2}\right)\mu\right] \\ \sigma_\theta(r) &= \frac{-3F}{4\pi H^2}\left[\left(2\ln R+2-C-\frac{C}{R^2}\right)\mu+\left(2\ln R-C+\frac{C}{R^2}\right)\right] \end{aligned}\right\} \tag{3}$$

where

$$R=\frac{r}{A},\quad C=\frac{-2K^2\ln K}{1-K^2},\quad K=\frac{B}{A}$$

It is obviously shown that the initial stress of the beam varies when it is attached to different parts of the diaphragm.

5 Mathematical Model of the Beam under a Varying Axial Load

Fig. 2 shows the mathematical model of the beam. The vibrating displacement of the beam at an arbitrary point can be written as

$$\left.\begin{aligned} u(s,z,t) &= -z\frac{\mathrm{d}w(s)}{\mathrm{d}s}\cos\omega t \\ w(s,t) &= w(s)\cos\omega t \end{aligned}\right\} \tag{4}$$

The axial strain and stress of the beam can be written as follows:

$$\left.\begin{aligned}\varepsilon_s(s,z,t) &= -z\frac{d^2w(s,t)}{ds^2}\\ \sigma_s(s,z,t) &= -zE\frac{d^2w(s,t)}{ds^2}\end{aligned}\right\} \quad (5)$$

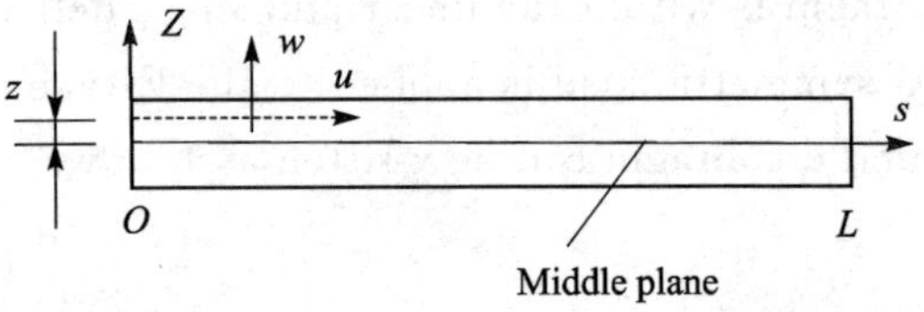

Fig. 2 Mathematical model of the beam

The potential energy of the beam is

$$U = \frac{1}{2}\iiint_V \sigma_s(s,z,t)\varepsilon_s(s,z,t)\mathrm{d}V = \frac{Ebh^3\cos^2\omega t}{24}\int_s\left[\frac{d^2w(s)}{ds^2}\right]^2 ds \quad (6)$$

The kinetic energy of the beam is

$$T = \frac{1}{2}\iiint_V\left\{\left[\frac{\partial u(s,z,t)}{\partial t}\right]^2 + \left[\frac{\partial w(s,t)}{\partial t}\right]^2\right\}\rho\mathrm{d}V \approx$$
$$\frac{\rho bh\omega^2\sin^2\omega t}{2}\int_s[w(s)]^2 ds \quad (7)$$

Because the beam'S principal vibration $w(s,t)$ is along its normal direction(direction of the beam's thickness)and the applied initial stress is along its axial direction(direction of the beam's length), the independent initial potential energy of the beam, which is caused by $\sigma_s^0(s)$, can be obtained:

$$U_0 = \frac{1}{2}\iiint_V \sigma_s^0(s)\left\{\left[\frac{\partial u(s,z,t)}{\partial s}\right]^2 + \left[\frac{\partial w(s,t)}{\partial s}\right]^2\right\}\mathrm{d}V \approx$$
$$-\frac{bh\cos^2\omega t}{2}\int_s\sigma_s^0(s)\left[\frac{dw(s)}{ds}\right]^2 ds \quad (8)$$

From Fig. 1 and Eq. (3)

$$\sigma_s^0(s) = \sigma_r(R_0+s) = \frac{-3F}{4\pi H^2}\left\{\left[2\ln\left(\frac{R_0+s}{A}\right)+2-C-\frac{CA^2}{(R_0+s)^2}\right]+\right.$$
$$\left.\left[2\ln\left(\frac{R_0+s}{A}\right)-C+\frac{CA^2}{(R_0+s)^2}\right]\mu\right\} \quad (9)$$

Then the total potential energy of the beam is

$$U_T = U - U_0 \quad (10)$$

In Eq. (8), if $\sigma_s^0(s)$ is a constant, we can directly obtain the analytical relationship between the frequency and the initial axial stress. In this paper. we present the FEM equations to solve the relationship between the frequency and the initial varying axial stress.

Based on the above equations we can divide the element along the axial direction of the beam, see Fig. 3. Introducing the dimensionless length $x=(s-S_j)/l$, $l=0.5(S_{j+1}-S_j)$, then a map of domain $s \in [S_j, S_{j+1}]$ is $x \in [-1, +1]$. For the displacement of the jth element, we introduce the second order Hermite interpolation

$$w_j(s) = w_j(x) = \boldsymbol{X}_2^0 \boldsymbol{G}_2 \boldsymbol{a}_j \quad (11)$$

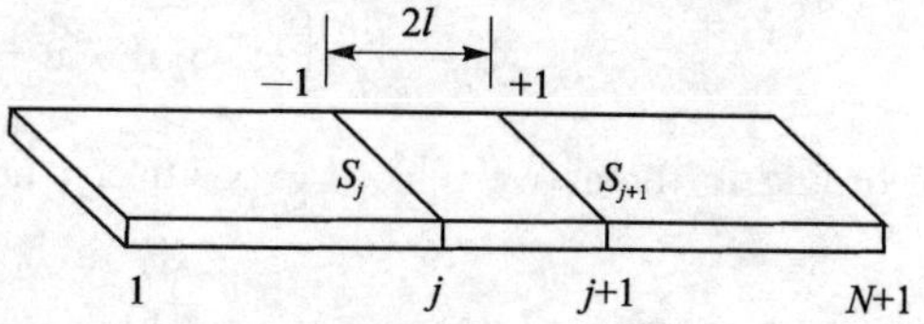

Fig. 3 Dividing element along the beam's axial direction

where

$$\boldsymbol{X}_2^0 = [1 \quad x \quad x^2 \quad x^3 \quad x^4 \quad x^5]$$

$$\boldsymbol{a}_j = [w(-1) \quad w'(-1) \quad w''(-1) \quad w(+1) \quad w'(+1) \quad w''(+1)]^{\mathrm{T}}$$

$$\boldsymbol{G}_2 = \frac{1}{16}\begin{bmatrix} 8 & 5l & l^2 & 8 & -5l & l^2 \\ -15 & -7l & -l^2 & 15 & -7l & l^2 \\ 0 & -6l & -2l^2 & 0 & 6l & -2l^2 \\ 10 & 10l & 2l^2 & -10 & 10l & -2l^2 \\ 0 & l & l^2 & 0 & -l & l^2 \\ -3 & -3l & -l^2 & 3 & -3l & l^2 \end{bmatrix}$$

From Eqs. (6)~(11), we can obtain the element potential energy, kinetic energy and initial potential energy caused by the initial stress, in the domain $s \in [S_j, S_{j+1}]$:

$$U^j = \frac{Ebh^3 \cos^2 \omega t}{24 l^3} \int_{-1}^{+1} \boldsymbol{a}_j^{\mathrm{T}} \boldsymbol{G}_2^{\mathrm{T}} \boldsymbol{X}_2^{2\mathrm{T}} \boldsymbol{X}_2^2 \boldsymbol{G}_2 \boldsymbol{a}_j \, \mathrm{d}x \quad (12)$$

$$\boldsymbol{X}_2^2 = \frac{\mathrm{d}^2 \boldsymbol{X}_2^0}{\mathrm{d}x^2}$$

$$T^j = \frac{\rho bhl\omega^2 \sin \omega t}{2} \int_{-1}^{+1} \boldsymbol{a}_j^{\mathrm{T}} \boldsymbol{G}_2^{\mathrm{T}} \boldsymbol{X}_2^{0\mathrm{T}} \boldsymbol{X}_2^0 \boldsymbol{G}_2 \boldsymbol{a}_j \, \mathrm{d}x \quad (13)$$

$$U_0^j = -\frac{bh \cos^2 \omega t}{2l} \int_{-1}^{+1} \sigma_s^0 [S_j + (x+1) l] \boldsymbol{a}_j^{\mathrm{T}} \boldsymbol{G}_2^{\mathrm{T}} \boldsymbol{X}_2^{1\mathrm{T}} \boldsymbol{X}_2^1 \boldsymbol{G}_2 \boldsymbol{a}_j \, \mathrm{d}x \quad (14)$$

$$\boldsymbol{X}_2^1 = \frac{\mathrm{d}}{\mathrm{d}x} \boldsymbol{X}_2^0$$

Then we can obtain the element stiffness matrix

$$\boldsymbol{K}^j = \frac{Ebh^3}{12 l^3} \int_{-1}^{+1} \boldsymbol{G}_2^{\mathrm{T}} \boldsymbol{X}_2^{2\mathrm{T}} \boldsymbol{X}_2^2 \boldsymbol{G}_2 \, \mathrm{d}x \quad (15)$$

the element mass matrix

$$\boldsymbol{M}^j = \rho bhl \int_{-1}^{+1} \boldsymbol{G}_2^{\mathrm{T}} \boldsymbol{X}_2^{0\mathrm{T}} \boldsymbol{X}_2^0 \boldsymbol{G}_2 \, \mathrm{d}x \quad (16)$$

and the element initial stiffness matrix

$$\boldsymbol{K}_0^j = \frac{bh}{l}\int_{-1}^{+1}\sigma_s^0[S_j + (x+1)/l]\boldsymbol{G}_2^{\mathrm{T}}\boldsymbol{X}_2^{1\mathrm{T}}\boldsymbol{X}_2^1\boldsymbol{G}_2\,\mathrm{d}x \tag{17}$$

According to the above relaced expositi on, the elementtotal stiffness matrix is as follows:

$$\boldsymbol{K}_{\mathrm{T}}^j = \boldsymbol{K}^j + \boldsymbol{K}_0^j \tag{18}$$

From Eqs. (18) and (16), we can obtain the assembly stiffness matrix $\boldsymbol{K}$ and assembly mass matrix $\boldsymbol{M}$ in the local domain $s\in[0,L]$. Then the finite-element equation can be written as folllows:

$$(\boldsymbol{K} - \omega^2\boldsymbol{M})a = \boldsymbol{0} \tag{19}$$

where the assembly nodal vector a consists of all a_j.

For the actual structural feature of Fig. 1. as the excitation of the sensor is selected to be contactless, such as electrostatically driven[4], and the detector of the sensor is selected as a piezoresistor[2], the boundary conditions of the beam are as follows:

$$\left.\begin{aligned} s = 0: \quad w(s) = w'(s) = 0 \\ s = L: \quad w(s) = w'(s) = 0 \end{aligned}\right\} \tag{20}$$

From Eqs. (19) and (20), we can obtain the natural frequencies and the corresponding vibrating shapes.

6 Calculation and Discussion of Concentrated Force Frequency Relationship

The sensor is made of silicon, $E=1.3\times10^{11}$ Pa. $\rho=2.33\times10^3$ kg/m^3, $\mu=0.17$. The outer radius and thickness of the E-type round diaphragm are $A=2.5\times10^{-3}$ m, $H=0.042\times10^{-3}$ m. In addition, the width and thickness of the beam are $b=50\times10^{-6}$ m and $h=5\times10^{-6}$ m. Considering the symmetry of the stress of the E-type round diaphragm, the results are the same if the beam is located at (R_0, R_0+L) and at $(-R_0-L, -R_0)$. Therefore, we shallonly discuss the half location of the beam on the E-type round diaphragm in this paper. In order to compare the concentrated force-frequency relationship under the different parameters of the E-type round diaphragm and the beam, we define a relative position R_L of the beam along the radial direction of the diaphragm as follows:

$$R_L = \frac{R_0 - B}{A - B - L} \tag{21}$$

Figs. 4 and 5 show the variation and the relative variation of the basic natural frequency of the beam within (0, 0.1)N for a beam located at different relative positions R_L along the radial direction of the diaphragm. corresponding to different lengths $L=400\times10^{-6}$, 500×10^{-6} and 600×10^{-6} m, respectively, when the inner radius of the E-type round diaphragm is

$B=0.9\times10^{-3}$ m.

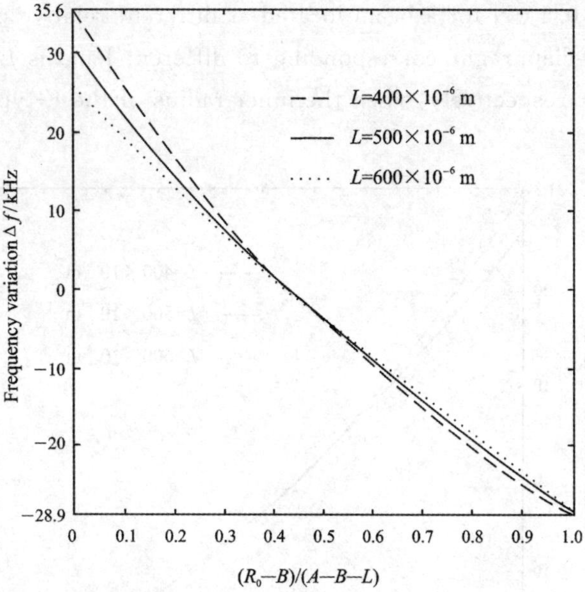

Fig. 4 The variation of the beam's basic frequency vs. its relative location along the radial direction of the diaphragm within (0, 0.1) N, for different beam lengths and $A=2.5\times10^{-3}$ m, $B=0.9\times10^{-3}$ m

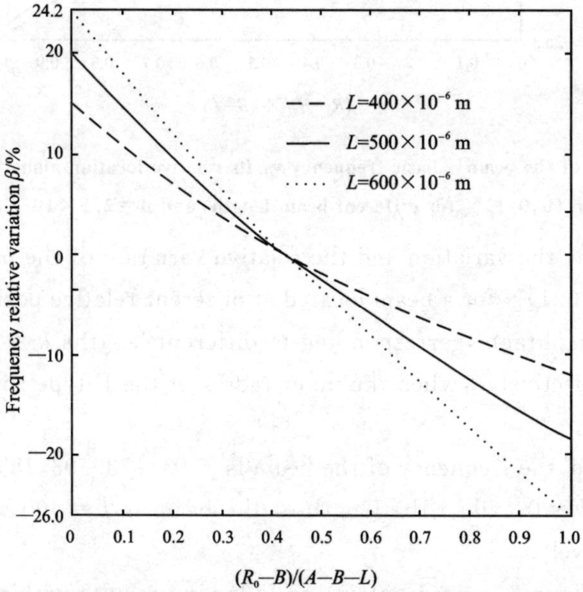

Fig. 5 The relative variation of the beam's basic frequency vs. its relative location along the radial direction of the diaphragm within (0, 0.1) N, for different beam lengths and $A=2.5\times10^{-3}$ m, $B=0.9\times10^{-3}$ m

Figs. 6 and 7 show the variation and the relative variation of the basic natural frequency of the beam within(0,0. 1) N for a beam located at different relative positions R_L along the radial direction of the diaphragm, corresponding to different lengths $L = 400 \times 10^{-6}$, 500×10^{-6} and 600×10^{-6} m, respectively, when the inner radius of the E-type round diaphragm is $B = 1.0 \times 10^{-3}$ m.

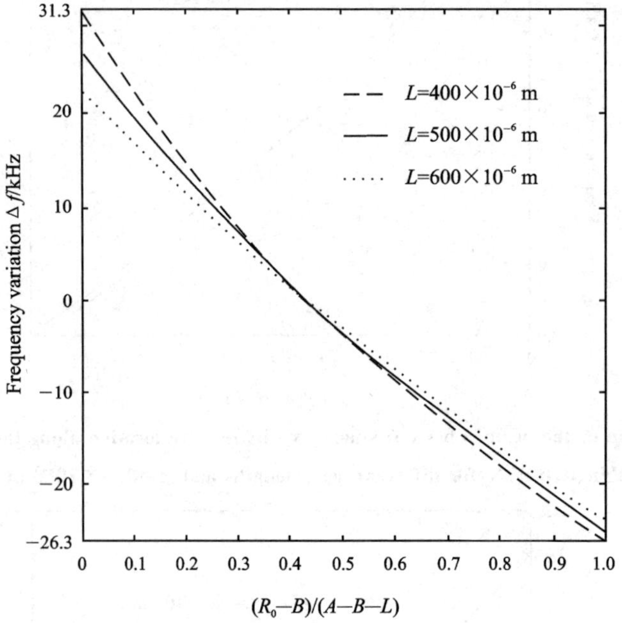

Fig. 6 The variation of the beam's basic frequency vs. its relative location along the radial direction of the diaphragm within (0,0. 1)N, for different beam lengths and $A = 2.5 \times 10^{-3}$ m, $B = 1.0 \times 10^{-3}$ m

Figs. 8 and 9 show the variation and the relative variation of the basic natural frequency of the beam within(0,0. 1)N for a beam located at different relative positions R_L along the radial direction of the diaphragm, corresponding to different lengths $L = 400 \times 10^{-6}$, 500×10^{-6} and 600×10^{-6} m, respectively. , when the inner radius of the E-type round diaphragm is $B = 1.1 \times 10^{-3}$ m.

On the other hand, the frequency of the beam is $f(0) = 239.93$, 153. 56 and 106. 65 kHz, for the applied force $F = 0$, when the length of the beam is $L = 400 \times 10^{-6}$, 500×10^{-6} and 600×10^{-6} m, respectively.

From the above figures, we can obtain the following results within the investigated ranges:

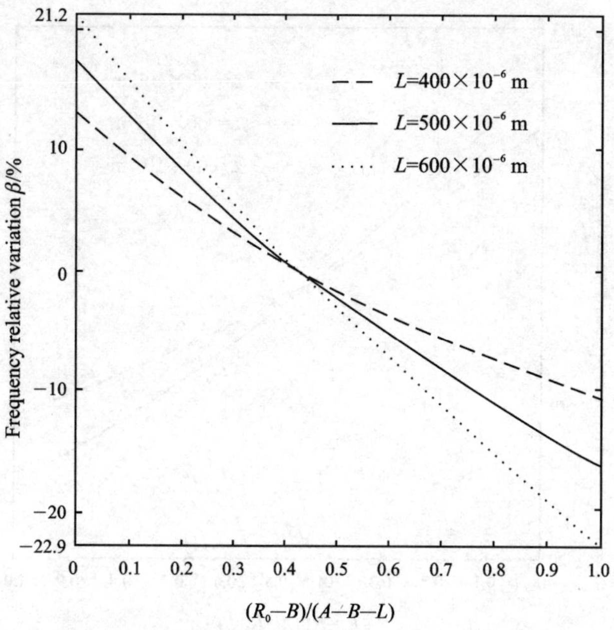

Fig. 7 The relative variation of the beam's basic frequency vs. its relative location along the radial direction of the diaphragm within (0, 0.1)N, for different beam lengths and $A=2.5\times10^{-3}$ m, $B=1.0\times10^{-3}$ m

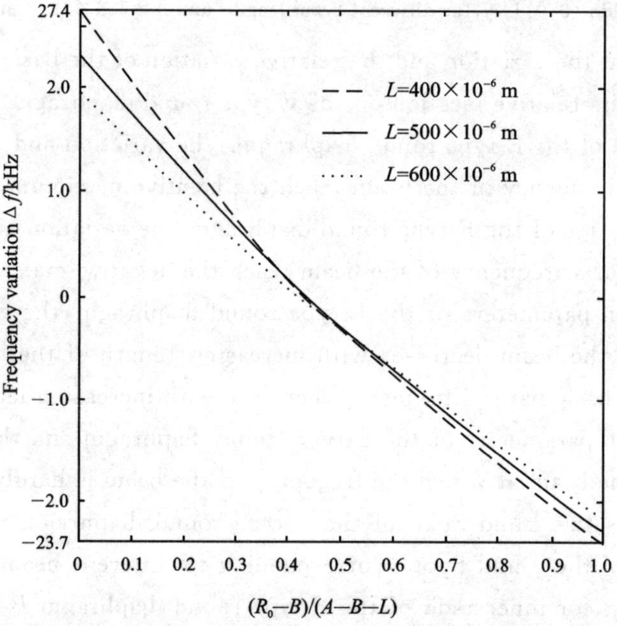

Fig. 8 The variation of the beam's basic frequency vs. its relative location along the radial direction of the diaphragm within (0, 0.1)N, for different beam lengths and $A=2.5\times10^{-3}$ m, $B=1.1\times10^{-3}$ m

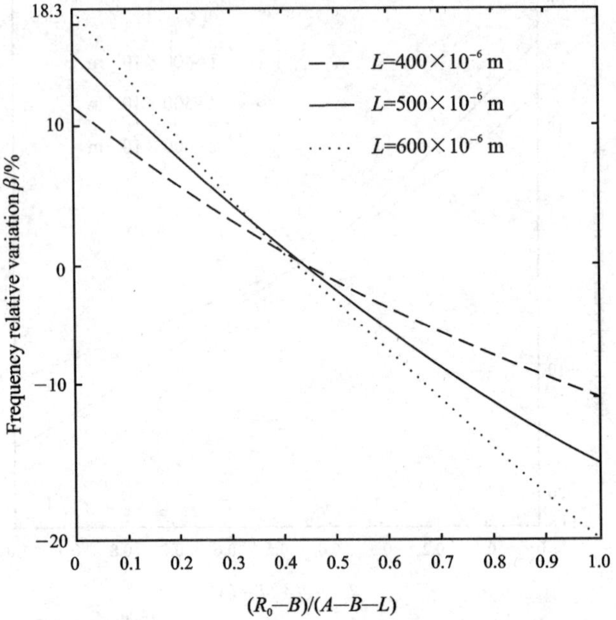

Fig. 9 The relative variation of the beam's basic frequency vs. its relative location along the radial direction of the diaphragm within (0,0.1)N, for different beam lengths and $A=2.5\times10^{-3}$ m, $B=1.1\times10^{-3}$ m

① It is clear that the variation and the relative variation of the basic natural frequency of the beam vary with its relative location on the E-type round diaphragm. If the beam is located at the inner edge of the E-type round diaphragm. the variation and the relative variation of the basic natural frequency of the beam reach the positive maximum; while if the beam is located at the outer edge of the E-type round diaphragm, the variation and the relative variation of the basic natural frequency of the beam reach the negative maximum.

② For the given parameters of the E-type round diaphragm, the variation of the basic natural frequency of the beam decreases with increasing length of the beam, while the relative variation of the basic natural frequency increases with increasing length of the beam.

③ For the given parameters of the E-type round diaphragm and the beam. there is one unique location of the beam at which the frequency of the beam is hardly varying. We define the above location as the 'Blind Spot' of the E-type round diaphragm in this paper. Fig. 10 shows the location of the 'Blind Spot', corresponding to different beam lengths from 300×10^{-6} to 700×10^{-6} m, for inner radii of the E-type round diaphragm $B=0.9\times10^{-3}$, 1.0×10^{-3} and 1.1×10^{-3} m, respectively. Designers should take note of this characteristic of the above complex sensing structure of Fig. 1. It is worthwhile for designers to notice that the relative location of the"Blind Spot"for the beam on the diaphragm is always the same. This result is valuable for designing sensing structures of different geometric parameters.

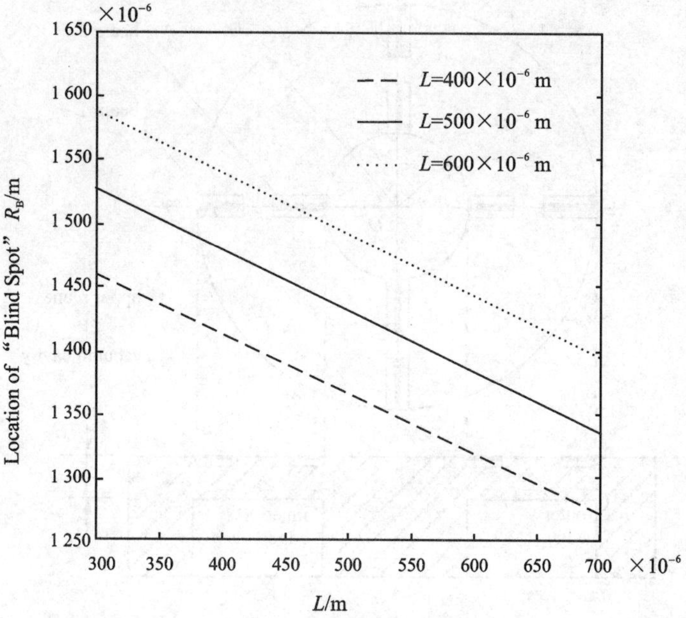

Fig. 10 Location of "Blind Spot" vs. beam length for different inner radii *B* of the E-type diaphragm and the same outer radius $A=2.5\times10^{-3}$ m

④ For the given parameters of the beam, the variation and the relative variation of the basic natural frequency of the beam increase with decreasing inner radius of the E-type round diaphragm when the outer radius of the E-type dinphragm has a given value.

Based on the above calculations and analyses, it is better for a pair of beams(beam 2 and beam 1) to be located at the inner and outer edge of the E-type round diaphragm, respectively, see Fig. 11, i. e., $r\in(B,B+L)$ for beam 2, while $r\in(A-L,A)$ for beam 1. Then the output of the sensor is the differential frequency of beam 2 and beam 1. Thus the sensitivity of the sensor will be twice that of the frequency output of only beam 2 or beam 1. The differential output mechanism can reduce some disturbances such as the influence of temperature. Finally, we would like to present a set of optimum parameters of the above sensing unit: for the E-type round diaphragm, $A=2.5\times10^{-3}$ m, $B=1.0\times10^{-3}$ m, $H=0.042\times10^{-3}$ m; for beam 2 and beam 1, $L=600\times10^{-6}$ m, $b=50\times10^{-6}$ m, $h=5\times10^{-6}$ m. The sensing structure is the same as in Fig. 11. It should be pointed out that the scheme with four pairs of beams Jocated at the upper plane of the E-type round diaphragm orthogonal can improve the productivity of the silicon sensor, and the production cost does not increase.

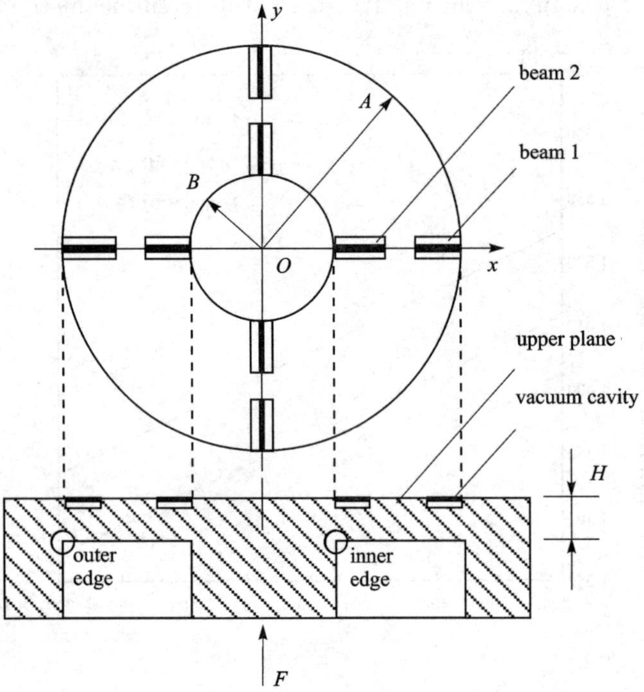

Fig. 11 Differential output scheme(four pairs of beams)

7 Conclusions

This paper establishes the FEM model of a practical silicon complex sensing structure for measuring the concentrated force. Based on the model, we calculate, analyse and investigate the relationship between the basic natural frequency of the beam resonator and the applied concentrated force. Moreover, we present some important results on designing the beam resonator. The authors suggest that a set of optimum parameters of the above sensing unit as follows: for the E-type round diaphragm, $A=2.5\times10^{-3}$ m, $B=1.0\times10^{-3}$ m, $H=0.042\times10^{-3}$ m; for the beams, $L=600\times10^{-6}$ m, $b=50\times10^{-6}$ m, $h=5\times10^{-6}$ m. A pair of beams is located at the inner and outer edges of the E-type round diaphragm. Then the simulation results are as follows: the output frequency is within (106.65, 129.21) kHz for beam 2 and (106.65, 82.20) kHz for beam 1 when the measured concentrated force is within (0, 0.1) N. In addition, $\Delta f=22.56$ kHz, $\beta=21.2\%$ for beam 2 and $\Delta f=-24.45$ kHz, $\beta=-22.9\%$ for beam 1. As the differential frequency of beam 2 and beam 1 is the output of the sensor, the output is within (0,47.01) kHz.

Acknowledgements

This work was supported by the Chinese Aeronautics Science Foundation, the Korea Science and Engineering Fonndation and research fund of POSCO Chair professor.

References

[1] Luo R C. Sensor technologies and microsensor issues for mechatronics systems(invited paper). IEEE/ASME Trans. Mechatronics I, 1996: 39-49.

[2] Parsons P, Glendinning A, Angelidis D. Resonator sensor for high accuracy pressure measurement using silicon technology. Schlumberger Industries Aerospace. Transducer Division, IEEE AES Magazine, July, 1992: 45-48.

[3] Beeby S P, Tudor M J. Modeling and optimization of micromachined silicon resonators. J. Micromech. Microeng., 1995, 5: 103-105.

[4] Tilmans H A C, Legtenberg R. Electrostatically driven vacuumencapsulated polysilicon resonators, Part II: Theory and performance. Sensors and Actuators A, 1994(45): 67-84.

[5] Hauptmann P. Resonant sensors and applications(invited paper). Sensors and Actuators A, 1991(27): 371-377.

[6] Timoshenko S P, Krieger S W. Theory of Plates and Shells, McGraw-Hill, New York, 1959.

Appendix E
Nonlinear Vibration Characteristics of Coriolis Mass Flowmeter①

Zheng Dezhi   Wang Shuai   Fan Shangchun

Abstract: Coriolis Mass Flowmeter (CMF) is generally regarded as a linear system when the vibration amplitude of the measuring tube is relatively small, but the spectrum of its output signals still show existence of some multiple frequency components. Taking into consideration the nonlinear factors and external excitation, the motion equations of the CMF straight measuring tube are established by properly simplifying the boundary conditions of the measuring tube and integrating the mathematical model of the CMF sensor. In addition, the motion equations are discretized with the Galerkin method, and from the discretized results, it is discovered that the system contains a group of nonlinear factors. By using a laser vibrometer, the state of the closed loop resonant vibration of the measuring tube and the forced vibration of the standard sine signal are measured. The results show that both contain nonlinear factors just as the nonlinear analysis had expected. Experiments have been undertaken to verify the measurement accuracy of the measuring tube with different vibration amplitudes, which evidences that nonlinear influences become more serious with the vibration amplitude rising. The theoretical analysis and experimental results are expected to be of great value for further improvement of CMF performance.

Keywords: flowmeters, Coriolis, closed loop control systems, nonlinear systems, vibrations

1 Introduction

As one of the high-precision devices to directly measure the mass flow of fluid, the Coriolis Mass Flowmeter (CMF) has found wide application in a variety of industries, such as, aeronautical industry, aerospace industry, petrochemical industry, food processing industry, and so on. CMF, through the mechanical vibration of the measuring tube, produces Coriolis

---

① Chinese Journal of Aeronautics, 2009, 22(2): 198-205.

effects that are sensitive to the mass flow of the fluid to be measured. It is generally regarded as a linear system when the vibration amplitude is relatively small. In their experimental researches on the influences of pulsation flow, in 1998, Britain scientists R. Cheesewright, et al.[1] found that there were excitation signals, with 2～4 frequency multiplication, in the spectrum from the measuring tube, which were suspected to be caused by nonlinear effects. Since then, few articles have been found on the nonlinear vibration of CMF sensor. This article uses the analytical method to perform a theoretical study on the straight CMF tube, its nonlinear vibration characteristics, and the influences on CMF, and then carries out verification through experiments. The results substantiate the correctness of the discovery in Ref. [1]. Moreover, this article also makes clear the relationship between vibration amplitude and sensitivity, through experiments.

2　Model Analysis with Nonlinear Factors Considered

2.1　Basic assumptions and approximations

The following assumptions are made for the CMF straight tube type as shown in Fig. 1[2-4], where V is the flow velocity, P is the pressure of fix end, $f(t)$ is the harmonic exciting force, and $w(x,t)$ is the offset, L is the length of the measuring tube, m_c is the mass of exciting unit; m_l and m_r are the mass of detecting units, this two detecting units' distance is s, the distance from each unit is m_c, m_l and m_r to the left fixed end is x_c, x_l and x_r.

Of the lateral and axial displacements of the measuring tube, only the former is taken into consideration.

The measuring tube is fully filled with the fluid to be measured, which has the same displacement and deflectionangle as the tube itself.

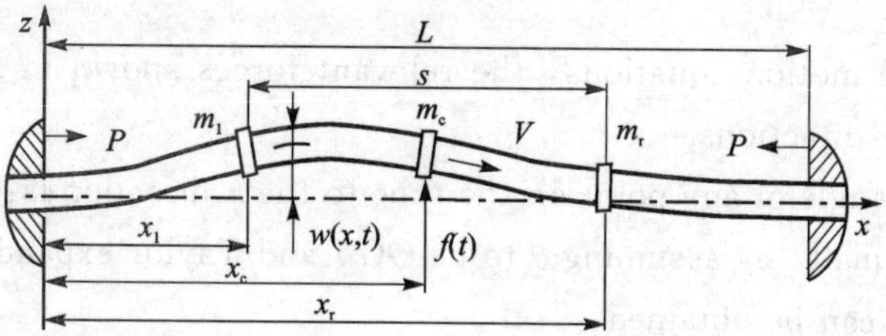

Fig. 1　Schematic construction diagram of the measuring tube of CMF straight shape

Generally, the measuring tube is 10 times as long as its diameter, and is assumed to vibrate with its inherent frequency. The measuring tube is treated as Euler beam by neglecting its rotational inertia and transversal shear deformation.

The measuring tube has a slight damping over the vibration period and an excitation frequency close to its inherent frequency, so the measuring tube is thought to vibrate free of damping.

The effects of the gravity of the measuring tube, and the exciting and detecting units on the measuring results are neglected.

The simple harmonic exciting force $f(t)$ is assumed to act on the midpoint of the tube length with the amplitude remaining unchanged: $f(t)=F_0 \cos \Omega t$.

The wavelength of the pulsating flow is far greater than the length of the measuring tube, and the fluid to be measured is supposed to evenly distribute along the pipe, changing in a simple harmonic manner with time. The velocity of the flow is

$$v = v_0(1+\eta \cos \omega_f t) \tag{1}$$

where η is the pulse coefficient.

2.2 Establishment of motion equation for the measuring tube

Based on the aforementioned assumptions, the motion equations of the measuring tube are formulated in the coordinates shown in Fig. 1 by means of a force balance method, in which the lateral displacement of the measuring tube is expressed by $z=z(x,t)$. Let dx be a lengthways differential element; then Fig. 2 shows the force conditions of the tube and fluid elements.

In Fig. 2, q_τ and q_n represent the tangential and normal forces acting on the unit length between the fluid and the tube wall, respectively; p is the fluid pressure; M and Q are the bending moment and shearing force acting on the cross-section of the tube; and T is the axial force on the cross-section of the tube, m_f the mass of fluid micro element, m_t the mass of tube micro element.

In deducting the motion equations, the relevant forces shown in Fig. 2 need to be resolved in the x and z directions.

The inclination angle of any point on the tube to the x direction is: $\theta=\partial z/\partial x=z'$. Since the angle θ is quite small, by assuming θ to be $O(e)$ and Taylor expanding $\sin\theta$ and $\cos\theta$ at $\theta=0$, the following can be obtained:

$$\left.\begin{aligned} \sin\theta &\approx \theta + O(e^3) \\ \cos\theta &\approx 1 + O(e^2) \end{aligned}\right\} \tag{2}$$

Neglecting the second and higher order terms leads to $\sin\theta\approx\theta$ and $\cos\theta\approx 1$.

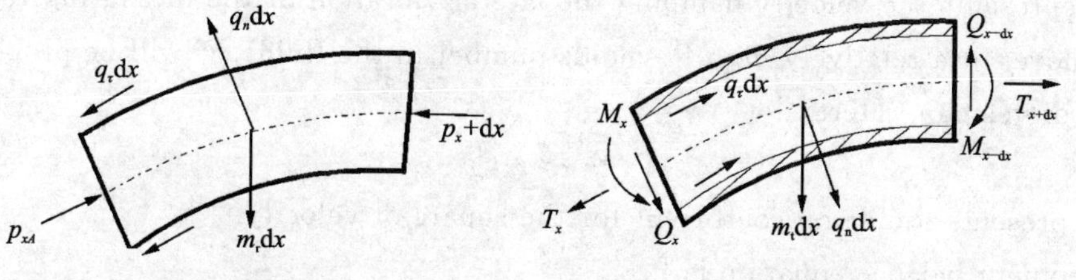

(a) Fluid element (b) Tube element

Fig. 2 Schematic diagram of force conditions of fluid and tube elements

By choosing a fluid particle randomly in the tube and neglecting the changes of velocity caused by the deformation of the tube, the fluid velocity can be described as

$$\boldsymbol{V}_{\mathrm{f}} = v\boldsymbol{i} + (\dot{z} + vz')\boldsymbol{j} \tag{3}$$

where $\boldsymbol{i}$ and $\boldsymbol{j}$ represent the unit direction vector in the x and z directions, respectively, and "·" and "′" are the derivatives with respect to time and space, respectively.

By partially differentiating Eq. (3) with respect to time (t), the fluid acceleration can be obtained as follows

$$\boldsymbol{a}_{\mathrm{f}} = \dot{v}\boldsymbol{i} + \left(\ddot{z} + 2v\frac{\partial^2 z}{\partial x \partial t} + \dot{v}z' + v^2 z''\right)\boldsymbol{j} \tag{4}$$

For a fluid element on the length of δx, the force balance equation can be expressed as follows:

In x direction:

$$-\frac{\partial}{\partial x}(pA\cos\theta) - q_n \sin\left(\theta + \frac{\partial\theta}{\partial x}\frac{\delta x}{2}\right) - q_\tau \cos\left(\theta + \frac{\partial\theta}{\partial x}\frac{\delta x}{2}\right) - m_{\mathrm{f}}\dot{v} = 0 \tag{5}$$

In z direction:

$$-\frac{\partial}{\partial x}(pA\sin\theta) + q_n \cos\left(\theta + \frac{\partial\theta}{\partial x}\frac{\delta x}{2}\right) - q_\tau \sin\left(\theta + \frac{\partial\theta}{\partial x}\frac{\delta x}{2}\right) - m_{\mathrm{f}} g -$$
$$m_{\mathrm{f}}\left(\ddot{z} + 2v\frac{\partial^2 z}{\partial x \partial t} + \dot{v}z' + v^2 z''\right) = 0 \tag{6}$$

For a tube element on the length of δx, the force balance equation can be expressed as follows:

In x direction:

$$\frac{\partial}{\partial x}(T\cos\theta) - \frac{\partial}{\partial x}(Q\sin\theta) + q_n \sin\left(\theta + \frac{\partial\theta}{\partial x}\frac{\delta x}{2}\right) + q_\tau \cos\left(\theta + \frac{\partial\theta}{\partial x}\frac{\delta x}{2}\right) = 0 \tag{7}$$

In z direction:

$$\frac{\partial}{\partial x}(T\sin\theta) + \frac{\partial}{\partial x}(Q\cos\theta) - q_n \cos\left(\theta + \frac{\partial\theta}{\partial x}\frac{\delta x}{2}\right) + q_\tau \sin\left(\theta + \frac{\partial\theta}{\partial x}\frac{\delta x}{2}\right) - m_t g - F_{\mathrm{d}} - m_t\ddot{z} = 0 \tag{8}$$

where F_d represents the velocity damping the lateral vibration of the measuring tube. When an object moves at a relatively high Reynolds number in the fluid, F_d will be proportional to the square of velocity. Here, let

$$F_d \approx c_d \dot{z}^2 \tag{9}$$

where c_d represents the damp coefficient for the square of velocity.

The moment balance equation is

$$Q + \frac{\partial M}{\partial x} = 0 \tag{10}$$

From the beam bending theory[5-6]:

$$Q = -EIz''' \tag{11}$$

where E and I represent the elastic modulus, and the cross-sectional moment of inertia respectively.

By omitting the second and higher order terms inclusive of $(\partial\theta/\partial x)\delta x$ and $Q \sin\theta$, Eqs. (5)~(8) could be rewritten as

$$\frac{\partial}{\partial x}(pA) + q_\tau + q_n z' + m_f \dot{v} = 0 \tag{12}$$

$$\frac{\partial}{\partial x}(pAz') + q_\tau z' - q_n + m_f g + m_f\left(\ddot{z} + 2v\frac{\partial^2 z}{\partial x \partial t} + \dot{v}z' + v^2 z''\right) = 0 \tag{13}$$

$$\frac{\partial T}{\partial x} + q_\tau + q_n z' = 0 \tag{14}$$

$$\frac{\partial}{\partial x}(Tz') + \frac{\partial Q}{\partial x} + q_\tau z' - q_n - m_t g - c_d \dot{z}^2 - m_t \ddot{z} = 0 \tag{15}$$

By subtracting Eq. (13) from Eq. (15), the following could be acquired:

$$\frac{\partial}{\partial x}[(T - PA)z'] + \frac{\partial\theta}{\partial x} - (m_f + m_t)g - C_d\dot{z}^2 - (m_f + m_t)\ddot{z} - m_f\left(2v\frac{\partial^2 z}{\partial x \partial t} + \dot{v}z' + v^2 z''\right) = 0 \tag{16}$$

By subtracting Eq. (12) from Eq. (14) and then integrating the result on $[x, L]$, the following could be obtained:

$$T - pA = (T - pA)\,|_{x=L} - m_f \dot{v}(L - x) \tag{17}$$

where L and A are the length of the measuring tube, and the cross—sectional area of flow.

When taking into account the axial elongation caused by lateral displacement, $(T - pA)|_{x=L}$ comprises the following three parts: the axial force caused by temperature changes and external factors, such as, manufacturing, installation, and so on, the additional axial forces[7-9] caused by the internal pressure of fluid in the measuring tube, and the additional axial forces obtained by the axial elongation due to lateral displacement. Thus, the following could be obtained:

$$(T-pA)\mid_{x=L}=T_L-p_LA(1-2\gamma)-\frac{EA_t}{2L}\int_0^L(z')^2\mathrm{d}x \tag{18}$$

where

A_t is the cross-sectional area of the tube wall;

T_L is the first kind of axial force at the end of the measuring tube;

$p_LA(1-2\gamma)$ is the second kind of axial force;

γ is the Poisson's ratio of the material; and the last term, the average value of the additional axial forces caused by lateral displacement.

By substituting Eqs. (17) and (18) into Eq. (16) and omitting the constant term, $(m_t+m_f)g$, the following could be obtained:

$$EIz^{(4)}-\left[T_L-p_LA(1-\gamma)-\frac{EA_t}{2L}\int_0^L(z')^2\mathrm{d}x-m_f\dot{v}(L-x)\right]z''+m_fv^2z''+$$
$$2m_fv\frac{\partial^2 z}{\partial x\partial t}+c_d\dot{z}^2+(m_f+m_t)\ddot{z}=0 \tag{19}$$

This is the motion equation of the straight CMF measuring tube in the absence of external excitation, although taking into account the nonlinear factors. Obviously, Eq. (19) contains the nonlinear square terms[10-11].

In a normal working condition, the CMF measuring tube is in the forced vibration state under the action of external excitation. The external exciting force $f(t)$ is the concentrated force acting on the middle part of the tube (see Fig. 1), which can be expressed in the following distribution form, with the δ function:

$$f(t)=F_0\cos(\Omega t)\delta(x-L/2) \tag{20}$$

Therefore, a differential equation of the vibration of the CMF measuring tube under the action of external excitation can be written as

$$EIz^{(4)}-\left[T_L-p_LA(1-2\gamma)-\frac{EA_t}{2L}\int_0^L(z')^2\mathrm{d}x-m_f\dot{v}(L-x)\right]z''+$$
$$m_fv^2z''+2m_fv\frac{\partial^2 z}{\partial x\partial t}+c_d\dot{z}^2+(m_f+m_t)\ddot{z}=F_0\cos(\Omega t)\delta\left(x-\frac{L}{2}\right) \tag{21}$$

Substituting Eq. (1) into Eq. (21), the following can be acquired:

$$EIz^{(4)}-\left\{T_L-p_LA(1-2\gamma)-\frac{EA_t}{2L}\int_0^L(z')^2\mathrm{d}x+m_fv_0\eta\omega_f(L-x)\sin(\omega_f t)-\right.$$
$$\left.m_fv_0^2[1+\eta\cdot\cos(\omega_f t)]^2\right\}z''+2m_fv_0[1+\eta\cos(\omega_f t)]\frac{\partial^2 z}{\partial x\partial t}+c_d\dot{z}^2+(m_f+m_t)\ddot{z}=$$
$$F_0\cos(\Omega t)\delta\left(x-\frac{L}{2}\right) \tag{22}$$

Eq. (22) is the motion equation of the CMF measuring tube taking nonlinearity, simple harmonic pulsation flow, and external excitation into consideration.

2.3 Approximate solution of motion equation

(1) Discretization of motion equation

To analyze the solution of the motion equation of the CMF measuring tube, the Galerkin method[12] is first used to discretize Eq. (22). Based on the first two orders of the vibration mode, the $z(x,t)$ can be expressed as

$$z(x,t) \approx \varphi_1(x)q_1(t) + \varphi_2(x)q_2(t) \tag{23}$$

where

$\varphi_1(x)$ and $\varphi_2(x)$ represent the first order and the second order undamped free vibration modes of the straight beam, which has the same geometrical parameters and boundary conditions as the measuring tube full of static fluid;

$q_1(t)$ and $q_2(t)$ represent the first order and the second order generalized coordinates.

According to the vibration theory of beams, $\varphi_1(x)$ and $\varphi_2(x)$ have orthogonality relations:

$$\left.\begin{aligned} &\int_0^L \varphi_1(x)\varphi_2(x)\mathrm{d}x = \delta_{12} \\ &\int_0^L \varphi_s(x)\frac{\mathrm{d}^4\varphi_r(x)}{\mathrm{d}x^4}\mathrm{d}x = \omega_{bs}^2\frac{m_t + m_f}{EI}\delta_{sr} \qquad (s,r = 1,2) \end{aligned}\right\} \tag{24}$$

where $\omega_{bs}(s=1,2)$ separately represent correspondingly the first order and the second order inherent frequencies.

Substitute Eqs. (23)~(24) into Eq. (22); multiply both sides of the result by $\varphi_s(x)(s=1,2)$; integrate the result on $[0, L]$ and, meanwhile, introduce the following symbols:

$$\begin{cases} b_{sr} = \int_0^L \varphi_s(x)\varphi_r'(x)\mathrm{d}x \\ c_{sr} = \int_0^L \varphi_s(x)\varphi_r''(x)\mathrm{d}x \qquad (s,r = 1,2) \\ d_{sr} = \int_0^L x\varphi_s(x)\varphi_r''(x)\mathrm{d}x \end{cases}$$

$$e_{11} = \int_0^L \varphi_1^3\mathrm{d}x, \quad e_{12} = \int_0^L \varphi_1^2\varphi_2\mathrm{d}x$$

$$e_{21} = \int_0^L \varphi_2^2\varphi_1\mathrm{d}x, \quad e_{22} = \int_0^L \varphi_2^3\mathrm{d}x$$

By taking advantage of the orthogonality of $\varphi_s(x)$ and the values of b_{sr}, c_{sr} and d_{sr}, and after rearrangement, the following expressions can be acquired:

$$(m_t+m_f)\ddot{q}_1+2m_f v_0 b_{12}\dot{q}_2+\left[p_L A(1-2\gamma)-T_L-m_f v_0^2\cdot\left(1+\frac{\eta^2}{2}\right)\right]c_{11}q_1+(m_t+m_f)\omega_{b1}^2 q_1+$$

$$2\eta m_f v_0 b_{12}\dot{q}_2\cdot\cos(\omega_f t)+\left[2\eta m_f v_0^2\cos(\omega_f t)+\frac{\eta^2}{2}m_f v_0^2\cos(2\omega_f t)-\eta m_f v_0\omega_f L\sin(\omega_f t)\right]c_{11}q_1+$$

$$\eta m_f v_0\omega_f(d_{11}q_1+d_{12}q_2)\cdot\sin(\omega_f t)-\frac{EA_t}{2L}(c_{11}^2 q_1^3+c_{11}c_{22}q_1 q_2^2)+c_d(e_{11}\dot{q}_1^2+2e_{12}\dot{q}_1\dot{q}_2+e_{21}\dot{q}_2^2)=$$

$$\varphi_1\left(\frac{L}{2}\right)F_0\cos(\Omega t) \tag{25a}$$

$$(m_t+m_f)\ddot{q}_2+2m_f v_0 b_{21}\dot{q}_1+\left[p_L A(1-2\gamma)-T_L+m_f v_0^2\cdot\left(1+\frac{\eta^2}{2}\right)\right]c_{22}q_2+(m_t+m_f)\omega_{b2}^2 q_2+$$

$$2\eta m_f v_0 b_{21}\dot{q}_1\cdot\cos(\omega_f t)+\left[2\eta m_f v_0^2\cos(\omega_f t)+\frac{\eta^2}{2}m_f v_0^2\cos(2\omega_f t)-\eta m_f v_0\omega_f L\sin(\omega_f t)\right]c_{22}q_2+$$

$$\eta m_f v_0\omega_f(d_{21}q_1+d_{22}q_2)\cdot\sin(\omega_f t)-\frac{EA_t}{2L}(c_{22}^2 q_2^3+c_{11}c_{22}q_1^2 q_2)+c_d(e_{12}\dot{q}_1^2+2e_{21}\dot{q}_1\dot{q}_2+e_{22}\dot{q}_2^2)=$$

$$\varphi_2\left(\frac{L}{2}\right)F_0\cos(\Omega t) \tag{25b}$$

By introducing a small parameter ε and letting

$$\eta=\varepsilon\tilde{\eta}$$

$$a_{11}=\frac{2m_f v_0}{m_t+m_f}b_{12}$$

$$a_{12}=\frac{m_f v_0^2\left(1+\frac{\eta^2}{2}\right)-T_L+p_L A(1-2\gamma)}{m_t+m_f}c_{11}+\omega_{b1}^2$$

$$a_{21}=\frac{2m_f v_0}{m_t+m_f}b_{21}$$

$$a_{22}=\frac{m_f v_0^2\left(1+\frac{\eta^2}{2}\right)-T_L+p_L A(1-2\gamma)}{m_t+m_f}c_{22}+\omega_{b2}^2$$

$$\begin{cases} h_{r2}=\tilde{\eta}\dfrac{m_f v_0\omega_f}{m_t+m_f}(Lc_{rr}-d_{rr}) \\ h_{r3}=-2\tilde{\eta}\dfrac{m_f v_0^2}{m_t+m_f}c_{rr} \qquad (r=1,2) \\ h_{r5}=-\tilde{\eta}^2\dfrac{m_f v_0^2}{2(m_t+m_f)}c_{rr} \end{cases}$$

$$h_{11}=-2\tilde{\eta}\frac{m_f v_0}{m_t+m_f}b_{12},\quad h_{21}=-2\tilde{\eta}\frac{m_f v_0}{m_t+m_f}b_{21}$$

$$h_{14} = -\tilde{\eta}\frac{m_f v_0 \omega_f}{m_t + m_f} d_{12}, \quad h_{24} = -\tilde{\eta}\frac{m_f v_0 \omega_f}{m_t + m_f} d_{21}$$

$$\varepsilon h_a = \frac{EA_t}{2L(m_t + m_f)} c_{11}^2, \quad \varepsilon h_b = \frac{EA_t}{2L(m_t + m_f)} c_{11} c_{22}$$

$$\varepsilon h_c = \frac{EA_t}{2L(m_t + m_f)} c_{22}^2, \quad \varepsilon h_d = -\frac{c_d e_{11}}{m_t + m_f}$$

$$\varepsilon h_e = -\frac{c_d e_{12}}{m_t + m_f}, \quad \varepsilon h_f = -\frac{c_d e_{21}}{m_t + m_f}$$

$$\varepsilon h_g = -\frac{c_d e_{22}}{m_t + m_f}$$

then, Eq. (25) can be rewritten as

$$\begin{aligned}\ddot{q}_1 + a_{11}\dot{q}_2 + a_{12}q_1 = {} & \varepsilon[h_{11}\dot{q}_2\cos(\omega_f t) + h_{12}q_1\sin(\omega_f t) + h_{13}q_1\cos(\omega_f t) + h_{14}q_2\sin(\omega_f t)] + \\ & \varepsilon^2 h_{15}q_1\cos(2\omega_f t) + \varepsilon[h_a q_1^3 + h_b q_1 q_2^2 + h_d \dot{q}_1^2 + 2h_e \dot{q}_1\dot{q}_2 + h_f \dot{q}_2^2] + \\ & \varphi_1\left(\frac{L}{2}\right)\frac{F_0}{m_t + m_f}\cos\Omega t\end{aligned} \tag{26a}$$

$$\begin{aligned}\ddot{q}_2 + a_{21}\dot{q}_2 + a_{22}q_2 = {} & \varepsilon[h_{21}\dot{q}_1\cos(\omega_f t) + h_{22}q_2\sin(\omega_f t) + h_{23}q_2\cos(\omega_f t) + h_{24}q_1\sin(\omega_f t)] + \\ & \varepsilon^2 h_{25}q_2\cos(2\omega_f t) + \varepsilon[h_c q_2^3 + h_b q_1^2 q_2 + h_e \dot{q}_1^2 + 2h_f \dot{q}_1\dot{q}_2 + h_g \dot{q}_2^2] + \\ & \varphi_2\left(\frac{L}{2}\right)\frac{F_0}{m_t + m_f}\cos\Omega t\end{aligned} \tag{26b}$$

It is obvious that after having been discretized by the Galerkin method, the motion equation of the CMF measuring tube is converted into a coupled 2 degrees of freedom (2-DOF) vibration problem, including nonlinear terms, under the combined action of parameter excitation and external excitation. Thus the vibration of the measuring tube, that is, the approximate solution of $z(x,t)$ can be obtained from Eq. (23) as long as the solutions of $q_1(t)$ and $q_2(t)$ have been found.

(2) Influences of nonlinear factors on the vibration characteristics of CMF measuring tube

Taking into account the action of steady flow ($\eta=0, v=v_0$), the following discrete differential equation can be achieved for the vibration of the CMF measuring tube under the action of steady flow from Eq. (26), taking into consideration the nonlinear factors:

$$\left.\begin{aligned}\ddot{q}_1 + a_{11}\dot{q}_2 + a_{12}q_1 = {} & \varepsilon(h_a q_1^3 + h_b q_1 q_2^2 + h_d \dot{q}_1^2 + 2h_e \dot{q}_1\dot{q}_2 + h_f \dot{q}_2^2) + \\ & \varphi_1\left(\frac{L}{2}\right)\frac{F_0}{m_t + m_f}\cos\Omega t \\ \ddot{q}_2 + a_{21}\dot{q}_1 + a_{22}q_2 = {} & \varepsilon(h_c q_2^3 + h_b q_1^2 q_2 + h_e \dot{q}_1^2 + 2h_f \dot{q}_1\dot{q}_2 + h_g \dot{q}_2^2) + \\ & \varphi_2\left(\frac{L}{2}\right)\frac{F_0}{m_t + m_f}\cos\Omega t\end{aligned}\right\} \tag{27}$$

In engineering applications of CMF, the flow velocity of the fluid to be measured is relatively low and thus has a relatively small influence on the inherent frequency of the measuring tube[13-14]. For the convenience of making a qualitative analysis using the multiple scale method, suppose the influences of fluid pressure and axial stress of the tube are so minor that they can be omitted, then it can be assumed that,

$$\alpha_{r2} \approx \omega_{br}^2 \qquad (r = 1,2) \tag{28}$$

Meanwhile, let a_{11} and a_{21} be small enough and have the same order as ε, and

$$a_{r1} = -\varepsilon\tilde{a}_{r1} \qquad (r = 1,2) \tag{29}$$

Since $\Omega \approx \omega_{b1}$, when the CMF measuring tube is under normal working conditions, another detuning parameter σ is introduced. Let

$$\Omega^2 = \omega_{b1}^2 + \varepsilon\sigma \tag{30}$$

By substituting Eqs. (28)~(30) into Eq. (27) and rearranging, the following can be obtained:

$$\left.\begin{aligned}\ddot{q}_1 + \Omega^2 q_1 &= \varepsilon(\sigma q_1 + \tilde{a}_{11}\dot{q}_2 + h_a q_1^3 + h_b q_1 q_2^2 + h_d \dot{q}_1^2 + \\ &\quad 2h_e \dot{q}_1 \dot{q}_2 + h_f \dot{q}_2^2) + \varphi_1\left(\frac{L}{2}\right)\frac{F_0}{m_t + m_f}\cos \Omega t \\ \ddot{q}_2 + \omega_{b2}^2 q_2 &= \varepsilon(\tilde{a}_{21}\dot{q}_1 + h_c q_2^3 + h_b q_1^2 q_2 + h_e \dot{q}_1^2 + \\ &\quad 2h_f \dot{q}_1 \dot{q}_2 + h_g \dot{q}_2^2) + \varphi_2\left(\frac{L}{2}\right)\frac{F_0}{m_t + m_f}\cos \Omega t\end{aligned}\right\} \tag{31}$$

Thus, the discretized differential equation for the vibration of the CMF measuring tube, taking into consideration the nonlinear factors and the action of steady flow, is simplified into a 2-DOF nonlinear vibration equation with the left side uncoupled.

Express $q_1(t)$ and $q_2(t)$ as the first order expansion of ε using the multiple scale method as follows

$$q_r(t) = q_{r0}(T_0, T_1) + \varepsilon q_{r1}(T_0, T_1) + \varepsilon^2 q_{r2}(T_0, T_1) + \cdots \qquad (r = 1,2) \tag{32}$$

and let

$$\left.\begin{aligned}\varepsilon\widetilde{F}_1 &= \varphi_1\left(\frac{L}{2}\right)\frac{F_0}{m_t + m_f} \\ F_2 &= \varphi_2\left(\frac{L}{2}\right)\frac{F_0}{m_t + m_f}\end{aligned}\right\} \tag{33}$$

By substituting Eqs. (32)~(33) into Eq. (31) and comparing the coefficients of ε with the same power on both sides of the equal mark, the ε^0 and ε^1 order expansion formula of Eq. (31) can be derived, respectively, as follows

$$\left.\begin{aligned}D_0^2 q_{10} + \Omega^2 q_{10} &= 0 \\ D_0^2 q_{20} + \omega_{b2}^2 q_{20} &= F_2 \cos \Omega T_0\end{aligned}\right\} \tag{34}$$

$$D_0^2 q_{11} + \Omega^2 q_{11} = -2D_0 D_1 q_{10} + \sigma q_{10} + \tilde{a}_{11} D_0 q_{20} + h_a q_{10}^3 + h_b q_{10} q_{20}^2 + h_d (D_0 q_{10})^2 + 2h_e D_0 q_{10} D_0 q_{20} + h_f (D_0 q_{20})^2 + \tilde{F}_1 \cos \Omega T_0 \quad (35a)$$

$$D_0^2 q_{21} + \omega_{b2}^2 q_{21} = -2D_0 D_1 q_{20} + \tilde{a}_{21} D_0 q_{10} + h_c q_{20}^3 + h_b q_{10}^2 q_{20} + h_e (D_0 q_{10})^2 + 2h_f D_0 q_{10} D_0 q_{20} + h_g (D_0 q_{20})^2 \quad (35b)$$

Assuming the solution of Eq. (34) is

$$\left.\begin{aligned} q_{10} &= A_1(T_1)\exp(\mathrm{j}\Omega T_0) + c \\ q_{20} &= A_2(T_1)\exp(\mathrm{j}\omega_{b2} T_0) + B_2 \exp(\mathrm{j}\Omega T_0) + c \end{aligned}\right\} \quad (36)$$

where c is the error of expansion formula, then

$$B_2 = \frac{F_2}{2(\omega_{b2}^2 - \Omega^2)} \quad (37)$$

By substituting Eq. (36) into Eq. (35), the frequency components contained by the square and cubic nonhomogeneous terms at the right side of the linear partial differential equations relevant to q_{11} and q_{21} (see Tab. 1) could be deduced.

Tab. 1 Frequency components contained in nonhomogeneous items

Nonhomogeneous items	$(D_0 q_{10})^2$	$(D_0 q_{20})^2$	$D_0 q_{10} D_0 q_{20}$	q_{10}^3	q_{20}^3	$q_{10} q_{20}^2$	$q_{10}^2 q_{20}$
Frequency components contained	2Ω	$2\Omega, 2\omega_{b2}$ $\omega_{b2} \pm \Omega$	2Ω $\omega_{b2} \pm \Omega$	Ω 3Ω	$\Omega, 3\Omega$ $\omega_{b2}, 3\omega_{b2}$ $2\Omega \pm \omega_{b2}$ $2\omega_{b2} \pm \Omega$	$\omega_{b2}, 3\Omega$ $2\Omega \pm \omega_{b2}$ $2\omega_{b2} \pm \Omega$	$\Omega, 3\Omega$ $2\Omega \pm \omega_{b2}$

According to the superposition law of the linear system, q_{11} and q_{21} have the following frequency components[15]:

$$\Omega, 2\Omega, 3\Omega, \omega_{b2}, 2\omega_{b2}, \omega_{b2} \pm \Omega, 2\Omega \pm \omega_{b2}, \omega_{b2} \pm \Omega$$

Moreover, q_{21} also contains the components of $3\omega_{b2}$ frequency. By further analyzing ε^2 and other higher order expansion formule using the multiple scale method, it is understood that q_{1r} and q_{2r} ($r=2,3,\cdots$) also contain other multiple frequencies and a combination of Ω and ω_{b2}.

From the above-mentioned analysis, it is clear that the vibration $z(x,t)$ of the CMF measuring tube contains many frequency components, such as Ω and its multiplication, ω_{b2} and its multiplication, and the combination frequencies formed with Ω and ω_{b2}. In this article, vibration components other than Ω are called additional vibration components, and are caused by nonlinear factors. Among them, ω_{b2} and its multiplication and combination frequencies formed by ω_{b2} and Ω are relevant to the free vibration terms contained in q_{20}. With there being no super-harmonic consonant, its amplitude will attenuate to zero, due to damping. Therefore, the main components of the vibration of a measuring tube are Ω and its multiplication and the vibration amplitude of the additional com-ponents decrease with an increase in the multiplication number, due to the effect of $\varepsilon^i (i=1,2,\cdots)$ in Eq. (32).

This analysis cannot show the difference in magnitude between the vibration amplitudes of frequency 2Ω and 3Ω appearing simultaneously in the expansion formula of ε^1 order, but the intensity of square and cubic nonlinear terms should be able to reflect it since the additional components of frequency 2Ω and 3Ω are caused by square and cubic nonlinear terms.

3 Experimental Research on Nonlinear Factors

From the above-mentioned analysis, it is understood that the nonlinear terms in the motion equations give birth to nonlinear vibration factors in a CMF measuring tube. To verify the existence of nonlinear vibration factors, the vibration state of the tube has been measured with a Polytec laser vibrometer in the laboratory and the corresponding spectrum analyzed[16]. Two experimental procedures are described as follows:

Experiment 1 The CMF measuring tube itself is actuated and made to work in the simple harmonic oscillation state with the closed loop oscillating circuit, and meanwhile, the vibration state of the measuring tube is measured with the laser vibrometer. Fig. 3 shows the results of spectrum analysis, with the amplitude of closed loop excitation signal being 1 V.

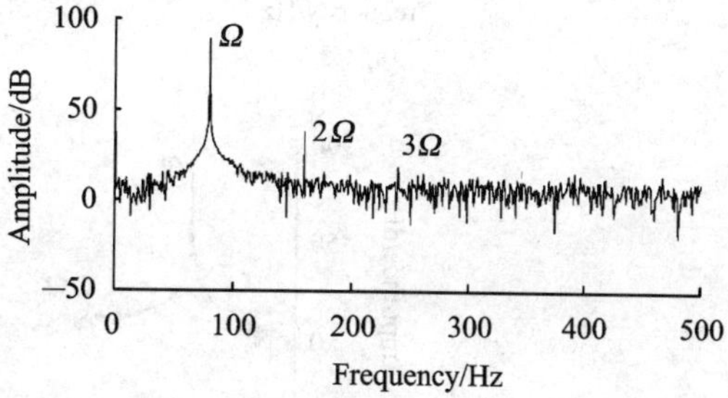

Fig. 3 Vibration spectrum of CMF measuring tube with closed loop working

Experiment 2 A signal generator is adopted to force the CMF measuring tube to work at the first order inherent frequency point and different amplitudes are set for the excitation signal, namely, the vibration amplitude of the measuring tube is changed by changing the exciting force. Fig. 4 shows the results from the comparison of vibration spectrums of the measuring tube determined at different amplitudes of excitation signals.

In summary, Experiment 1 evidences the existence of nonlinear factors, such as frequency doubling and tripling, in the vibration of the measuring tube of the CMF sensor, and the amplitude of the signal decreasing with the multiplication order increasing, which in turn verifies the theoretical analysis. Experiment 2 changes the vibration amplitude of the measuring tube by means of the exciting voltage with different amplitudes. From the comparison of the three diagrams (a), (b) and (c) shown in Fig. 4, it can be seen that with the change in amplitude of the exciting voltage, that is, the change in vibration amplitude of the measuring tube, the components of frequency multiplication in the vibration of the measuring tube also change. When the amplitude of the exciting voltage is relatively high (i. e., the vibrating amplitude of the measuring tube is relatively large), the amplitudes of nonlinear factors such as frequency multiplication and so on are also relatively large and vice versa. Consequently, it can be seen that the vibration amplitude of the measuring tube exerts a direct influence on the amplitude of the nonlinear factors.

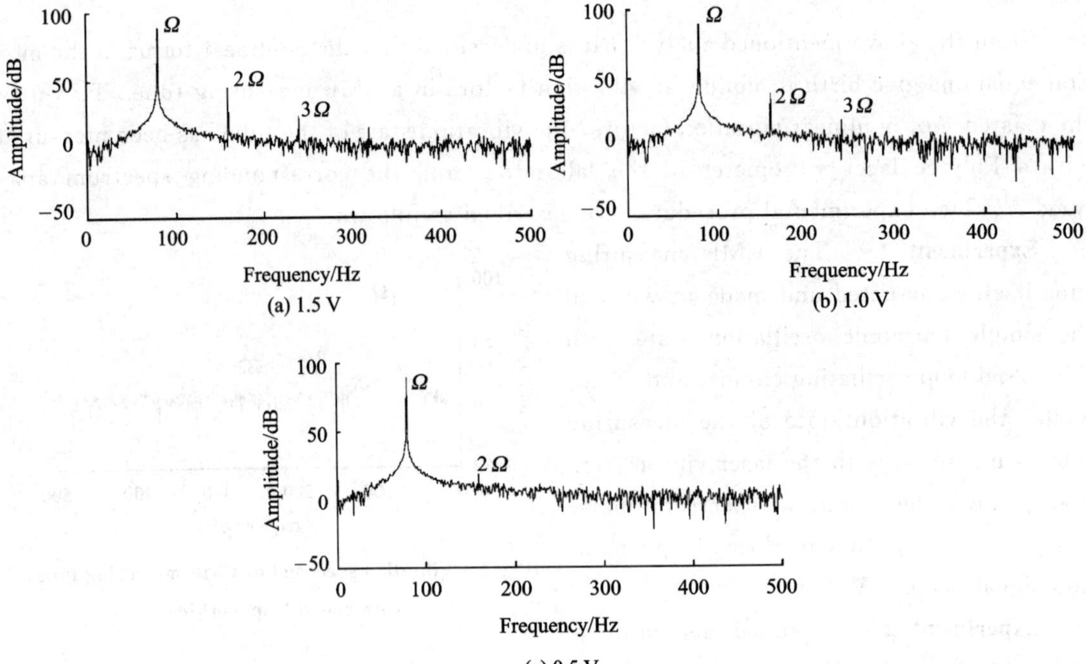

Fig. 4 Vibration spectrum of CMF measuring tube at different amplitudes of exciting signals

However, the vibration amplitude of the measuring tube has a direct relevance to the signal to noise ratio (SNR) of the CMF; the larger the vibration amplitude of the measuring tube is, the higher the SNR of the CMF is[17]. Therefore, there is a contradiction between the SNR and the anti-interference ability of CMF with respect to the nonlinear factors of vibration. In practice, both SNR and anti-interference ability of CMF are required to be as high as possible. This means it is necessary to increase the vibration amplitude of the measuring tube, and meanwhile, lower the nonlinear factors of vibration, namely, reduce the vibration amplitude of the measuring tube. Regrettably, so far, there is no way out for it both at home and abroad. In the CMF design, the optimal vibration amplitude of the measuring tube is often well determined through tests. Fig. 5 shows the curves of the CMF measuring accuracy of a commercial ϕ25 mm analogical straight tube, with different vibration amplitudes.

When the amplitude of the output signal of the measurement coil is 0. 28 V, meaning the measuring tube has a moderate amplitude, the curve of the CMF measurement error appears quite smooth, and indicates a higher accuracy; when the amplitude is 0. 20 V, meaning the vibration amplitude of the tube is rather small, the curve looks to have a remarkable fluctuation; when the amplitude is 0. 35 V, meaning the tube is under serious vibration, the curve is subjected to sharp fluctuations beyond the range of ± 0. 2%. Thus, it can be seen that the vibration amplitude of the CMF measuring tube requires to be tested for possibly the best working properties, under concrete conditions, because of the need for increasing the vibration amplitude of the CMF measuring tube to improve its SNR as well as lower the nonlinearity of CMF.

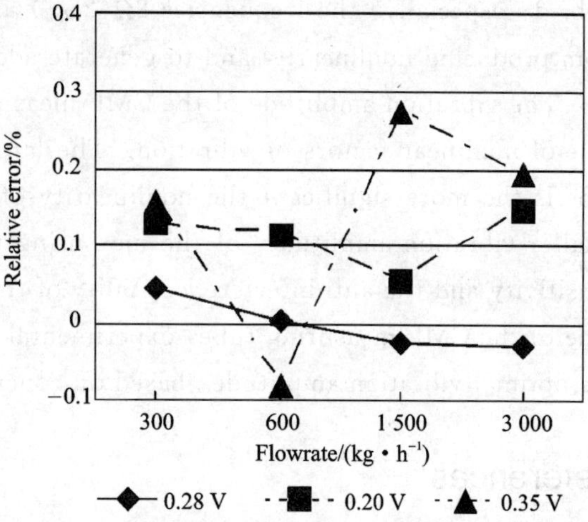

Fig. 5 Curves showing measurement accuracy of CMF tube with different vibration amplitudes

4 Conclusions

On the basis of the proposed motion equations for the CMF measuring tube and through theoretical analysis and experiments, the following conclusions could be drawn:

The influences that nonlinearity exerts on the vibration of the CMF measuring tube find chief expression in generating in the tube additional vibration components dominated by the multiplication of the excitation frequency. Its vibration amplitude decreases with the increase in the frequency multiplication number.

When the vibration frequency of the additional components equals to ω_{b2}, q_2 generates a super-harmonic resonance and the nonlinearity cannot make the free vibration solution attenuate to zero[18], thus inducing the additional vibration components of the measuring tube onto ω_{b2}. This would significantly distort the relationship between the output signal of CMF calibrated on frequency ω_{b1} and the measured mass flow, thereby producing a relatively large measurement error. Therefore, during designing the structure of the measuring tube, attention should be paid to keep the ω_{b2} clear of the frequencies of additional components shown in Tab. 1, especially, the frequencies $2\Omega(2\omega_{b1})$ and $3\Omega(3\omega_{b1})$, to prevent the measuring tube from producing nonlinearity and to generate additional vibration components on ω_{b2}.

The vibration amplitude of the CMF measuring tube has a direct influence on the amplitude of nonlinear factors of vibration. The larger the vibration amplitude of the measuring tube is the more significant the nonlinearity of vibration thus caused. On the other hand, smaller vibration amplitudes of the measuring tube would have a negative influence on the sensitivity and the anti-interference ability of the CMF. When designing the vibration amplitude of the CMF measuring tube, experimental verification should be carried out, to find out the optimal vibration amplitude, based on concrete conditions.

References

[1] Cheesewright R, Clark C. The effect of flow pulsations on Coriolis mass flow meters. Journal of Fluids and Structures,1998,12(8): 1025-1039.

[2] Ji A M. Modeling research & simulation analysis of CMF. Ph. D. thesis. Hefei: University of Science and Technology of China, 2001.

[3] Cheesewright R, Clark C, Belhadj A, et al. The dynamic response of Coriolis mass flow meters. Journal of Fluids and Structures, 2003, 18(2): 165-178.

[4] Clark C, Cheesewright R. The influence upon Coriolis mass flow meters of external vibrations at selected frequencies. Flow Measurement and Instrumentation, 2003, 14(1-2): 33-42.

[5] Weng Z Y. Collections of translations on static-dynamics of beams, plates and shells. Shanghai: Tongji University Press, 1986: 24-46.

[6] Xu Y Y. Dynamics of mechanical system. Beijing: China Machine Press, 1991: 10-35.

[7] Cascetta F. Effect of fluid pressure on Coriolis mass flowmeter's performance. ISA Transactions, 1996, 35(4): 365-370.

[8] Cascetta F. Experimental investigations upon the pressure effects on two Coriolis mass flowmeters of different generation. ISA Transactions, 1999, 38(2): 149-156.

[9] Belhadj A, Cheesewright R, Clark C. The simulation of Coriolis meter response to pulsating flow using a general purpose F. E. code. Journal of Fluids and Structures, 2000,14(5): 613-634.

[10] Zhang X. Research on nonlinear problems in resonant sensors. M. S. thesis. Beijing University of Aeronautics and Astronautics, 2005.

[11] Kutin J, Bajsic I. Stability-boundary effect in Coriolis meters. Flow Measurement and Instrumentation, 2001,12(1): 65-73.

[12] Lu W D. Variational method in differential equations. Beijing: Science Press, 2003: 98-109.

[13] Chen Q, Chen M, Fan D Y. Design and interference analysis of measurement tube of single and straight tube CMF. Collection of Papers of the Sixth National Symposium on Sensitive Elements and Sensors. 1999:447-450.

[14] Henry M P, Clark C, Duta M, et al. Response of a Coriolis mass flow meter to step changes in flow rate. Flow Measurement and Instrumentation, 2003, 14(3): 109-118.

[15] Zheng D Z. Linear system theory. Beijing: Tsinghua University Press, 1990: 41-216.

[16] Kutin J, Smercnik A, Bajsic I. Phase-locking control of the Coriolis meter's resonance frequency based on virtual instrumentation. Sensors and Actuators A, 2003,104(1): 86-93.

[17] Zhu Y P, Gao X B, Chen J R. Analysis and calculations of sensitive character on Coriolis mass flowmeter. Journal of Nanjing University of Science and Technology, 1997, 21(6): 498-501.

[18] Huang J L, Huang H Y, Cheng H, et al. Superharmonic and subharmonic resonances of strongly nonlinear vibration of beams. Journal of Vibration and Shock, 2004,23(1): 1-4.

Appendix F
Modeling of Coriolis Mass Flow Meter of a General Plane-shape Pipe①

Samer Guirguis Fan Shangchun

Abstract: The vibration of a general plane tube with a flowing fluid, which is the measuring element in a Coriolis mass flow meter (CMF), is studied. The dynamic stiffness matrix method is used to model such a tube. The tube is divided into straight and circular elements. The elements dynamic stiffness matrices are derived from the equations of motion. By assembling the elements matrices into a global matrix the natural frequencies are obtained. The mode shapes are obtained by applying the boundary conditions at the supports and the compatibility conditions at the nodes. The effects of the flow velocity on the natural frequency and the relative phase difference are modeled. The method is applied to different tube shapes. The results are compared to the published data which reveals good agreement.

Keywords: Tubes conveying fluid, Coriolis mass flow meter, Dynamic stiffness matrix

1 Introduction

1.1 Measuring principle

The first application of the Coriolis effect for mass flow measurement was proposed by Li and Lee in 1953[1]. Since the early 1980's[2] the interest in Coriolis mass flow meters has been increasingly and widely accepted in many industries.

The main advantage of the Coriolis mass flow meter (CMF) is that it measures the true mass flow rate directly, unlike some other instruments that measure the volumetric flow rate.

The basic measurement principle[3] is that a flow tube is caused to vibrate sinusoidally at

① Flow Measurement and Instrumentation. 2010, 21(1): 40-47.

a resonant frequency (the fundamental natural frequency in most cases[4]) by one or more drivers while two sensors monitor the vibration. The flow tube geometry and sensor placement are arranged so that the frequency of oscillation can be used to calculate the density of the process fluid, while the phase difference between the two sensor signals provides the mass flow rate. The flowing fluid passing through the vibrating tube produces Coriolis forces acting asymmetrically on the tube. These forces, which are proportional to the mass flow rate, produce the phase difference mentioned above.

1.2 Modeling of CMF

The bending vibration of straight tubes with flowing fluid was studied in [5] by Ashley and Haviland. The model was a simply supported pipe that was treated like a beam. In [6] Housner derived the equations of motion for both free and forced vibration. Chen in [7] derived the out-of-plane equations of motion, of a curved circular tube with flowing fluid, from Hamilton's principle. Mathematical modeling of U-shape tube CMF was presented by Sultan and Hemp in [8]. Using perturbation theory, Raszillier et al. introduced a mathematical model for a straight pipe CMF in [4,9]. Another approach was given by the same authors in [10] where they assumed that the driving mode of the pipe is mixed with its two spectral neighbors. In [11] they studied the effect of a concentrated mass, attached at the midpoint of a straight pipe segment conveying fluid, on the CMF performance. Kutin and Bajsic in [12] included the axial force, added masses, damping and excitation in their model.

The dynamic response of a straight tube CMF was studied extensively by Cheesewright, Clark and others. In [13] they studied the effect of flow pulsations on a straight tube CMF. The external vibration excitation effect was studied in [14,15]. The step response was studied in [3].

According to literature, modeling of pipes conveying fluid of a general shape depends on the finite element method. [16] gives a finite element analysis including the effect of axial tension in tubes and demonstrates the importance of including this axial term. The influence of mass flow rate on the change in the natural frequency is well captured. In [17] a finite element model for a straight pipe conveying fluid is introduced assuming the pipe as a Timoshenko beam. The model has the advantage of including the damping effect of the structure and the fluid which is estimated via experiments. In [18] a numerical model is realized by coupling of a finite volume code for fluid flow analysis with a finite element code for structural analysis using the conventional staggered solution procedure. In [19] a finite volume model for the fluid and a finite element of the tube are coupled to find the natural frequencies and phase differences in a straight Coriolis mass flow meter. In [20] the errors in the finite element modeling of CMF are studied. It was found that errors are not due to the method of modeling but indeed due to the round off errors in the numerical solver of the eigenvalue problem. In [21] a procedure that depends on the work of Stack[22] and Paidoussis and Issid[23] is used for modeling the pulsating flow in Coriolis mass flow meter using a standard FE code (ANSYS). The procedure has been validated by comparison with the published analytical solution for a simple straight tube meter and a comparison with the experimental data from a commercially available meter.

Compared to the finite element method, the dynamic stiffness matrix (DSM) method that is used in this work, is considered exact in nature[24]. This is because it is based on exact shape function obtained from the exact solution of the differential equations. Independent of the number of elements used in modeling, the DSM method can provide the analys with much better accuracy with less computation time compared to the finite element method whose accuracy degrades with decreasing the number of elements used. There is no need to make nodes except at points of application of forces or points of change of structure or material properties. The DSM method is easy to apply as the geometry of a tube which can be the input to the model just by the input of a few numbers defining the lengths of the straight elements and the angles and the radii of curvature of the curved elements.

1.3 Assumptions and limitations

- The CMF tube has a general plane shape formed of straight and curved segments with the same slope at the points of conjunction.
- All the tube sections are treated as a Euler-Bernoulli beam (the effect of rotary inertia and transverse shear are neglected).
- Structural damping and damping due to fluid are both neglected.
- The fluid is modeled using the plug-flow model so the effects of velocity profile, fluid viscosity and compressibility are not considered.
- The fluid has a constant velocity. The dynamic response due to changes in the flow velocity is not modeled.

In the following sections a general method to model a general shape pipe CMF is introduced. Section 2 gives a description of the dynamic stiffness matrix method that is used in modeling. Section 3 applies the proposed algorithm to straight and curved portions of tubes conveying fluid. Section 4 contains numerical results of the given method applied to different tube shapes.

2 Dynamic Stiffness Matrix Method

In [24] Banerjee gives a general approach to model the vibrating structures. The dynamic stiffness matrix method is often referred to as an exact method as it is based on exact shape functions obtained from the exact solution of the element differential equations. Compared to the finite element method, this approach gives better accuracy with a smaller number of elements. The dynamic stiffness matrix method combined with the Wittrick-Williams algorithm[25] is used to model the vibration of structures with high coupling between bending and torsion modes of vibration.

The first step towards developing the dynamic stiffness matrix of a structural element is to derive its governing differential equations of motion in free vibration which is of the form:

$$L(\boldsymbol{u}) = 0 \tag{1}$$

where

L is a differential operator and $\boldsymbol{u}$ is a displacement vector whose elements are the degrees of freedom of the structure. For harmonically varying $\boldsymbol{u}$

$$\boldsymbol{u} = \boldsymbol{U}e^{i\omega t} \tag{2}$$

where

$\boldsymbol{U}$ is the vector of amplitudes of displacements,

ω is the frequency and t is the time.

Substituting Eq. (2) in Eq. (1) we get:

$$L_1(\boldsymbol{U}) = 0 \tag{3}$$

The solution of which will be of the form:

$$\boldsymbol{U} = \boldsymbol{AC} \tag{4}$$

where

$\boldsymbol{C}$ is a constant vector and $\boldsymbol{A}$ is a frequency dependent square matrix.

The boundary conditions of forces and displacements are applied to eliminate the constant vector $\boldsymbol{C}$.

$$\boldsymbol{V} = \boldsymbol{DC} \tag{5}$$

$$\boldsymbol{F} = \boldsymbol{BC} \tag{6}$$

$\boldsymbol{V}$ is the vector of nodal displacements while $\boldsymbol{F}$ is the vector of nodal forces.

From Eqs. (5) and (6) we get:

$$\boldsymbol{F} = \boldsymbol{KV} \tag{7}$$

$$\boldsymbol{K} = \boldsymbol{BD}^{-1} \tag{8}$$

Matrix **K** is the dynamic stiffness matrix relating the nodal forces to the nodal displacements.

The dynamic stiffness matrices of all elements in a structure are assembled in the usual way as is done in the finite element method. A suitable *eigen* solution is then used to find the natural frequencies.

Due to the exact nature of formulation, nodes are only required at points of support, changes of material or cross sectional properties, and points of application of external forces. In the case of plane tubes with a flowing fluid, used in the CMF, the nodes will be at the points of conjunction between straight and curved tube portions, and in the position of sensors and actuators.

3 Element Dynamic Stiffness Matrix Formulation

3.1 Curved tube equations

The shear force, bending moment and torsion moment in a curved tube can be described by the following partial differential equations as mentioned in [26]:

$$\frac{\partial S}{\partial s}=m\frac{\partial^2 v}{\partial t^2} \tag{9}$$

$$\frac{\partial M}{\partial s}+\frac{T}{a}-s=0 \tag{10}$$

$$\frac{\partial T}{\partial s}-\frac{M}{a}=0 \tag{11}$$

where

S, M and T are the shear force, bending moment and torsion moment per unit length respectively;

s is the position at any point of a curved beam span;

v is the out-of-plane linear displacement;

a is the radius of curvature;

t is the time;

m is the mass per unit length of the pipe.

The bending moment and torsion moment are given by the following equations:

$$M=EI\left(\frac{-\partial^2 v}{\partial s^2}+\frac{\phi}{a}\right) \tag{12}$$

$$T = GJ\left(\frac{\partial \phi}{\partial s} + \frac{1}{a}\frac{\partial v}{\partial s}\right) \tag{13}$$

where EI is the bending rigidity and GJ is the torsion rigidity.

Substituting Eqs. (12) and (13) in Eq. (10), the shear force can be found by the relation:

$$S = \frac{1}{a}(EI + GJ)\frac{\partial^2 v}{\partial s^2} + \left(\frac{GJ}{EI}\frac{\partial^2}{\partial s^2} - \frac{1}{a^2}\right)\phi \tag{14}$$

As described in [8] the inertia term in the right hand side of Eq. (9) for a tube with a flowing fluid will be of the form:

$$m\frac{d^2 v}{dt^2} = m_f\left(\frac{\partial^2 v}{\partial s^2}U^2 + 2\frac{\partial^2 v}{\partial s \partial t}U\right) + (m_f + m_t)\frac{\partial^2 v}{\partial t^2} \tag{15}$$

where

m_f and m_t are the mass per unit length of the fluid and tube respectively;

U is the flow velocity assuming

$$v(s, t) = V(s)e^{i\omega t} \tag{16}$$

$$\phi(s, t) = \Phi(s)e^{i\omega t} \tag{17}$$

Differentiating Eq. (14) with respect to s and substituting in Eq. (9) considering Eqs. (15), (16) and (17) we get:

$$\left[D^4 - \left(\frac{GJ}{a^2 EI} - \frac{m_f}{EI}U^2\right)D^2 + 2i\omega\frac{m_f}{EI}UD - \frac{m_f + m_t}{EI}\omega^2\right] \times V - \left[\frac{1}{a}\left(1 + \frac{GJ}{EI}\right)D^2\right]\Phi = 0 \tag{18}$$

where

$$D = \frac{\partial}{\partial s}$$

Differentiating Eq. (13) with respect to s and substituting the result and Eq. (12) in Eq. (11) considering Eq. (16) and Eq. (17) we get:

$$\left(D^2 - \frac{EI}{a^2 GJ}\right)\Phi + \frac{1}{a}\left(1 + \frac{EI}{GJ}\right)D^2 V = 0 \tag{19}$$

Eqs. (18) and (19) are the partial differential equations governing the free vibration of a curved tube with flowing fluid. These equations show the coupling between the bending and torsion modes of vibration. Multiplying Eq. (18) by $\left(D^2 - \frac{EI}{a^2 GJ}\right)$ and Eq. (19) by $\frac{1}{a} \cdot \left(1 + \frac{GJ}{EI}\right)D^2$ and adding the products will eliminate the twist angle Φ. Elimination of the displacement V will result in the same differential equation.

$$\left[D^6 + \left(\frac{2}{a^2} + \frac{m_f}{EI}U^2\right)D^4 + 2i\omega\frac{m_f}{EI}UD^3 + \left(\frac{1}{a^4} - \frac{m_f}{a^2GJ}U^2 - \frac{m_f + m_t}{EI}\omega^2\right)D^2 - \right.$$

$$\left. 2i\omega\frac{m_f}{a^2GJ}UD + \frac{m_f + m_t}{a^2GJ}\omega^2\right]W = 0 \tag{20}$$

where W can be replaced by V or Φ. The solution of this equation will be on the form:

$$V = \sum_{i=1}^{6} C_i e^{\lambda_i s} \tag{21}$$

$$\Phi = \sum_{i=1}^{6} K_i e^{\lambda_i s} \tag{22}$$

Substituting Eqs. (21) or (22) in Eq. (20), by putting $W=V$ or $W=\Phi$, we get the sixth order polynomial equation in λ.

$$\lambda^6 + \left(\frac{2}{a^2} + \frac{m_f}{EI}U^2\right)\lambda^4 + 2i\omega\frac{m_f}{EI}U\lambda^3 + \left(\frac{1}{a^4} - \frac{m_f}{a^2GJ}U^2 - \frac{m_f + m_t}{EI}\omega^2\right)\lambda^2 -$$

$$2i\omega - \frac{m_f}{a^2GJ}U\lambda + \frac{m_f + m_t}{a^2GJ}\omega^2 = 0 \tag{23}$$

Solving the above equation for λ and substituting the result in Eqs. (21) and (22) gives the displacements at any given frequency.

Substituting for the linear displacement, twist angle and their derivatives in Eqs. (14), (12) and (13), the shear force, bending moment and torsion moment can be determined at any given frequency.

$$S = \sum_{i=1}^{6} \lambda_i e^{\lambda_i s}\left[\frac{EI + GJ}{a}K_i + \left(\frac{GJ}{a^2} - EI\lambda_i^2\right)C_i\right] \tag{24}$$

$$M = EI\sum_{i=1}^{6} e^{\lambda_i s}\left(-C_i\lambda_i^2 + \frac{K_i}{a}\right) \tag{25}$$

$$T = GJ\sum_{i=1}^{6} \lambda_i e^{\lambda_i s}\left(K_i + \frac{C_i}{a}\right) \tag{26}$$

Substituting Eqs. (21) and (22) in Eq. (19) we can find the relation between K_i and C_i.

$$K_i = C_i\frac{\lambda_i^2}{a}\left(\frac{k+1}{\frac{k}{a^2} - \lambda_i^2}\right) \tag{27}$$

$$k = \frac{EI}{GJ}$$

and in matrix form:

$$\mathbf{K} = C_k\mathbf{C} \tag{28}$$

where C_k is a diagonal matrix with the ith element:

$$C_{k_i} = \frac{\lambda_i^2}{a}\left(\frac{k+1}{\frac{k}{a^2}-\lambda_i^2}\right) \tag{29}$$

3.2 Curved tube dynamic stiffness matrix

For a curved tube of total length l, the end point forces can be determined by substituting s by 0 and l in Eqs. (24)~(26).

Fig. 1 shows the nodal forces of a curved tube element. S, M and T are shear force, bending moment and twist moment respectively. Suffix (1) is for the start point and suffix (2) is for end point. The nodal forces can be put in matrix form as follows:

$$\boldsymbol{F} = \boldsymbol{A}_1\boldsymbol{C} + \boldsymbol{A}_2\boldsymbol{K} = (\boldsymbol{A}_1 + \boldsymbol{A}_2 C_k)\boldsymbol{C} = \boldsymbol{D}_c\boldsymbol{C} \tag{30}$$

$$\boldsymbol{F} = (S_1 \quad M_1 \quad T_1 \quad S_2 \quad M_2 \quad T_2)^{\mathrm{T}}$$

$$\boldsymbol{A}_1 = \begin{pmatrix} -\frac{1}{a}GJ\lambda_1 + EI\lambda_1^3 & \cdots & -\frac{1}{a}GJ\lambda_6 + EI\lambda_6^3 \\ EI\lambda_1^2 & \cdots & EI\lambda_6^2 \\ -\frac{1}{a}\lambda_1 GJ & \cdots & -\frac{1}{a}\lambda_6 + GJ \\ \left(\frac{1}{a}GJ\lambda_1 - EI\lambda_1^3\right)e^{\lambda_1 l} & \cdots & \left(\frac{1}{a}GJ\lambda_6 - EI\lambda_1^3\right)e^{\lambda_6 l} \\ -EI\lambda_1^2 e^{\lambda_1 l} & \cdots & EI\lambda_6^2 e^{\lambda_6 l} \\ \frac{1}{a}\lambda_1 GJ\, e^{\lambda_1 l} & \cdots & -\frac{1}{a}\lambda_6 GJ\, e^{\lambda_1 l} \end{pmatrix}$$

$$\boldsymbol{A}_2 = \begin{pmatrix} -\frac{1}{a}(EI+GJ)\lambda_1 & \cdots & -\frac{1}{a}(EI+GJ)\lambda_6 \\ -\frac{1}{a}EI & \cdots & -\frac{1}{a}EI \\ -GJ\lambda_1 & \cdots & -GJ\lambda_6 \\ \frac{1}{a}(EI+GJ)\lambda_1 e^{\lambda_1 l} & \cdots & \frac{1}{a}(EI+GJ)\lambda_6 e^{\lambda_6 l} \\ \frac{1}{a}EI\, e^{\lambda_1 l} & \cdots & -\frac{1}{a}EI\, e^{\lambda_6 l} \\ GJ\lambda_1 e^{\lambda_1 l} & \cdots & GJ\lambda_6 e^{\lambda_6 l} \end{pmatrix}$$

For a curved tube of total length l, the nodal displacements can be determined by substituting s by 0 and l in Eq. (21) and its first derivative for out-of-plane linear displacement and slope, and in Eq. (22) for twist angle. Fig. 2 shows the nodal forces of a curved tube element. V, Θ and Φ are the out-of-plane displacement, slope and twist angle respectively. Suffix (1) is for the start point and suffix (2) is for the end point. The nodal displacements can be put in matrix form as follows:

$$\boldsymbol{V} = \boldsymbol{A}_3\boldsymbol{C} + \boldsymbol{A}_4\boldsymbol{K} = (\boldsymbol{A}_3 + \boldsymbol{A}_4\boldsymbol{C}_k)\boldsymbol{C} = \boldsymbol{B}_c\boldsymbol{C} \tag{31}$$

$$\boldsymbol{V} = (V_1 \quad \Theta_1 \quad \Phi_1 \quad V_2 \quad \Theta_2 \quad \Phi_2)^{\mathrm{T}}$$

$$\boldsymbol{A}_3 = \begin{pmatrix} 1 & \cdots & 1 \\ \lambda_1 & \cdots & \lambda_6 \\ 0 & \cdots & 0 \\ \mathrm{e}^{\lambda_1 l} & \cdots & \mathrm{e}^{\lambda_6 l} \\ \lambda_1 \mathrm{e}^{\lambda_1 l} & \cdots & \lambda_6 \mathrm{e}^{\lambda_6 l} \\ 0 & \cdots & 0 \end{pmatrix}$$

$$\boldsymbol{A}_4 = \begin{pmatrix} 0 & \cdots & 0 \\ 0 & \cdots & 0 \\ 1 & \cdots & 1 \\ 0 & \cdots & 0 \\ 0 & \cdots & 0 \\ \mathrm{e}^{\lambda_1 l} & \cdots & \mathrm{e}^{\lambda_6 l} \end{pmatrix}$$

$$\boldsymbol{F} = \boldsymbol{B}_c\boldsymbol{D}_c^{-1}\boldsymbol{V} = \boldsymbol{K}_c\boldsymbol{V}$$

where $\boldsymbol{K}_c$ is the dynamic stiffness matrix relating the forces and displacements at the end points of a curved tube with a flowing fluid.

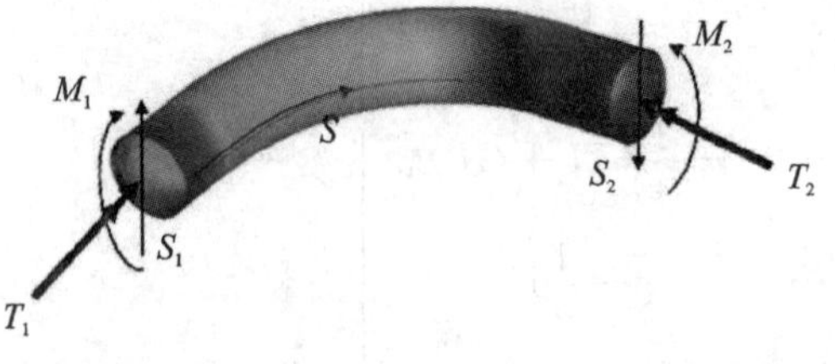

Fig. 1 Forces and moments at the tube ends

Fig. 2 Displacements at the tube ends

3.3 Straight tube equations and dynamic stiffness matrix

The equations of free vibration of a straight tube with a flowing fluid can be determined by taking the limit of Eqs. (18) and (19) as $a \to \infty$.

$$\left(D^4 - \frac{m_{\mathrm{f}}}{EI}U^2 D^2 + 2\mathrm{i}\omega \frac{m_{\mathrm{f}}}{EI}UD - \frac{m_{\mathrm{f}} + m_{\mathrm{t}}}{EI}\omega^2\right)\boldsymbol{V} = 0 \tag{32}$$

$$D^2 \Phi = 0 \tag{33}$$

In this case the bending and torsion are not coupled. The solution of the above two equations will be of the form:

$$V = \sum_{i=1}^{4} E_i e^{\lambda_i s} \tag{34}$$

$$\Phi = B_0 + B_1 s \tag{35}$$

Following the same procedure as in the curved tube formulation we can find the dynamic stiffness matrix that relates the forces and displacements at the end points of a straight tube.

$$\boldsymbol{F} = \boldsymbol{B}_s \boldsymbol{D}_s^{-1} \boldsymbol{V} = \boldsymbol{K}_s \boldsymbol{V}$$

$$\boldsymbol{B}_s = \begin{bmatrix} EI\lambda_1^3 & \cdots & EI\lambda_4^3 & 0 & 0 \\ EI\lambda_1^2 & \cdots & EI\lambda_4^2 & 0 & 0 \\ 0 & 0 & 0 & 0 & -GJ \\ -EI\lambda_1^3 e^{\lambda_1 s} & \cdots & -EI\lambda_4^3 e^{\lambda_4 s} & 0 & 0 \\ -EI\lambda_1^2 e^{\lambda_1 s} & \cdots & -EI\lambda_4^2 e^{\lambda_4 s} & 0 & 0 \\ 0 & 0 \quad 0 & 0 & 0 & GJ \end{bmatrix}$$

$$\boldsymbol{D}_s = \begin{bmatrix} 1 & 1 & 1 & 1 & 0 & 0 \\ \lambda_1 & \lambda_2 & \lambda_3 & \lambda_4 & 0 & 0 \\ 0 & 0 & 0 & 0 & 1 & 0 \\ e^{\lambda_1 l} & e^{\lambda_2 l} & e^{\lambda_3 l} & e^{\lambda_4 l} & 0 & 0 \\ \lambda_1 e^{\lambda_1 l} & \lambda_2 e^{\lambda_2 l} & \lambda_3 e^{\lambda_3 l} & \lambda_4 e^{\lambda_4 l} & 0 & 0 \\ 0 & 0 & 0 & 0 & 1 & l \end{bmatrix}$$

4 Numerical Application

The U-tube Coriolis mass flow meter described in [8] is chosen here to compare the results of applying the proposed work to the published data.

The U-tube CMF is made from stainless steel 316L with properties: Young's modulus E=208 GN/m^2, shear modulus G=80 GN/m^2 and density ρ=8 027 kg/m^3. The outer diameter of the tube is 50.8 mm, the tube thickness t=1.8 mm, the radius of curvature of the circular portion a=15 cm and the length of the straight tubes b=3a (for natural frequency comparison b=2a).

4.1 Natural frequencies

The U-tube is divided into two straight tubes and one curved tube with four nodes as shown in Fig. 3. The dynamic stiffness matrices of the individual elements are calculated as in the previous section.

The global dynamic stiffness matrix is assembled in the same pattern that is used in the finite element method. This matrix relates the forces and displacements at the nodes of the whole tube which are the end points of individual elements.

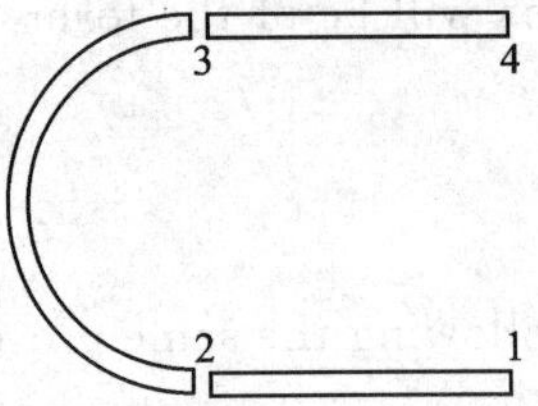

Fig. 3 U-tube elements and nodes

There is no need for coordinate transformation because the variable s, which defines the position at any point on the tube, has a continuous slope at the points of conjunction of individual tube portions. Thus, the forces and displacements of the nodes are related by:

$$\boldsymbol{F} = \boldsymbol{K}_g \boldsymbol{V} \tag{36}$$

In the case of a U-shape tube, the global stiffness matrix $\boldsymbol{K}_g$ is of order 12. The order of the matrix is computed by $3(n+1)$ where n is the number of elements.

To study the free vibration we set $\boldsymbol{K}_g \boldsymbol{V} = 0$. The next step is to apply the boundary conditions. For fixed end conditions, the displacements at the end points of the tube are nullified. This results in reducing the order of the global matrix by removing the columns and rows corresponding to the nullified displacements. The determinant of the reduced dynamic stiffness matrix $\boldsymbol{K}_{gr}$ is evaluated at different frequencies. The frequencies at which $|\boldsymbol{K}_{gr}|$ vanishes are the natural frequencies of the tube.

4.2 Mode shapes

We use the results of the previous subsection to find the mode shapes corresponding to the natural frequencies found. The number of nodes in the model is too small to be used to find the mode shapes by finding the eigenvectors of the global stiffness matrix. In the case of a U-shape tube, there are four nodes two of which have zero displacements. Instead we apply the boundary conditions at the end nodes and the compatibility conditions at the conjunction nodes. Thus, at nodes (1) and (4) the out of plane displacement, slope and twist angle are zeros. At nodes (2) and (3) the displacements and forces are both equated at either side of each node. This procedure is done to find the constants in Eqs. (21), (22), (34) and (35). We can put all the above conditions in one matrix equation of the form:

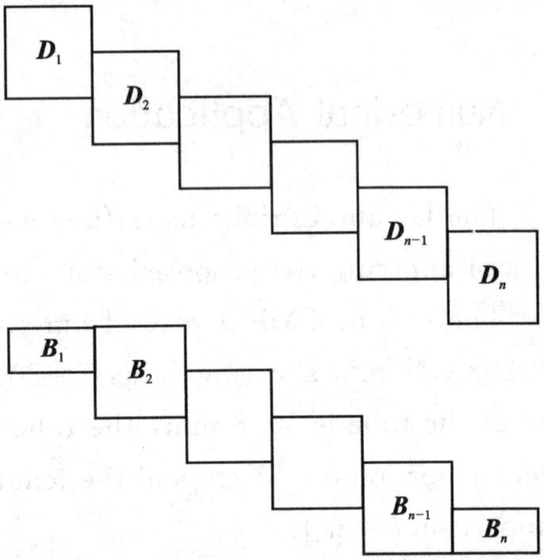

Fig. 4 *G* matrix assembly pattern

$$GA = 0 \tag{37}$$

where vector $\boldsymbol{A}$ contains all the constants to be determined and matrix $\boldsymbol{G}$ can be constructed (as shown in Fig. 4) from the element matrices $\boldsymbol{B}$ and $\boldsymbol{D}$ that are evaluated at the computed natural frequencies. The first three rows in each $\boldsymbol{D}$ matrix should be multiplied by (-1). Matrix $\boldsymbol{B}_1$ consists of the last three rows in the first element $\boldsymbol{B}$ matrix. Matrix $\boldsymbol{B}_n$ consists of the first three rows in the nth element $\boldsymbol{B}$ matrix.

For a non-trivial solution of Eq. (37) we use the method of back substitution to determine the values of elements in $\boldsymbol{A}$. This method involves assuming the value of one element in $\boldsymbol{A}$ and reducing the order of matrix $\boldsymbol{G}$. As $\boldsymbol{G}$ is generally complex, the assumed element should be complex. Eq. (37) will be on the form:

$$\boldsymbol{G}^* \boldsymbol{A}^* = -(1+i)g \tag{38}$$

where vector g consists of the column in matrix $\boldsymbol{G}$ corresponding to the assumed element in $\boldsymbol{A}$.

$$\boldsymbol{A}^* = -(1+i)(\boldsymbol{G}^*)^{-1}g \tag{39}$$

Substituting the values of $\boldsymbol{A}$ in Eqs. (21), (22), (34) and (35) we can find the mode shapes. Fig. 5 shows the first three normalized mode shapes of the U-shape tube without fluid.

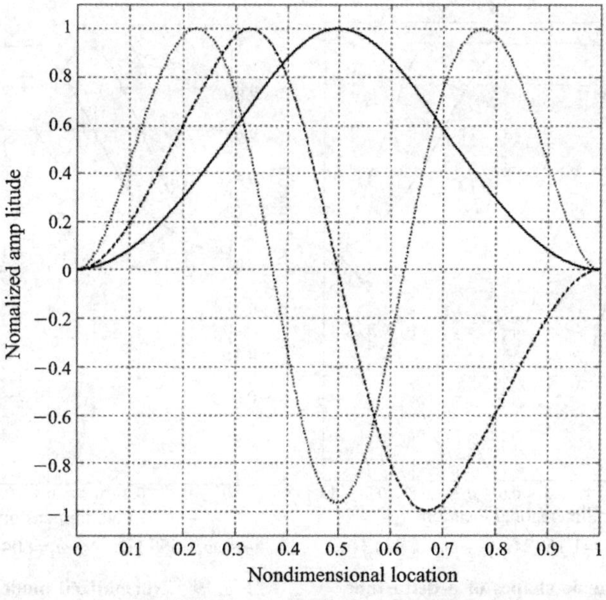

Fig. 5　Normalized mode shapes of a U-tube without fluid

The same procedure is done with different shapes of tubes. The delta-shape and S-shape tubes shown in Figs. 6 and 7 have the same material and cross section properties of the U-tube discussed above. All the straight elements are of length 30 cm and all the curved elements are of radius 15 cm. The curved elements of the delta-tube have an angle 135°. Figs. 8 and 9 show the first three mode shapes of a delta-shape and S-shape tubes respectively, without fluid. These shapes are included as examples but any tube that consists of a straight and circular elements can be modeled the same way.

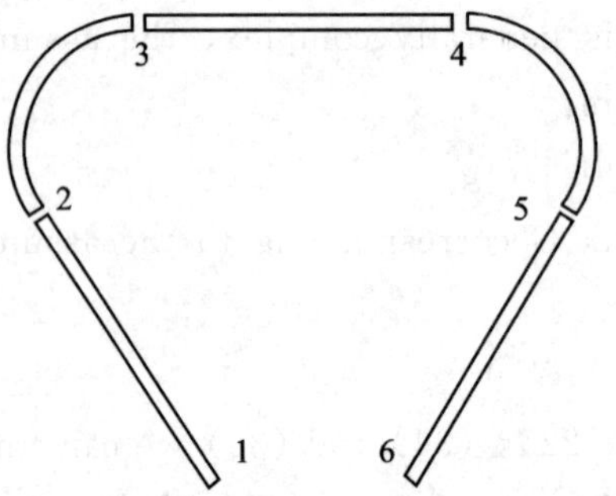

Fig. 6 Delta-tube elements and nodes

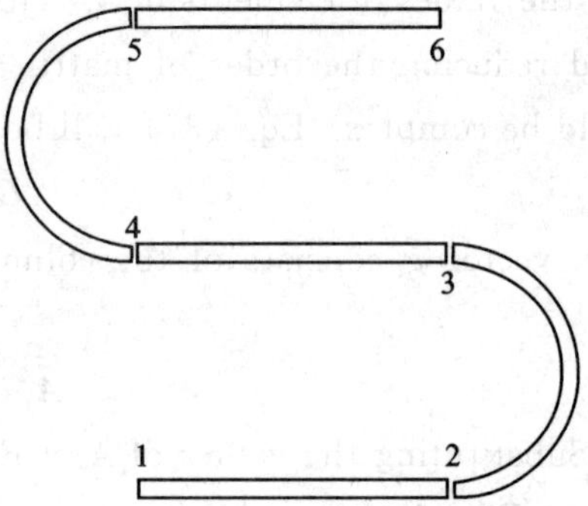

Fig. 7 S-tube elements and nodes

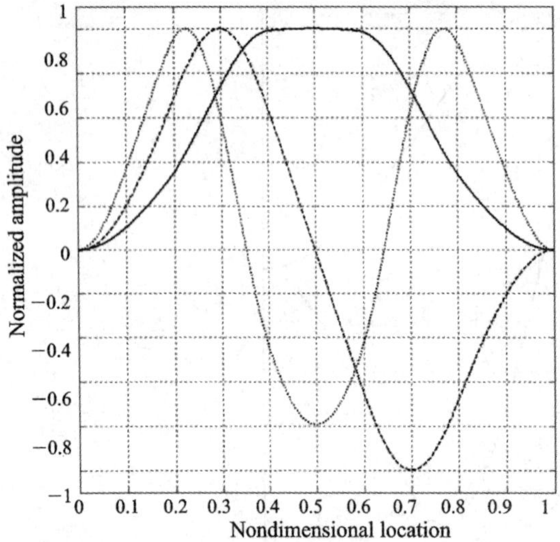

Fig. 8 Normalized mode shapes of a delta-tube without fluid

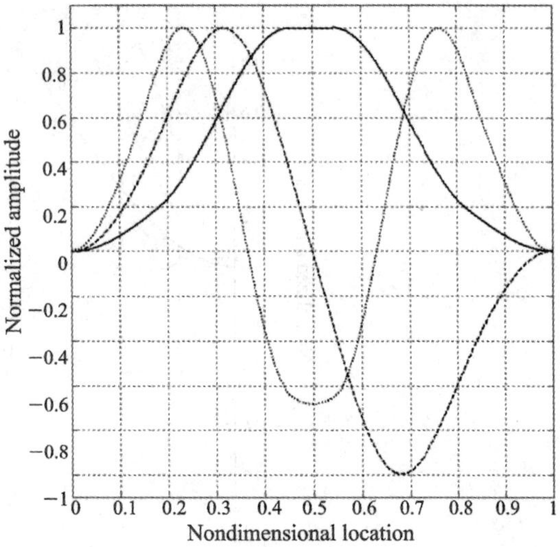

Fig. 9 Normalized mode shapes of an S-tube without fluid

4.3 Effect of flow velocity

In [8] the effects of flow velocity on the values of the first natural frequency as well as the phase difference between two points are studied. The curves given in this reference are regenerated using the given algorithm to verify it. Non-dimensional quantities are used in comparison.

$$\bar{u} = Ua\left(\frac{m_f}{EI}\right)^{\frac{1}{2}}$$

$$\bar{\omega} = \omega a^2\left(\frac{m_f + m_t}{EI}\right)^{\frac{1}{2}}$$

$$\beta = \frac{m_f}{m_f + m_t}$$

where $\bar{u}$ is the non-dimensional flow velocity, $\bar{\omega}$ is the nondimensional frequency and β is the mass ratio.

Fig. 10 shows the relation between the flow velocity and the first natural frequency for different fluid masses. Fig. 11 shows the relation between the flow velocity and the phase difference between two points of a U-tube, for different fluid masses. The phase difference is computed by computing the values of complex out of plane displacements at the points of interest, then computing the phase at each point and subtracting the phases.

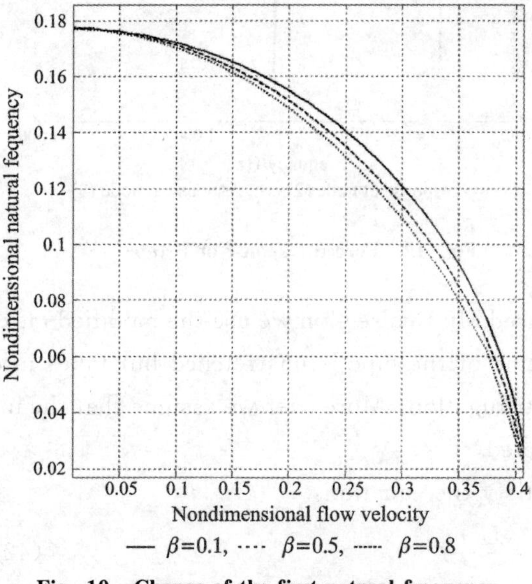

Fig. 10 Change of the first natural frequency with flow velocity change

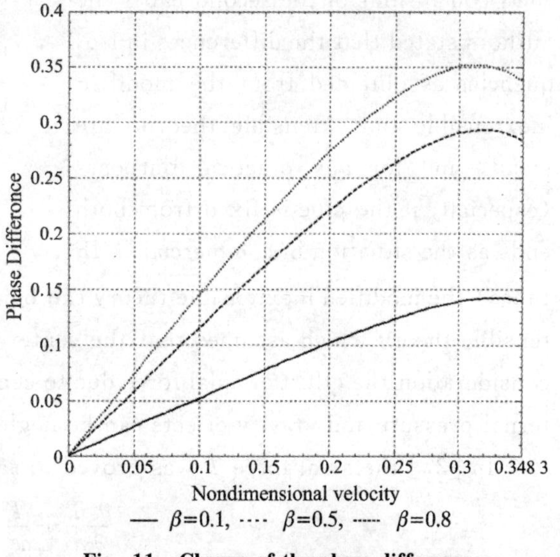

Fig. 11 Change of the phase difference with flow velocity change

4.4 Forced response

In order to study the control law of the CMF, it is needed to find a relation between the input actuator force and the output sensor signals. Applying the boundary conditions, Eq. (36) will be reduced to:

$$\boldsymbol{F}_r = \boldsymbol{K}_{gr}\boldsymbol{V}_r \tag{40}$$

Vector $\boldsymbol{F}_r$ is set such that it contains elements that correspond to the actuator forces. The displacement due to a point force can be determined by:

$$\boldsymbol{V}_r = \boldsymbol{K}_{gr}^{-1}\boldsymbol{F}_r \tag{41}$$

For a U-tube with an actuator force at node (2), the response at nodes (2) and (3) is given in Fig. 12.

4.5 Effect of axial force

In [27,28] the authors studied the effect of axial force on the natural frequencies of a curved pipe conveying a fluid. Three theories were used, namely, inextensible theory, extensible theory and modified inextensible theory. In their conclusions of the second paper the authors stated that the differences in frequencies as obtained from the modified inextensible and extensible theories are small, and for any practical purposes (especially if the pipe is fixed from both ends as the situation in commercial CMF tubes) the modified inextensible theory can be used. In this section we use the modified inextensible theory which assumes that the center line of the pipe is unstretched but takes into consideration the effect of axial force due to centrifugation. More over we assume that the internal pressure and gravity effects can be neglected.

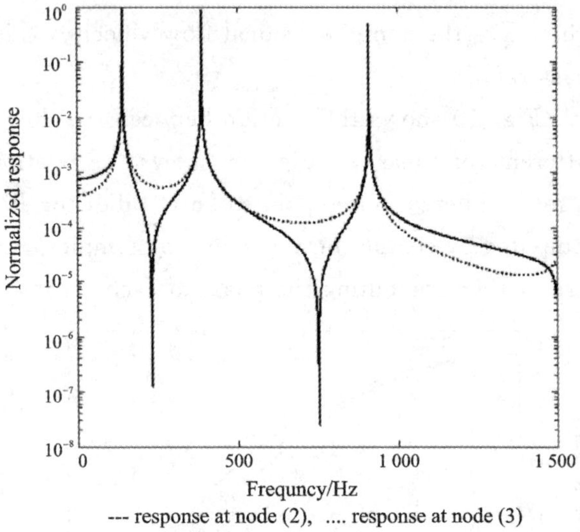

Fig. 12 Forced response of U-tube

In [27] the axial force T was proved to satisfy the equation:

$$\frac{\partial^2 T}{\partial s^2} + \frac{T}{a^2} = \frac{U^2 m_f}{a^2} \tag{42}$$

For fixed-fixed tubes the solution of the above equation is

$$T = m_f U^2 \tag{43}$$

In order to account for the axial forces, the differential equation describing the out-of-plane vibration of a curved pipe conveying fluid should be modified to be

$$\left[D^6 + \left(\frac{2}{a^2} + \frac{m_f}{EI}U^2 - \frac{T}{EI}\right)D^4 + 2\mathrm{i}\omega\frac{m_f}{EI}UD^3 + \left(\frac{1}{a^4} - \frac{m_f}{a^2GJ}U^2 - \frac{m_f + m_t}{EI}\omega^2 + \frac{T}{a^2GJ}\right)D^2 - 2\mathrm{i}\omega\frac{m_f}{a^2GJ}UD + \frac{m_f + m_t}{a^2GJ}\omega^2\right]V = 0 \tag{44}$$

The straight pipe equation will be of the form:

$$\left[D^4 - \left(\frac{m_f}{EI}U^2 - \frac{T}{EI}\right)D^2 + 2\mathrm{i}\omega\frac{m_f}{EI}UD - \frac{m_f + m_t}{EI}\omega^2\right]V = 0 \tag{45}$$

Applying the same procedure of Section 3 the dynamic stiffness matrices of the straight and curved tube elements can be obtained. Consequently all the above numerical applications can be performed again to account for the axial force.

Fig. 13 shows the change of the first three natural frequencies with flow velocity with and without considering the axial force effect for a U-tube, delta-tube and S-tube. The fluid used is water.

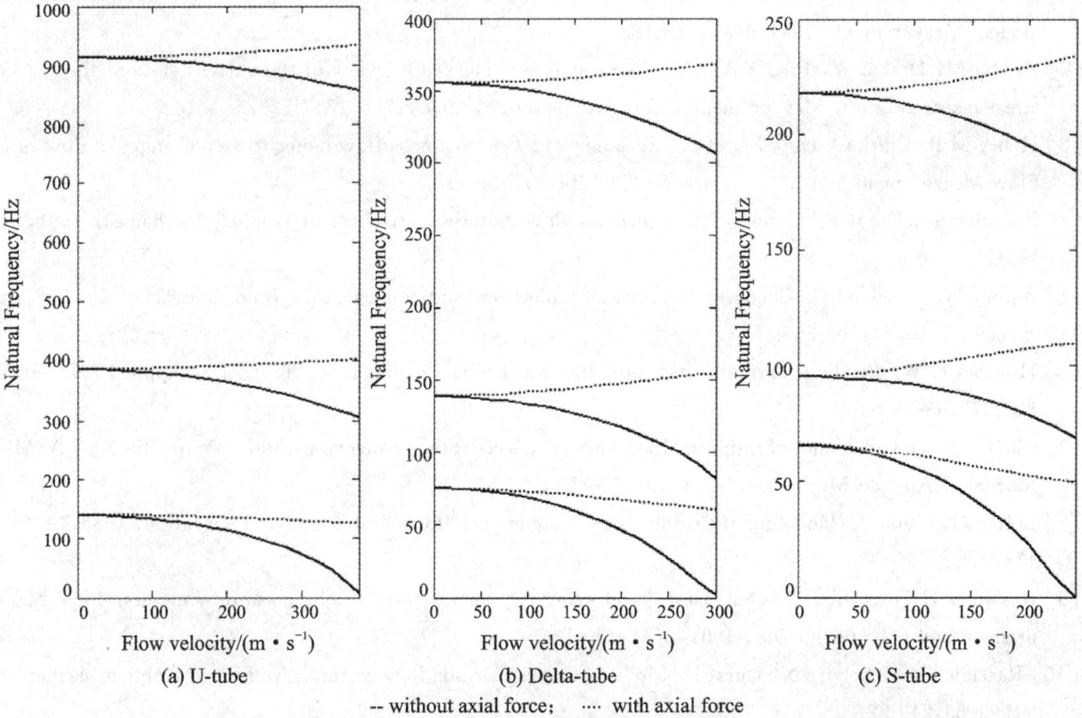

Fig. 13　Variation of the first three natural frequencies with flow velocity

5 Conclusion

The dynamic stiffness matrix method is capable of modeling a general plane tube with a flowing fluid. This method can be used to model Coriolis mass flow meters of different shapes. Damping of the tube fluid system, which is not accounted for, can be added to the same model. The tube end supports are assumed here to be infinitely rigid, which eliminates the effect of environmental noise. Both topics can be studied in a future work. The easiness of obtaining the response of the tube due to an actuator point force, gives a step forward in solving the problem of optimizing the actuator's and sensor's positions. Also this facilitates the study of the control law between the actuator(s) and sensors which is essential in maintaining the CMF tuned to its variable natural frequency. An experimental verification for different shapes of CMF tubes is essential to validate and enhance the given model.

References

[1] Cascetta F, Valle S D, Guido A R, et al. A new straight-pipe Coriolis mass flow meter: The mathematical model. Measurement, 1991,9(3):115-123.

[2] Anklin M, Drahm W, Rieder A. Coriolis mass flow meters: Overview of the current state of the art and latest research. Flow Measurement and Instrumentation, 2006,17(6):317-323.

[3] Henry M P, Clark C, Duta M, et al. Response of a Coriolis mass flow meter to step changes in flow rate. Flow Measurement and Instrumentation 2003,14(3):109-118.

[4] Raszillier H, Durst F. Coriolis effect in mass flow metering. Archive of Applied Mechanics, 1991,61: 192-214.

[5] Ashley H, Haviland G. Bending vibrations of a pipe line containing flowing fluid. Journal of Applied Mechanics, 1950,74:205-208.

[6] Housner G W. Bending vibrations of a pipe line containing flowing fluid. Journal of Applied Mechanics, 1952,77:65-68.

[7] Chen S S. Out of plane vibration and stability of curved tubes conveying fluid. Transactions of ASME. Journal of Applied Mechanics, 1973,40:362-368.

[8] Sultan G, Hemp J. Modeling of Coriolis mass flow meter. Journal of Sound and Vibration, 1989,132(3): 473-489.

[9] Raszillier H, Raszillier V. Dimensional and symmetry analysis of Coriolis mass flow meters. Flow Measurement and Instrumentation, 1991,2(3):180-184.

[10] Raszillier H, Alleborn N, Durst F. Mode mixing in Coriolis flow meters. Archive of Applied Mechanics, 1993,63:219-227.

[11] Raszillier H, Alleborn N, Durst F. Effect of a concentrated mass on Coriolis flow metering. Archive of Applied Mechanics, 1994,64:373-382.

[12] Kutin J, Bajsic I. An analytical estimation of the Coriolis meter's characteristics based on modal superposition. Flow Measurement and Instrumentation, 2002,12(5):345-351.

[13] Cheesewright R, Clark C. The effect of flow pulsation on Coriolis mass flow meters. Journal of Fluids and Structures, 1998,12(8):1025-1039.

[14] Cheesewright R, Clark C, Bisset D. The identification of external factors which influence the calibration of Coriolis mass flow meters. Flow Measurement and Instrumentation, 2000,11(1):1-10.

[15] Clark C, Cheesewright R. The influence upon Coriolis mass flow meters of external vibrations at selected frequencies. Flow Measurement and Instrumentation,2003,14(1-2):33-42.

[16] Hulbert G M, Darnell I. Numerical and experimental analysis of Coriolis mass flow meters. (AIAA-95-1384-CP), 1995.

[17] Wang T, Baker R, Hussain Y. An advanced numerical model for single straight tube Coriolis flow meters. Transactions of the ASME 2006;127:1346-1350.

[18] Mole N, Bobovnik G, Kutin J, et al. An improved three-dimensional coupled fluid-structure model for Coriolis flow meters. Journal of Fluids and Structures, 2008,24(4):559-575.

[19] Bobovnik G, Mole N, Kutin J, et al. Coupled finite-volume/finite-element modeling of the straight-tube Coriolis flow meter. Journal of Fluids and Structures,2005,20(6):785-800.

[20] Cheesewright R, Shaw S. Uncertainties associated with finite element modeling of Coriolis mass flow meters. Flow Measurement and Instrumentation,2006,17(6):335-347.

[21] Belhadj A, Cheesewright R, Clark C. The simulation of Coriolis meter response to pulsating flow using a general purpose f. e. code. Journal of Fluids and Structures, 2000,14(5):613-634.

[22] Stack C P, Garnett R B, Pawlas G E. A finite element for the vibration analysis of a fluid-conveying Timoshenko beam. AIAA (AIAA-93-1552-CP), 93.

[23] Paidoussis M, Issid N. Dynamic stability of pipes conveying fluid. Journal of Sound and Vibration, 1974,33(3):267-294.

[24] Banerjee J R. Dynamic stiffness formulation for structural elements: A general approach. Computers and Structures, 1997,63(1):101-103.

[25] Wittrick W H, Williams F W. General algorithm for computing natural frequencies of elastic structures. Quarterly Journal of Mechanics and Applied Mathematics, 1971,24(3):263-284.

[26] Clark C, Cheesewright R, Wang S, et al. A radically new dynamic response capability for Coriolis flow meters. Sensors and Actuators A: Physical, 2005,123-124:54-62.

[27] Misra A K, Paidoussis M P, Van K S. On the dynamics of curved pipes transporting fluid. Part I: Inextensible theory. Journal of Fluids and Structures, 1988,2:221-244.

[28] Misra A K, Paidoussis M P, Van K S. On the dynamics of curved pipes transporting fluid. Part II: Extensible theory. Journal of Fluids and Structures,1988,2:221-244.

Appendix G Frequency Measurement Study of Resonant Vibratory Gyroscopes[①]

Li Yan Fan Shangchun Guo Zhanshe Li Jing Cao Le

Abstract: In this paper, frequency measurement methods of resonant vibratory gyroscopes are studied. First, the working principle and dynamic output characteristics of the resonant vibratory gyroscope are introduced, which provide a theoretical analysis base to analyze the dynamic frequency output characteristic of the resonant vibratory gyroscope. Moreover, some relevant frequency measurement methods are introduced, a frequency measurement method based on dynamic output characteristics has been proposed for the purpose of investigating the modulated output signal of resonant vibratory gyroscopes. Finally, in order to verify the feasibility of the proposed frequency measurement method, resonator oscillator and dynamic frequency measurement experiments of the resonant vibratory gyroscopes are built. It is concluded from the resonator oscillator result that the designed and observed resonant frequency of the resonator have a good match, which shows the design of resonator oscillator is feasible. It is concluded from the dynamic frequency measurement result that correlation coefficient of output fitted curve is 99.95%, and the output characteristic of the gyroscope is linear frequency output, which shows the effectiveness of the proposed frequency measurement method.

1 Introduction

Vibratory gyroscope senses the angular motion by detecting the Coriolis force acting on a vibratory structure when the supporting frame is undergoing angular motion. Compared to the conventional vibratory gyroscope, resonant vibratory gyroscope is not only advantageous for micromechanical actuators but also for resonant sensing, resonant sensing often involves the detection of input measurand by means of resonant frequency shift in the sensing device. Detection of frequency shift is more advantageous compared to detecting changes in ampli-

① Journal of Sound and Vibration, 2012,331(20): 4417-4424.

tude, as it is less sensitive to effects such as feedthrough coupling from undesired sources and parasitic passive elements. Therefore, it is very important and necessary to study the resonant frequency measurement method of resonant vibratory gyroscopes.

There are some resonant frequency measurement methods in recent papers, but some of them are using frequency measurement to achieve an especial function, such as crack detection[1], dental implant stability[2], Micro-electromechanical Systems (MEMS) actuators[3]. Some of them are using specialized method to measure resonant frequency, such as via magnetic[4] or dielectric[5] method to measure the resonant frequency of microwave resonators, via a simple setup to measure the resonant frequency of quartz crystal microbalance[6], and via hard contact readout to measure the resonant frequency of nano-scale cantilevers[7]. There are also some papers reported the resonant frequency measurement about resonant sensors, they are all concentrated on the resonant sensors about strain resonant sensor [8], mass resonant sensor[9], accelerator resonant sensor[10], and pressure resonant sensor[11], but there are not reports about the resonant frequency measurement methods about resonant vibratory gyroscopes. The reason maybe that the frequency output characteristics of other resonant sensors is linear output, but the dynamic output characteristics of resonant vibratory gyroscope is time varying.

This paper studies the frequency measurement method of resonant vibratory gyroscopes. Theory analysis and mathematical model are built. Experimental result on a resonant vibratory gyroscope indicates the theory analysis is right and feasible for the measurement of frequency shift.

2 Mathematical Model

2.1 Working principle of the resonant vibratory gyroscope

A schematic of the z-axis resonant vibratory gyroscope is shown in Fig. 1. The proposed gyroscope dynamical system consists of a sense-mode oscillator and a drive-mode oscillator. The proof mass is free to oscillate both in the drive and the sense directions and is excited in the drive direction. The two tines of each tuning fork vibrate anti-phase to each other. When an angular velocity is applied to the resonant gyroscope about the z-axis, the resonant frequencies of the two double-ended tuning fork (DETF) resonators vary with the applied angular velocity due to the Coriolis effect. The Coriolis force will cause axial push and pull loading on the resonators to shift their resonant frequency. One resonator is subjected to tension due to pull axial loading, thus the resonant frequency will increase. The other will be under compression and the frequency will decrease. As a result, the difference between the resonance frequencies of the two modes is a function of the applied angular velocity, the frequency shifts of the two resonators are proportional to the applied angular velocity input. The periodic compression and tension of the tuning fork tines by the Coriolis force at the proof mass drive frequency modulates the resonant frequency of these force sensors. Each force sensor comprises of the tuning fork mechanical structure embedded in the feedback loop of an oscillator circuit. Therefore, by demodulating the oscillation frequency, the rotation rate applied to the device can be estimated.

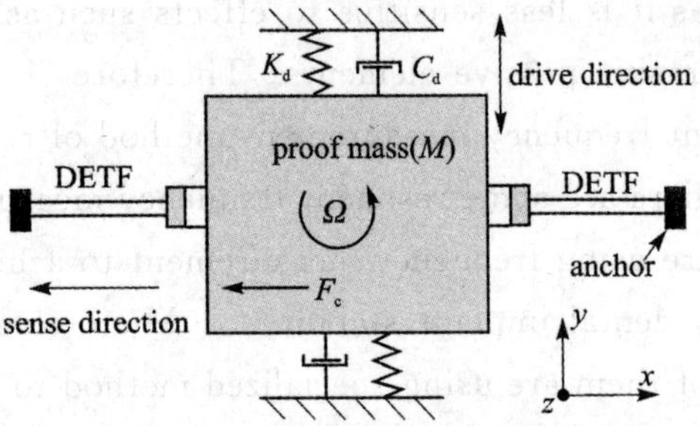

Fig. 1 Simplified model of a resonant vibratory gyroscope

2.2 Dynamic output characteristics of the resonant vibratory gyroscope

As mentioned above, the drive mode of the resonant vibratory gyroscope is a 1-DOF dynamic system where the operational frequency can be achieved by adjusting the drive mode stiffness, which the drive mode is independent of the sense mode. Therefore, the major focus of the resonant vibratory gyroscope is on the DETF resonators sense mode, specifically on demodulating the oscillator output frequency.

The dynamic equation of DETF could be described as[12,13]

$$\int_0^L \rho A\phi^2 \mathrm{d}x\ddot{q} + \int_0^L \xi\phi^2 \mathrm{d}x\dot{q} + \left[\int_0^L EI\left(\frac{\partial^2\phi}{\partial x^2}\right)^2 \mathrm{d}x + \int_0^L F\left(\frac{\partial\phi}{\partial x}\right)^2 \mathrm{d}x\right]q = F_\mathrm{d} \tag{1}$$

It is assumed that Young's modulus E, beam mass density ρ, the moment of inertia I, damped coefficient ξ. The function of driving force F_d keeps the resonator working in resonant frequencies, so

$$\int_0^L \xi\phi^2 \mathrm{d}x\dot{q} = F_\mathrm{d} \tag{2}$$

$$\ddot{q}(t) + \omega_0^2\left(1 + \frac{0.293L^3F_0\cos\omega_\mathrm{d}t}{EI}\right)q(t) = 0 \tag{3}$$

If the transformations are chosen suitably, the Eq. (3) can be transformed to a standard form of Mathieu equation[14,15].

$$q''(z) + (\delta + \varepsilon\cos 2z)q(z) = 0 \tag{4}$$

From the Eq. (4), the resonator vibration stiffness $k=\delta+\varepsilon\cos 2z$ changes the changing of the Coriolis force applied on the sensing masses of the resonant vibratory gyroscope. While the equivalent mass is the same, which makes the resonator resonant frequencies change. The parameter δ represents the natural resonant frequency of the resonator. The significant perturbation term $\varepsilon\cos 2z$ represents modulation of the spring constant of the resonant sensor at the proof masses driving frequency, which is directly contributed by the Coriolis force impinging axially on the resonator tines.

Since the input into the resonant vibratory gyroscope is modulated by a periodic function, its frequency output must be demodulated into stability-state. If the parameters δ and ε that we choose are in the stable regions of Mathieu equation, the solution of the parametric excitement Eq. (4) must in its stable regions. The approximate solution could be derived in its stable regions and achieved by small parameter perturbation method. The approximate solution could be described as

$$q(z) = \mathrm{e}^{\sqrt{\delta}\int w(z)\mathrm{d}z} = \exp\left(\ln\mu^{-1/2} \pm \mathrm{i}\int\sqrt{\delta}\mu\,\mathrm{d}z\right) = A\mu^{-1/2}\exp\left(\pm\mathrm{i}\sqrt{\delta}\int\mu\mathrm{d}z\right) = \frac{B}{(\delta+\varepsilon\cos 2z)^{1/4}}\cos\left[\int(\delta+\varepsilon\cos 2z)^{1/2}\mathrm{d}z+\theta\right] \tag{5}$$

where

$\mu^2 = 1+\beta\cos 2z$,

A, B and θ are arbitrary constants, which are determined by primal conditions.

Form Eq. (5), it is easy to see that there are amplitude modulation at the same time of frequency modulation in the output of the resonant vibratory gyroscope. Just because the resonant vibratory gyroscope we researched is by means of resonant frequency shift of the resonator to sense angular velocity, so the frequency of output signal of the resonant vibratory gyroscope is:

$$\omega(z) = \frac{\mathrm{d}\left[\int(\delta+\varepsilon\cos 2z)^{1/2}\mathrm{d}z+\theta\right]}{\mathrm{d}z} = (\delta+\varepsilon\cos 2z)^{1/2} \tag{6}$$

$$\omega(t) = \frac{\omega_\mathrm{d}}{2}\sqrt{\delta+\varepsilon\cos\omega_\mathrm{d}t} = \omega_0\sqrt{1+\beta\cos\omega_\mathrm{d}t} = \omega_0 + \frac{\varepsilon'}{\omega_0}\cos\omega_\mathrm{d}t \tag{7}$$

where $z=\omega_\mathrm{d}t/2$, $\omega_0=\sqrt{\delta'}$, ω_0 is the no-load resonant angular frequency of resonator. Form Eq. (7), it is easy to see that the change in tension along the resonator result in change in stiffness, shifting the resonant frequency, at the same time, the modulated power of output resonant frequency is not only determined by modulated stiffness ε', but also determined by static modulated stiffness δ'.

The relation between modulated frequency output and input angular velocity is

$$\hat{\omega}(t) = \frac{\varepsilon'}{\omega_0}\cos\omega_d t = \omega_0 \cdot S \cdot F(t) = \omega_0 \cdot S \cdot \frac{m}{4} \cdot Q \cdot \omega_d \cdot \Omega \cdot y_m(t) \tag{8}$$

where

$y_m(t) = y_0 \cos\omega_d t$ is the vibrating amplitude of proof mass.

The dynamic output characteristics of resonator based on Mathieu equation presents the instant output frequency of the resonant vibratory gyroscope, the most important part of the resonant vibratory gyroscope is how to measure the resonant output frequency.

3 Theory Analysis for the Frequency Measurement Method

3.1 Review of relevant frequency measurement methods

There are two methods for detecting the sensor resonant frequency[16,17]. The first is a direct detection method, it is the simplest way to measure frequency of a periodic signal, which could be obtain the signal frequency by counting zero crossings of the signal amplitude. A precise (low frequency) reference oscillator is used to gate the counting scheme. It is clear that the longer the interval for which the counting scheme is continued, the better is the resolution. However, there is a trade-off in terms of the bandwidth of the counting scheme. This imposes a limitation on the drive frequency of the resonant output gyroscope. The other common method for frequency measurement is indirect method, which is to measure the frequency of the input signal in terms of a precise high frequency reference signal.

In the case of the resonant vibratory gyroscope or a FM signal, the frequency of the input signal must be measured at the driven frequency of the proof mass. It is known that the phase of the frequency is dependent on the Coriolis force (in phase with the velocity of the driven structure). In this case the proposed counting technique is planned in the two phases of the drive signal. We can use either a two-phase direct counting method or a two-phase indirect counting method. The indirect counting method is better suited for the resonant output gyroscope with the quantization noise scaling inversely as the frequency of the reference signal. But in this case, the bandwidth of the measurement is again based on the averaging time. The quantization error has its roots in phase synchronization between the reference and input signals.

A phase-locked loop can be used for application in frequency demodulation. This method uses an oscillator (feedback loop) to force the sensor to resonate at its resonant frequency. The oscillator output frequency tracks the resonant frequency of the sensor by forcing the phase shift through the loop to be zero. As a result, this methods bandwidth is only limited by the bandwidth of the oscillator electronics. Because of the improvement in sensor bandwidth, this is the method used in resonant sensor measurement system developed in this research.

3.2 Proposed frequency measurement method

Resonant vibratory gyroscope frequency measurement system is composed of two components (Fig. 2). The first component is a resonant vibratory gyroscope combined with electronics to create an oscillator, referred to as a resonant vibratory gyroscope oscillator. This oscillator output is equal to the resonant frequency of the gyroscope and hence its output is frequency modulated (FM) by the measurand. This modulated frequency is demodulated by the second component of the measurement system to output a signal proportional to the measurand.

In the other resonant sensors but resonant vibratory gyroscope, frequency demodulation could be done with phase locked loops (Fig. 3), just because the relation between output frequency and measurand are linear. In the case of resonant vibratory gyroscope, as the Eq. (8) shown, the relation between modulated frequency output and input angular velocity is concluded a time-varying parameter $y(m)$, therefore the output characteristics of the resonant vibratory gyroscope is dynamic, and it is necessary to propose frequency measurement method to make the frequency output characteristic is linear.

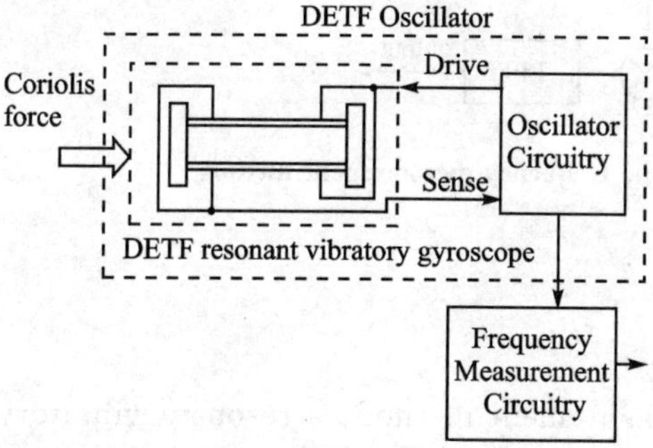

Fig. 2 Frequency measurement system

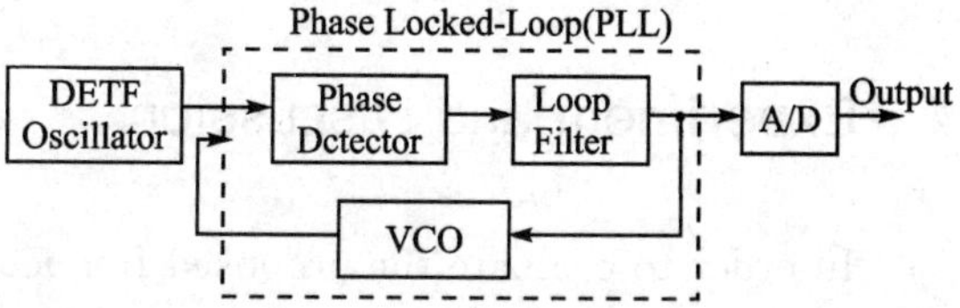

Fig. 3 Frequency demodulation system

In order to obtain a linear output of modulated frequency, some transformations in one period transform the Eq. (8) to

$$y_{\text{output}} = \int_T \hat{\omega}(t) \cdot y_{\text{m}}(t)\,\text{d}t \Big/ \int_T y_{\text{m}}^2(t)\,\text{d}t =$$

$$\int_T \frac{SmQ}{4}\omega_0\omega_d\Omega y_m(t)\cdot y_m(t)\mathrm{d}t \Big/ \int_T y^2 m(t)\mathrm{d}t = \frac{SmQ}{4}\omega_0\omega_d\Omega \tag{9}$$

From the Eq. (8), the dynamic output characteristics of resonator based on Mathieu equation presents the instant output frequency of the resonant vibratory gyroscope, the most important part of the resonant vibratory gyroscope is how to measure the resonant output frequency. The proposed frequency measurement method is as Fig. 4 shown, the task of the frequency measurement method is the frequency demodulation of the respective oscillator outputs followed by an amplitude demodulation with the proof mass oscillator output, which followed the theory analysis of the Eq. (9). To obtain a frequency output proportional to rotation rate, a Low Pass Filter (LPF) is added after the frequency.

No matter what the dynamic output characteristics from the equation or the proposed frequency measurement method, they are all just theory analysis, which will be evaluated by the experiment date in the next section.

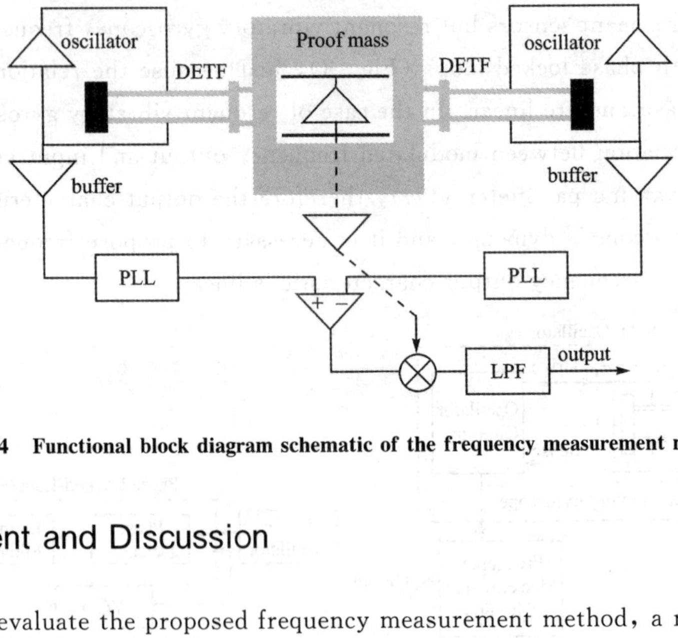

Fig. 4 Functional block diagram schematic of the frequency measurement method

4 Experiment and Discussion

In order to evaluate the proposed frequency measurement method, a resonant vibratory gyroscope sample was produced by TAIYUAN AEROINSTRUMENTS CO. LTD, which has the same design, working principle and signal modulation of a MEMS resonant vibratory gyroscope. Just because the resonant vibratory gyroscope frequency measurement systems are composed of two components, the experiment was carried out by the two parts of resonator oscillator and frequency measurement.

4.1 Resonator oscillator

Resonator oscillator experiment is designed to test the resonant frequency of the double-ended tuning forks without Coriolis force, and evaluate the closed-loop pierce oscillator circuits for the double-ended tuning fork resonators. The experimented DETF resonator as shown in Fig. 5, the experiment for the resonator is carried out using laser vibrometer and mechanical testing device. Detailed experimental sketch is shown in Fig. 6. The DETF was fixed on the mechanical testing device. During the course of the experiment, DETF vibrating in its resonant frequency, the laser was lighted on beam, the laser vibrometer testing the mechanical vibrating and recorded the vibratory signal. The piezoelectric driving was used to drive the DETF vibrating in its resonant frequency, but it is difficult to find the resonant frequency, so it is necessary to give different drive voltage and frequency to find the resonant frequency. Mechanical method is more advantageous compared to electrical method to test resonant frequency of resonator, as it is less sensitive to effects such as feedthrough coupling from undesired sources and parasitic passive elements. The double-ended tuning forks were designed for a resonant frequency of 1.735 0 kHz, and the observed resonant frequency of the double-ended tuning forks is 1.734 4 kHz, which lie slightly under designed values. It is concluded from the testing result that theory and experiment have a good match, and the design of DETF oscillator is feasible.

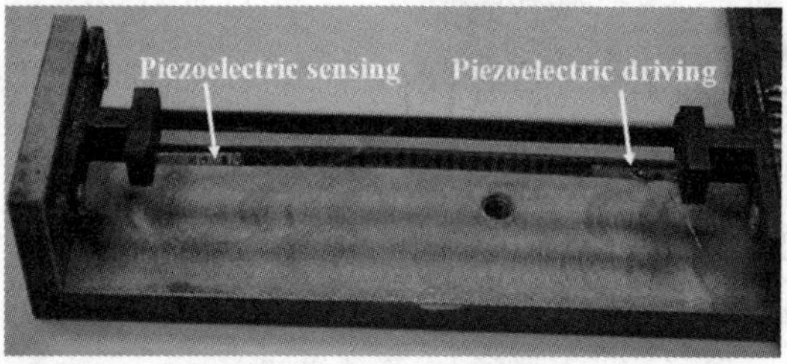

Fig. 5 The resonator in the resonant vibratory gyroscope

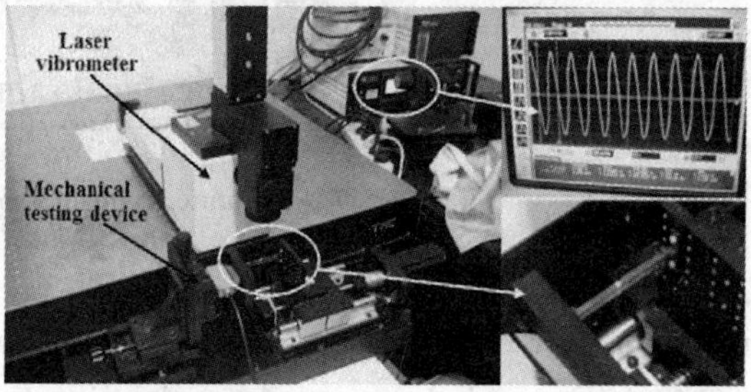

Fig. 6 The resonator oscillator assembly

4.2 Dynamic frequency measurement

Actually, the dynamic frequency measurement experiment is detecting the resonant frequency shift of resonant vibratory gyroscope, which is carried out by analyzing the dynamic characteristic of resonator.

The experiment as what is photographed in Fig. 7. The system is composed of 3 parts which are the resonant vibratory gyroscope, rate table and signal sampling device, the resonant vibratory gyroscope can be used to detect the rotations produced by the rate table. During the course of the experiment a standard sinusoidal driving voltage is firstly applied on the resonator to make it vibrate in its resonant frequency. At the same time, another sinusoidal driving voltage is applied on the proof mass to also make it vibrate in its resonant frequency. An off-device phase-locked loop is used to demodulate the signal down to the drive frequency. Further conventional amplitude demodulation with the proof mass vibrating signal, followed by a low-pass filter directly results in a rotation rate voltage output.

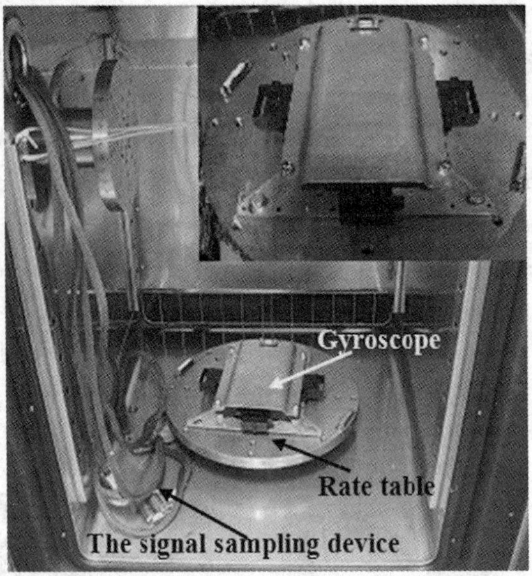

Fig. 7 Dynamic frequency measurement experiment assembly

Due to the requirement of a real time feedthrough, the device could not be subjected to a fixed constant rotation rate. The rate table sends out a series of angular velocities ($-200\sim+200(°)/s$), and these angular velocities are applied on the resonant vibratory gyroscope. The direct output signal of the resonant vibratory gyroscope sampled by the signal sampling device with every 20 s. The sampled crude signal at a fixed angular velocity (100(°)/s) is drawn in Fig. 8, the proof mass vibrating signal is Fig. 8(a), and the beam original output signal is Fig. 8(b).

When the resonant vibratory gyroscope is working, the proof mass as well as the two resonators is driven into motion, a component of the actuation force couples along the sense direction, resulting in a frequency modulated oscillator output. This coupling signal is in-

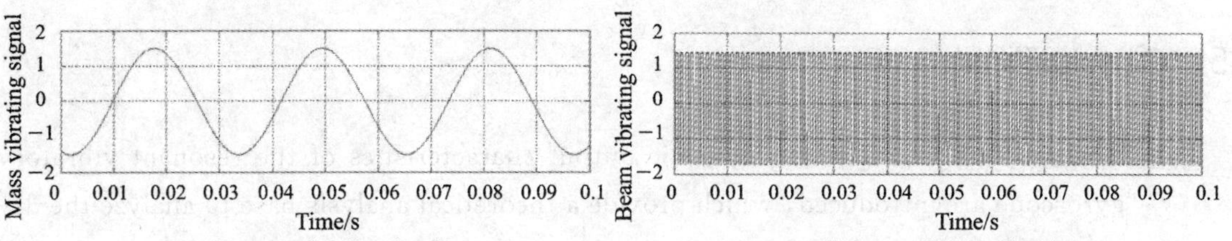

(a) mass vibrating signal (b) beam vibrating signal

Fig. 8 The original output signals of the mass and beam in resonant vibratory gyroscopes

phase with the Coriolis force and serves to introduce an offset in the output on the order of several hundred degrees per second, which indicates there is angular velocity trace in the resonator beam vibrating output signal. However, in Fig. 8, the output signal of the beam vibrating cannot follow the linear frequency output with angular velocity. On the condition of proposed measurement method, the proof mass vibrating signal (Fig. 8(a)) is using to demodulate different output signal. The ratio between alternating part of the frequency output and the changing angular velocity is plotted in Fig. 9, their relation can be expressed as

$$y_{out} = k_1 \times \omega_0^3 + k_2 \times \omega_0^2 + k_3 \times \omega_0 + k_4 \tag{10}$$

$k_4 = 0.198\,851$, $k_3 = -0.211\,636$, $k_2 = -3.491\,62E-5$, $k_1 = 6.577\,18E-7$

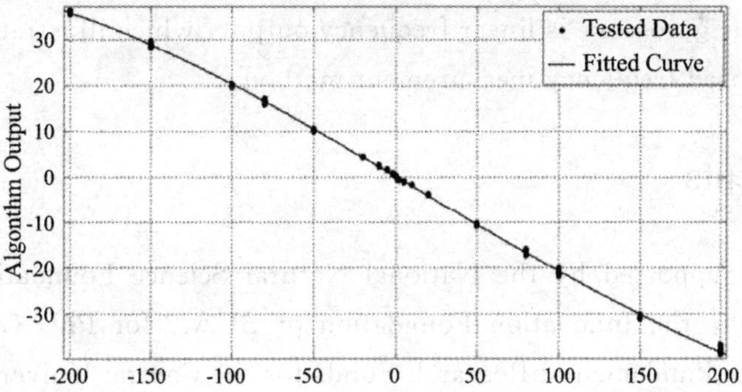

Fig. 9 The curve fitting result of the frequency output signal

The correlation coefficient of fitted curve is 99.95%, and in the linear space, the algorithm frequency output of the gyroscope is 0.211 636. Therefore, the resonant vibratory gyroscope beam output frequency is achieved linear output, which is so close to the theoretical analysis with the transformation result from the foregoing section. The correlation of the theoretical curve and the experimental data processing result is so close, which means the frequency measurement method is feasible in resonant vibratory gyroscope.

5 Conclusions

- The working principle and dynamic output characteristics of the resonant vibratory gyroscope are introduced, which provide a theoretical analysis base to analyze the dynamic frequency output characteristic of resonant vibratory gyroscopes.
- Relevant frequency measurement methods are introduced and analyzed. A proposed frequency measurement method based on dynamic output characteristics has been employed for the purpose of investigating the modulated output signal of the resonant vibratory gyroscope.
- Resonator oscillator experiment is made by using the measuring device laser vibrometer and mechanical testing device. Experiment result indicates the designed and observed resonant frequency of the resonator have a good match, which shows the design of the resonator oscillator is feasible.
- Dynamic frequency measurement experiment is performed to verify the feasibility of the proposed frequency measurement method. It is concluded from the tested result that correlation coefficient of output fitted curve is 99.95%, and the output characteristic of the gyroscope is linear frequency output, which illustrated the effectiveness of the proposed frequency measurement method.

Acknowledgments

This study is supported by the National Natural Science Foundation of China under Grant No. 60927005, the Innovation Foundation of BUAA for PhD Graduates (YWF-12-RBYJ-026), and the Fundamental Research Funds for the Central Universities (YWF-10-01-A17).

References

[1] McGuire Shaun M, Fine Morris E, Jan D. Achenbach Crack, Detection by resonant frequency measurements. Metallurgical and Materials Transactions A, 1995,26:1123-1127.

[2] Quesada-García Pilar, Prados-Sánchez Estrella, Olmedo-Gaya Victoria, et al. Measurement of dental implant stability by resonance frequency analysis: A review of the literature. Medicine Oral, Patol Oral Cirugía Bucal,2009,14:538-546.

[3] Smithl Norman F, Tanner Danelle M, Swansonl Scot E, et al. Non-Destructive Resonant Frequency Measurement on MEMS Actuators// 39th Annual International Reliability Physics Symposium. Florida. 2001: 99-105.

[4] Amigó R, Hernandez J M, García-Santiago A, et al. High-resolution detection of resonant frequencies of microwave resonators via magnetic measurements. Applied Physics Letter, 2003,82: 4528-4530.

[5] Petersan Paul J, Anlage Steven M. Measurement of resonant frequency and quality factor of Microwave resonators: comparison of methods. Journal of Applied Physic, 1998,84: 3392-3402.

[6] Rodahl Michael, Kasemo Bengt. A simple setup to simultaneously measure the resonant frequency and the absolute dissipation factor of a quartz crystal microbalance. Review of Scientific Instruments, 1996, 67: 3238-3241.

[7] Dohn S, Hansen O, Boisen A. Measurement of the resonant frequency of nano-scale cantilevers by hard-contact readout. Microelectronic Engineering, 2008, 85: 1390-1394.

[8] Ahn Y. Resonant Microbeam Electronic Oscillators for Strain Sensing, P h. D. thesis. University of Wisconson, Madison. 1999:8-22.

[9] Hagleitner C, Lange D, Hierlemann A, et al. CMOS single-chip gas detection system comprising capacitive, calorimetric and mass sensitive microsensors. IEEE Journal of Solid-State Circuits, 2002, 37: 1867-1878.

[10] Burrer C, Esteve J, Lora-Tamayo E. Resonant silicon accelerometers in bulk micromachining technology-an approach. IEEE Journal of Microelectromechanical Systems, 1996, 5:122-130.

[11] Clayton L D, EerNisse E P. Quartz thickness-shear mode pressure sensor design for enhanced sensitivity. IEEE Transactions on Ultrasonics, Ferroelectrics and Frequency Control, 1998, 45:1196-1203.

[12] Younesian D, Esmailzadeh E. Vibration suppression of rotating beams using time-varying internal tensile force, Journal of Sound and Vibration, 2011, 330: 308-320.

[13] Ghommem M, Nayfeh A H, Choura S, et al. Modeling and performance study of a beam microgyroscope. Journal of Sound and Vibration, 2010, 329: 4970-4979.

[14] Zhuang Haihan, Fan Shang chun, Guo Zhanshe, et al. Dynamic characteristics analysis of cylinder pressure transducers (VCPT). Sensors and Actuators A, 2010, 157: 219-227.

[15] McLachlan N W. Theory and Application of Mathieu Function. Dover Publications, Inc. , 1964:90-95.

[16] Seshia A A. Integrated Micromechanical Resonant Inertial Sensors for Inertial Measurement Systems. Ph. D. thesis. University of California, Berkeley, USA, 2002: 88-109.

[17] Wojciechowski K E. Electronics for Resonant Sensors. Ph. D. thesis. University of California, Berkeley, USA, 2005: 74-84.

Appendix H
An Electrothermally Excited Dual Beams Silicon Resonant Pressure Sensor with Temperature Compensation①

Tang Zhangyang   Fan Shangchun   Xing Weiwei   Guo Zhanshe   Zhang Zhengyuan

Abstract: An electrothermally excited dual beams silicon resonant pressure sensor with temperature compensation is proposed in this paper. One beam locates upon the diaphragm, whose resonant frequency changes with the measured pressure; the other beam is on the fixed edge, isolated from the pressure effect but sensitive to the temperature variation. Taking the difference of the dual beams' resonant frequencies, temperature influence can be corrected. Compensation algorithm with consideration of the fabrication errors is deducted theoretically and implemented by a homemade readout system. Experimental results of the test sample indicates that the maxim pressure measurement residual errors without compensation is up to 53.9 kPa in the working temperature range from −40 to 60℃; while with compensation the maxim residual errors decreased to 1.8 kPa from −40 to 60℃, which is only 3.3% of the uncompensated sensor. The experimental results confirm that the new designed sensor has good temperature compensation ability.

1 Introduction

Silicon resonant pressure sensors with frequency output have been successfully used for the barometer in aircrafts, because of their small volume, high accuracy, outstanding stability and negligible hysteresis. The resonator of this sensor must be excited into vibration and the vibration must be detected when measuring its resonant frequency. Several excitation methods have been utilized for silicon resonators in practice, and the most prevalent methods

① Microsystem Technologies-Micro-and Nanosystems-Information Storage and Processing Systems, 2011, 17(9): 1481-1490.

are electrostatic and thermoelastic (including photothermal and electrothermal) excitation (Stemme 1991).

Electrothermal excitation is much simpler for fabrication than electrostatic excitation, as it requires only a power-dissipating resistance on the resonator which is compatible to the standard IC process and there is no small gap requirement. Since Wilfinger et al. (1968) firstly utilized electrothermal excitation for the microcantilever resonance, it has been widely used in the resonant sensors until now (Ferrari et al. 2005; Seo and Brand 2008).

Because of its inherent heating effects, it is more necessary to utilize compensation techniques to correct the additional temperature drift of the resonant frequency when used electrothermal excitation. Greenwood and Wray (1993) mounted a temperature sensing diode on the carrier to correct the temperature drift. In this method, there were some errors caused by temperature difference as the temperature sensor was far away from the resonator. Věříš (1996) utilized different temperature coefficients of two vibration modes to correct the temperature drift of the resonant frequency. This method overcame the temperature difference but the readout rate would slow down. Han (2007) integrated a microcantilever as a thermometer, measuring its deformation by Wheatstone bridge; however, this way needs transformation between the voltage and the frequency.

In this paper, an electrothermally excited dual beams resonant pressure sensor was proposed to compensate the temperature drift. Firstly, the working principle was analyzed theoretically. Then, test samples of the sensor were fabricated following the detailed process steps. Finally, a series of experiments were carried out to verify the performance of the sensor. Experimental results indicate that the new designed sensor has good temperature self-compensation ability.

2 Sensor Operation and Theoretical Analysis

Fig. 1 illustrates the new designed sensor structure schematically. It consists of a square diaphragm, a working beam, a compensating beam and the surrounding fixed edge. Both beams are electrothermally excited and piezoresistively detected, with the excitation resistors and the detection resistors at the center and the root of the beams, respectively.

The working beam locates at the center of the diaphragm on its surface, sensitive to the deformation of the diaphragm caused by the measured pressure, as well as temperature. The compensating beam is on the surrounding fixed edge, parallel to the working beam, isolated from the effects of changes in pressure but sensitive to temperature changes. Both beams have the same geometrical size and are fabricated simultaneously with same materials by identical process. Consequently, the actual pressure with temperature compensation can be obtained from these two resonant frequencies.

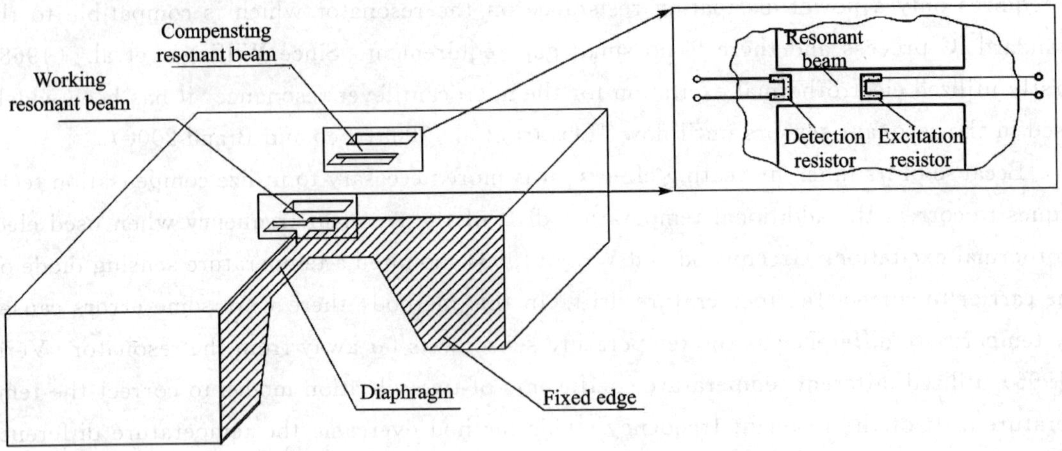

Fig. 1 A schematic drawing of the dual beams resonant pressure sensor

2.1 Resonant frequency dependence on the pressure, temperature, and static thermal power

The first vibrating mode of the beam is usually chosen as the working mode, to maintain the lowest working energy. Resonant frequency of the first mode is a function of its axial stress.

$$f^2(\sigma_x) = f_0^2 + k_b \cdot \sigma_x \tag{1}$$

where

f is the first mode resonant frequency of the beam;

$f_0 = 1.028h/l^2(E/\rho)^{1/2}$ is the nature frequency without stress in the beam, with h and l (the thickness and the length) of the beam, E (Young's modulus) and ρ (the density);

$k_b = 0.312/\rho l^2$ is the stress coefficient of the beam;

σ_x is the axial stress in the beam.

There are mainly four factors change the axial stress of the beam, the measured pressure, the ambient temperature, the static thermal power and the residual stress by fabricating deformation.

For the beam clamped at the center of a square diaphragm on its surface, the relation between the measured pressure and the axial stress is

$$\sigma_p = k_p \cdot P \tag{2}$$

where

σ_p is the pressure induced axial stress of the beam;

$k_p = (\beta_x + v\beta_y)(a^2/H^2)/(1-m^2)$ is the pressure coefficient of the square diaphragm, with β_x and β_y the location coefficients, depending on the beam's location; a and H the side length and the thickness of the square diaphragm, v the Poisson ratio;

P is the measured pressure.

Since the whole sensor structure expands with the ambient temperature elevation, the clamped ends of the beam moves accordingly, reducing its thermal strain. The relation between the thermal stress and the ambient temperature is assumed as

$$\sigma_{\Delta T} = -k_{\Delta T} \cdot \Delta T \tag{3}$$

where

$\sigma_{\Delta T}$ is the thermal stress in the beam of the new designed sensor;

$k_{\Delta T} = \chi E\alpha$ is the thermal stress coefficient, with $\chi \in (0,1)$ the adjusting coefficient which can be obtained by calibration;

ΔT is the temperature elevation of the current temperature to the reference temperature (20℃ in this paper).

For electrothermal excitation, the temperature elevation ΔT is not only caused by the ambient temperature variation, but also by the static thermal power heating effect. Lammerink (1990) deducted the relation between the static thermal power and the average temperature elevation of the doubly clamped beam. As the heat generated by the exciting resistor concentrates at the center of the beam, the movement at the clamped ends is ignored. Hence, the thermal stress caused by static thermal power is

$$\sigma_w = -k_w \cdot P_{stat} \tag{4}$$

where

σ_w is the static thermal power induced axial stress of the beam;

$k_w = E\alpha l/8bh\lambda$ is the thermal power stress coefficient of the beam, with λ the thermal conductivity;

P_{stat} is the static thermal power produced by the exciting resistor.

In addition, there is some residual stress produced by the silicon wafer deformation during the fabrication process. Usually, the residual stress is unknown and is determined by calibration.

2.2 Compensation mechanism with the dual beams

For the new designed sensor, the resonant frequency of the working beam is determined by all of the above factors, which can be expressed as

$$f_1^2 = f_{01}^2 + k_{b1}(\sigma_{p1} + \sigma_{\Delta T1} + \sigma_{w1} + \sigma_{res1}) \tag{5}$$

where

f_1 is the resonant frequency of the working beam;

f_{01} is the nature frequency of the working beam without stress;

k_{b1} is the stress coefficient of the working beam;

σ_{p1}, $\sigma_{\Delta T1}$, σ_{w1} and σ_{res1} are the pressure induced stress, the ambient temperature induced stress, the static thermal power induced stress and the residual stress in the working beam, respectively.

Substitute Eqs. (2), (3), (4) into Eq. (5), the resonant frequency of the working beam as a function of the four factors is

$$f_1^2 = f_{01}^2 + k_{b1} \cdot k_{p1} \cdot P - k_{b1} \cdot k_{\Delta T1} \cdot \Delta T - k_{b1} \cdot k_{w1} \cdot P_{stat} + k_{b1} \cdot \sigma_{res1} \tag{6}$$

where

f_1 is the resonant frequency of the working beam;

f_{01} is the nature frequency of the working beam without stress;

k_{b1} is the stress coefficient of the working beam;

k_{p1} is the pressure coefficient of square diaphragm at the location of the working beam;

$k_{\Delta T1}$, $k_{\Delta T1}$ and k_{w1} are the thermal stress coefficient of the beam related to the ambient temperature and the static thermal power, respectively;

P, ΔT, P_{stat} and σ_{res1} are the measured pressure, the ambient temperature elevation, the static thermal power and the residual stress in the working beam, respectively.

Although the dual beams are designed to have the same geometrical size and are fabricated simultaneously with same materials by identical process, there are inevitably some differences between them due to the fabrication errors. Similar to Eq. (6), the resonant frequency of the compensating beam is expressed as

$$f_2^2 = f_{02}^2 + k_{b2} \cdot k_{p2} \cdot P - k_{b2} \cdot k_{\Delta T2} \cdot \Delta T - k_{b2} \cdot k_{w2} \cdot P_{stat} + k_{b2} \cdot \sigma_{res2} \tag{7}$$

In practice, the exciting thermal power is optimized to a constant to maintain the linear vibration of the beam, and the residual stress will not change after fabrication. Thus, we can define the parameters as following

$$\left.\begin{aligned} \bar{f}_{0n}^2 &= f_{0n}^2 - k_{bn} \cdot k_{wn} \cdot P_{stat} + k_{bn} \cdot \sigma_{resn} & (n = 1,2) \\ k_n &= k_{bn} \cdot k_{pn} = 0.312(\beta_{xn} + v\beta_{yn})(a^2/H^2)/\rho l^2(1 - v^2) & (n = 1,2) \\ s_n &= k_{bn} \cdot k_{\Delta Tn} = 0.312\chi_n E\alpha/\rho l^2 & (n = 1,2) \end{aligned}\right\} \tag{8}$$

where

$\bar{f}_{01}^2$ and $\bar{f}_{02}^2$ are the resonant frequencies including the static thermal power and the residual stress influence, which is taken as the initial frequencies of the dual beams;

k_1 and k_2 are the pressure sensitivity of the resonant frequencies;

s_1 and s_2 are the temperature sensitivity of the resonant frequencies, respectively.

Hence, the resonant frequency of the working beam and the compensating beam can be written as

$$\left.\begin{aligned} f_1^2 &= \bar{f}_{01}^2 + k_1 \cdot P - s_1 \cdot \Delta T \\ f_2^2 &= \bar{f}_{02}^2 + k_2 \cdot P - s_2 \cdot \Delta T \end{aligned}\right\} \tag{9}$$

The temperature sensitivities of the working beam and the compensating beam are of the same order in magnitude and not zero, $s_1 \neq 0$, $s_2 \neq 0$. The pressure sensitivity of the working beam is much larger than that of the compensating beam, $k_1 s_2 > k_2 s_1$. We can calculate the measured pressure by

$$P = \frac{s_2(f_1^2 - \bar{f}_{01}^2) - s_1(f_2^2 - \bar{f}_{02}^2)}{k_1 s_2 - k_2 s_1} \tag{10}$$

Compensation algorithm should be implemented according to the calibration of pressure and temperature characteristics of the dual beams. The apparent pressure has nothing to do with the ambient temperature induced stress but still depends on Young's modulus temperature coefficient, because the temperature sensitivities are related to the Young's modulus.

3 Sensor Fabrication and Experimental Setup

To verify the feasibility of the new design, test samples were fabricated by silicon micromachining technology, and an automatic open-loop test system was built up to calibrate the resonant frequencies characteristics of the dual beams. In addition, a readout system for the new designed sensor was designed to interpret the resonant frequencies into pressure data.

3.1 Fabrication process of the sensor

The fabrication process starts from the silicon wafer (⟨100⟩ p-type) with resistivity of 7～10 Ω · cm.

Firstly, a 1 μm thick layer of thermal oxide is grown onto the wafer surface, and followed by a 0.1 μm thick Si_3N_4 layer (Fig. 2(a));

Then, a cavity is etched using anisotropically etching with KOH from the wafer backside, and the bottom of the cavity forms the diaphragm (Fig. 2(b)). A timed etch stop is used to achieve a predefined diaphragm thickness.

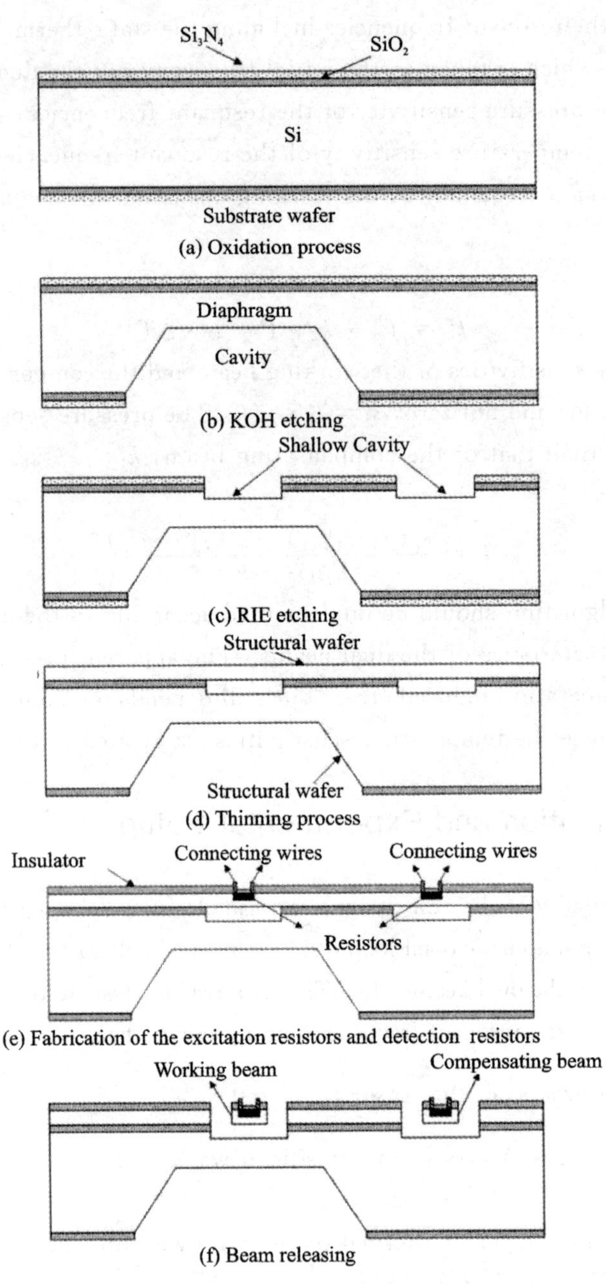

Fig. 2 Detailed fabrication process steps for the sensor

The third step is to etch two shallow cavities under the dual beams for their vibration by deep RIE (Reactive IonEtching) process from the wafer front-side (Fig. 2(c)).

Next, to fusion bond the etched substrate wafer to an 〈100〉 n-type structural wafer, from which the dual beams will be formed. The bonded wafers are immediately annealed at 1 150℃ for 3 h, and followed by thinning, grinding and polishing process to obtain a 15 μm thickness film (Fig. 2(d)).

The fifth step is to make the excitation resistors and detection resistors by boron diffusion, and the contact holes, the connecting wires and electrode pads are made by Al sputtering and wet etching (Fig. 2(e)).

Finally, the two beams are released by deep RIE process (Fig. 2(f)).

Fig. 3(a) shows a micrograph of the fabricated dual beams structure. The sensor chip is mounted onto a glass base by silicon-glass bonding, with the diaphragm side face to the pressure hole of the glass base. The glass base is fastened on an 8-pined metallic seat with a central pressure inlet. The pads on silicon wafer are connected to the pins of the seat through wire bonding (Fig. 3(b)). Finally, vacuum encapsulation of the sensor is achieved by jointing a stainless steel cap on the metallic seat by vacuum electron beam welding, and the quality factor of the fabricated sensor is about 3 000 after encapsulation. This sensor is an absolute pressure sensor.

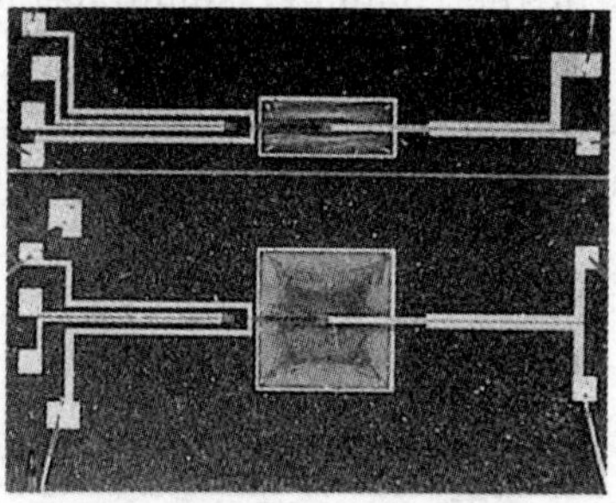

(a) Micrograph of the working beam (below) and the compensating beam (above)

(b) Sensor chip assemble on a 8-pined metallic seat

Fig. 3 Sketch of the fabricated sensor

3.2 Experimental setups for the sensor test

The experimental setup for pressure and temperature characteristic calibration is shown in Fig. 4. Standard pressure is supplied by PPC3 type standard pressure calibrator from DHI Corp. with regulating accuracy of 10 ppm; while the constant temperature environment is supplied by a GDS-50L type temperature controller with regulating accuracy of ±0.1℃. Frequency response is measured by a homemade open-loop test instrument. It mainly consists of the detecting unit and the frequency sweeping control unit. The amplitude and phase-shift of the beam vibration were abstracted from the very low signal-to-noise ratio (SNR) pickup signal based on lock-in principle. Three separate frequency sweeps were carried out with each sweep iterating to a narrower frequency range by a finer frequency step. The last sweep spanned 200 Hz with 1 000 data points, which yielded 0.2 Hz resolution for the measurement. The results are recorded, processed and plotted with a program code using Matlab software.

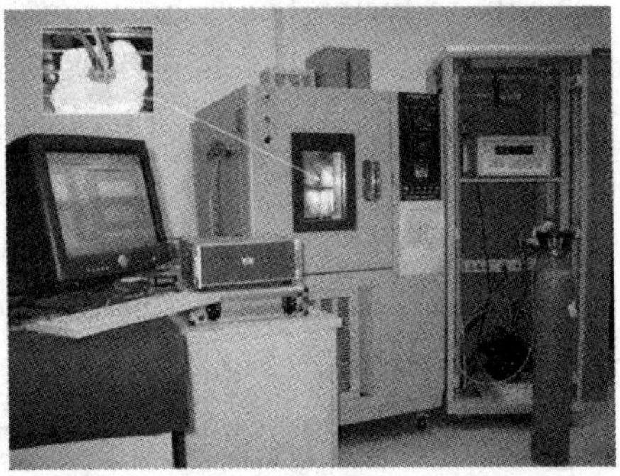

Fig. 4 A photograph of the automatic open-loop test system

Fig. 5 shows the typical results of the working beam and the compensating beam, respectively.

The readout system for the new designed sensor consists of the closed-loop circuits and the calculating unit, as shown in Fig. 6. The resonant frequencies of the dual beams are detected and tracked by the phase-locked loop (PLL)-based closed-loops at the same time, respectively. Compensation algorithm is implemented to give out pressure data with improved accuracy. The apparent pressure data are recorded in the computer simultaneously for the next processing step.

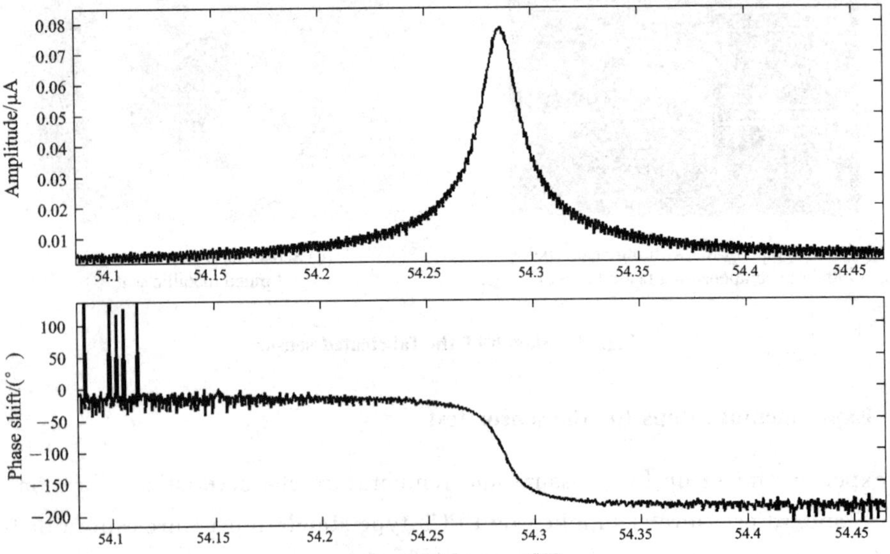

(a) Frequency characteristic of the working beam with static power of 300 μW at 100 kPa and 20℃

Fig. 5 Typical results of the working beam and the compensating beam

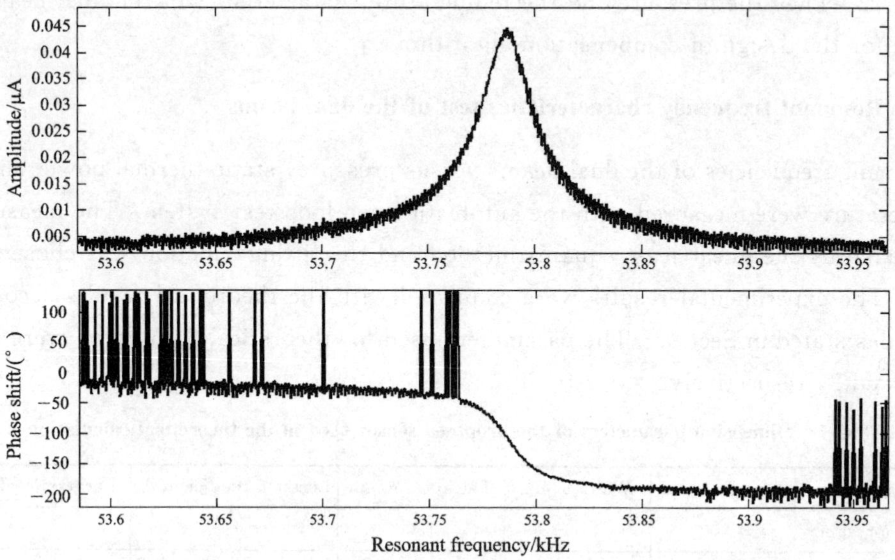

(b) Frequency characteristic of the compensating beam with static power of 300 μW at 100 kPa and 20℃

Fig. 5 Typical results of the working beam and the compensating beam(continued)

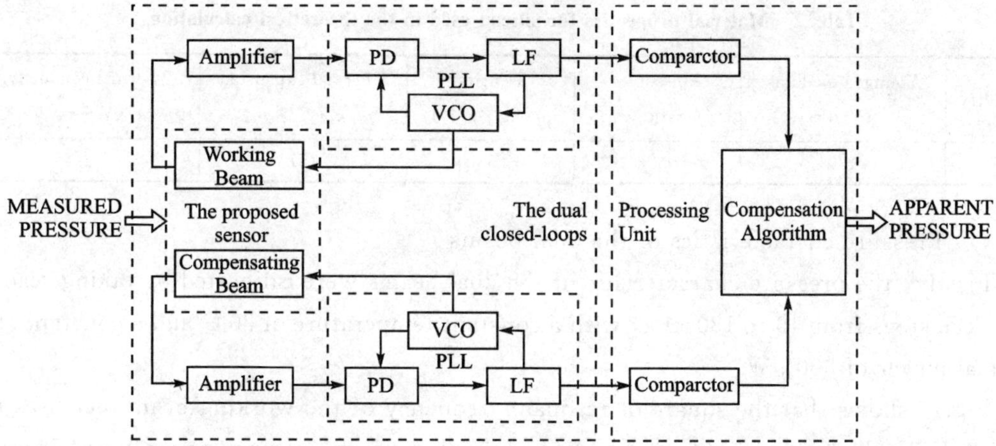

Fig. 6 Block diagram of the readout system for the dual beams resonant pressure sensor

4 Experimental Results and Discussion

A series of experiments were carried out to verify the resonant frequency dependency on the measured pressure, the ambient temperature and the static thermal power, which is stated in Sect. 2. Then the pressure and the temperature characteristics of the dual beams were calibrated for the design of compensation algorithm.

4.1 Resonant frequency characteristics test of the dual beams

Resonant frequencies of the dual beams versus pressure, static thermal power and ambient temperature were measured with the automatic open-loop test system. The measured data were fitted by the linear least squares method and the fitting function was chosen as $y = a_1 x + a_0$. The experimental results were compared with the theoretical results according to the formulas stated in Sect. 2. The parameters used for theoretical calculation were listed in Tables 1 and 2, respectively.

Tab. 1 Dimension parameters of the proposed sensor used in the theoretical calculation

Items	Length $l/\mu m$	Width $b/\mu m$	Thickness $h/\mu m$	Location coefficient β_x	Location coefficient β_y
Square diaphragm	4 000	4 000	150	—	—
Working beam	1 500	90	15	0.085 2	0.084 3
Compensating beam	1 500	90	15	0	0

Tab. 2 Material properties for silicon used in the theoretical calculation

Property	Young's modulus E/GPa	Poisson's ratio ν	Density $\rho/(kg \cdot m^{-3})$	Thermal expansion $\alpha/(℃^{-1})$	Thermal conductivity $10^6 \cdot \lambda/(W \cdot m^{-1} \cdot ℃^{-1})$
Si	130	0.22	2 330	157	2.5

(1) Pressure characteristics of the dual beams

Firstly, the pressure characteristic of the dual beams were calibrated by taking readings at 20 kPa steps from 10 to 130 kPa, with a constant temperature of 20℃ and a constant static thermal power of 300 μW.

Fig. 7 shows that the square of resonant frequency of the working beam increases proportionally with the increasing pressure; while that of the compensating beam increases slightly. The linear fitting for the working beam is $f_1^2 = 4.075 \times 10^6 P + 2.548 \times 10^9$; while for the compensating beam is $f_2^2 = 0.155 \times 10^6 P + 2.877 \times 10^9$. According to Eq. (9), the equivalent nature frequencies of the two beams are 50.469 and 53.642 kHz, differing from the theoretical value of 51.186 kHz by −1.4% and 4.8%, respectively. The measured pressure sensitivity of the working beam is 4.074×10^6 Hz^2/kPa, lower than the theoretical value of 4.608×10^6 Hz^2/kPa by 11.6%. The pressure sensitivity of the compensating beam is actually 0.155×10^6 Hz^2/kPa and is 3.8% of the working beam. These differences are mainly due to the fabrication error on the beams' thickness.

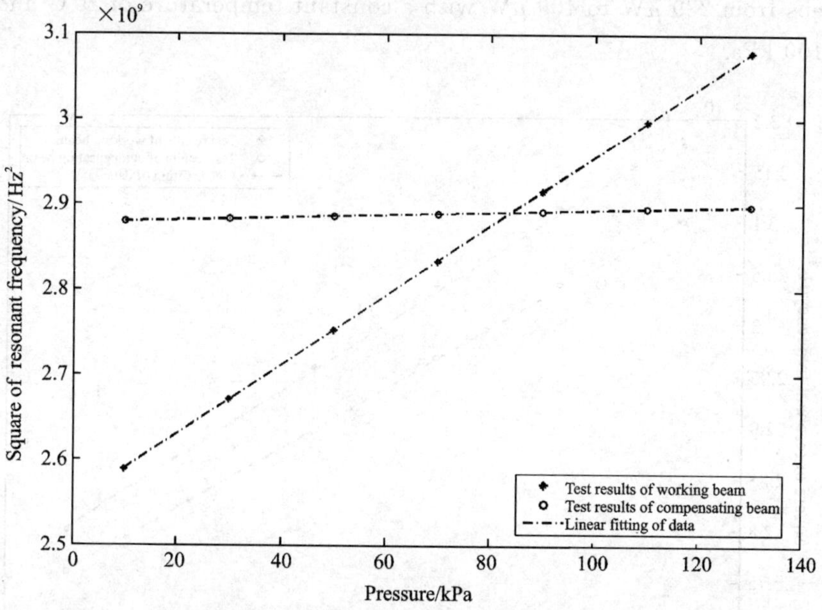

Fig. 7 Square of resonant frequencies as a function of the pressure for the dual beams

(2) Temperature characteristics of the dual beams

Secondly, the ambient temperature characteristic of the dual beams were determined by taking readings at 20℃ steps from −40℃ to 60℃ with a constant pressure of 100 kPa and a constant exciting power of 300 μW.

Fig. 8 shows that the square of resonant frequencies of the dual beams decreases linearly as the temperature increases with the approximately same slopes. The linear fitting for the working beam is $f_1^2=-3.919\times10^6\Delta T+2.954\times10^9$ and for the compensating beam is $f_2^2=-3.452\times10^6\Delta T+2.895\times10^9$. The linear correlation coefficients of the two fittings are both 0.999 6, which validated the good linearity assumed in Eq. (3). The measured temperature coefficient for the working beam is -3.919×10^6 $Hz^2/℃$ and for the compensating beam is -3.452×10^6 $Hz^2/℃$. Compared to the theoretical value of -1.932×10^7 $Hz^2/℃$ for the ideally doubly clamped beam, the measured coefficients decrease by 79.7% and 82.1%, thus the adjusting coefficients should be 0.203 and 0.179, respectively.

(3) Static thermal power characteristics of the dual beams

Thirdly, the resonant frequency as a function of the static thermal power were measured at 20 μW steps from 229 μW to 409 μW with a constant temperature of 20℃ and a constant pressure of 100 kPa.

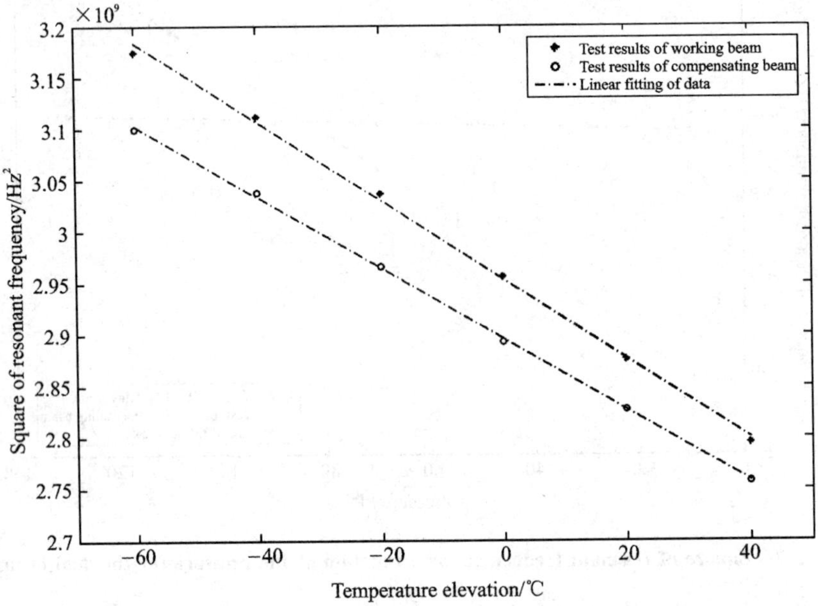

Fig. 8 Square of resonant frequencies as a function of the ambient temperature for the dual beams

Fig. 9 shows that the square of resonant frequencies of the dual beams decreases linearly as the static thermal power increases with the approximately same slopes. The linear fitting for the working beam is $f_1^2=-0.012\times10^6\ P_{stat}+2.959\times10^9$ and for the compensating beam is $f_2^2=-0.013\times10^6 P_{stat}+2.904\times10^9$. The measured thermal power coefficient for the working beam and the compensating beam are -1.2×10^4 and -1.3×10^4 $Hz^2/\mu W$, differing from the theoretical value of -1.7×10^4 $Hz^2/\mu W$ by -28.0% and -22.8%, respectively. This decrease is mainly because the concentrated thermal power actually causes some displacements of the clamped ends, which is ignored in the theoretical analysis.

The above results verify the linear dependence of the resonant frequency on the measured pressure, the ambient temperature and the static thermal power. They also indicate the differences between the dual beams caused by fabrication errors.

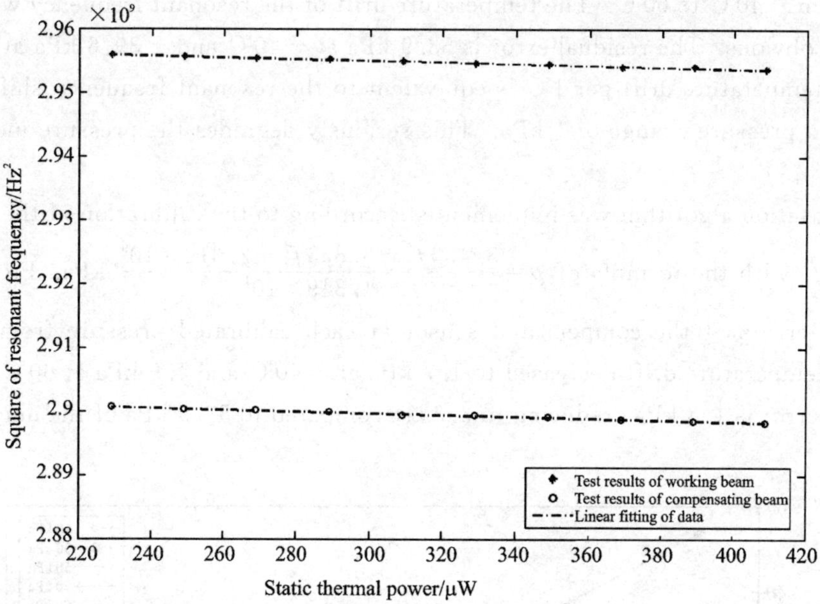

Fig. 9 Square of resonant frequencies as a function of the static thermal power for the dual beams

4.2 Compensation characteristic test of the new designed sensor

In the next experimental step, the performances of the pressure sensors without and with compensation are tested and compared. The apparent pressures of the uncompensated sensor and compensated sensor were obtained with the readout system. Both experiments were carried out at 20 kPa steps from 10 kPa to 130 kPa and at −40℃, −20℃, 0℃, 20℃, 40℃, 60℃ with a constant static thermal power of 300 μW.

The diaphragm and the working beam of the new designed sensor form a resonant pressure sensor without compensation; while with the compensating beam form a temperature compensated pressure sensor. As the closedloops for the dual beams are independent to each other, the apparent pressures without and with compensation can be calculated simultaneously by the readout system. The residual error is defined as the difference between the apparent pressure and the calibrated pressure.

The apparent pressure was calculated from the calibrated pressure characteristic of the working beam measured at 20℃ and 300 μW, by the formula of $p=\dfrac{f_1^2-2.548\times10^9}{4.075\times10^6}$ kPa. Fig. 10 shows the residual errors of the uncompensated pressure sensor at each calibrated pressure from -40℃ to 60℃. The temperature drift of the resonant frequency without compensation is obvious. The residual error is 53.9 kPa at -40℃ and -39.6 kPa at 60℃. This implies the temperature drift per 1℃ is equivalent to the resonant frequency shift caused by the measured pressure change of 1 kPa. This seriously degrades the pressure measuring accuracy.

Compensation algorithm was implemented according to the calibration of the dual beams stated above, with the formula of $p=\dfrac{3.453f_1^2-3.853f_2^2+2.312\times10^9}{1.349\times10^7}$ kPa. Fig. 11 shows the residual errors of the compensated sensor at each calibrated pressure from -40℃ to 60℃. The temperature drift decreased to 1.7 kPa at -40℃ and 1.5 kPa at 60℃. The maxim residual error is 1.8 kPa, reducing to 3.3%, compared to 53.9 kPa of the uncompensated sensor.

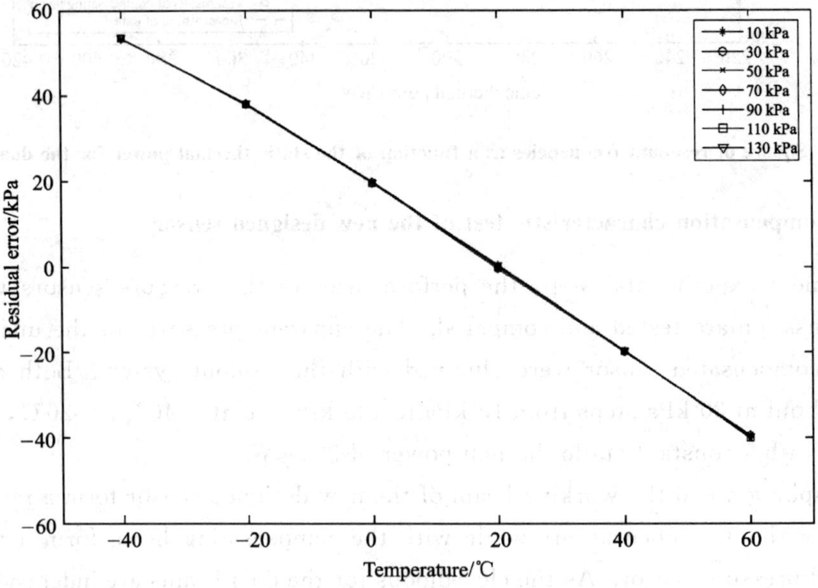

Fig. 10 Pressure measuring residual errors of the uncompensated sensor from -40℃ to 60℃

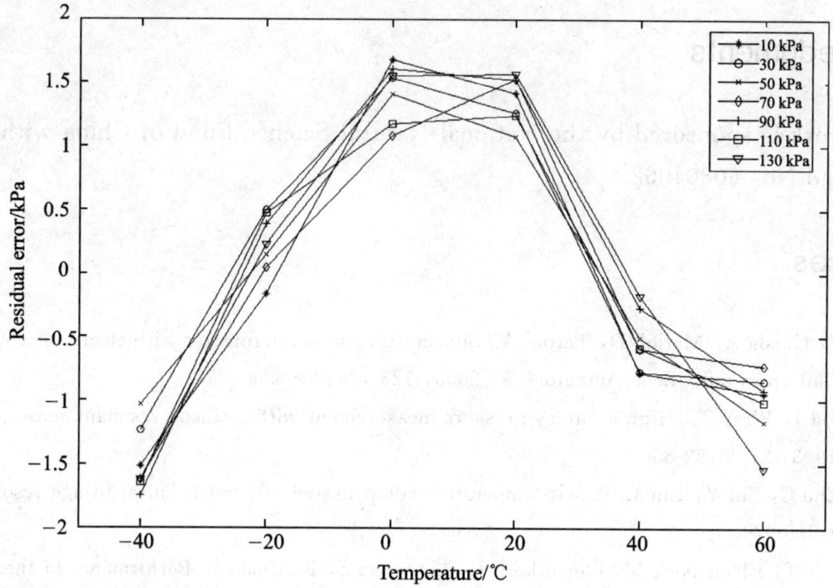

Fig. 11 Pressure measuring residual errors of the compensated sensor from −40℃ to 60℃

5 Conclusions

We have designed, fabricated and tested an electrothermally excited dual beams silicon resonant pressure sensor in this paper. This pressure sensor bases on a silicon diaphragm interrogated by a vibrating beam attached to it with temperature compensation from another vibrating beam, isolated from the effects of changes in pressure. The new designed sensor has two significant advantages over the resonant pressure sensor compensated with temperature sensing diode and the dual-mode method. Firstly, there is no need for an analog-to-digital (A/D) converter, because the compensation implementation deals with frequency signals. Secondly, the temperature drift is compensated with a faster readout than the dual-mode method, as the dual beams resonant frequencies are interrogated simultaneously.

Experimental results show that the maxim pressure measurement residual errors without compensation is up to 53.9 kPa in the working temperature range from −40℃ to 60℃; while with compensation the maxim residual errors decreased to 1.8 kPa from −40℃ to 60℃, only 3.3% of the uncompensated one. It proves that the new designed sensor has good temperature compensation ability.

Acknowledgments

This work is sponsored by the National Natural Science Fund of China with grant No. 60927005 and No. 60804062.

References

[1] Ferrari V, Ghisla A, Marioli D, Taroni A. Silicon resonant accelerometer with electronic compensation of input-output cross-talk. Sens Actuators A, 2005, 123-124:258-266.

[2] Greenwood J, Wray T. High accuracy pressure measurement with a silicon resonant sensor. Sens Actuators A, 1993, 37-38:82-85.

[3] Han J, Zhu C, Shi Y, Liu J. A self-temperature-compensated micromechanical bridge resonator. Meas, 2007, 40:975-979.

[4] Lammerink T, Elvenspoek M, Ouwerkerk R, Bouwstra S, Fluitman J. Performance of thermally excited resonators. Sens Actuators A, 1990, 21-23:352-356.

[5] Seo J, Brand O. High Q-factor in-plane-mode resonant microsensor platform for gaseous/liquid environment. J. Microelectromech Syst, 2008, 17:483-493.

[6] Stemme G. Resonant silicon sensors. J. Micromech Microeng, 1991, 1:113-125.

[7] Věříš. Temperature compensation of silicon resonant pressure sensor. Sens Actuators A, 1996, 57: 179-182.

[8] Wilfinger R, Bardell P, Chhabra D. The resonistor: a frequency selective device utilizing the mechanical resonance of a silicon substrate. IBM J. Res. Dev., 1968, 12:113-118.

Appendix I
Non-linear Dynamics of an Electrothermally Excited Resonant Pressure Sensor[①]

Li Qingfeng Fan Shangchun Tang Zhangyang Xing Weiwei

Abstract: Non-linear dynamics of a resonant silicon bridge pressure sensor with electrothermal excitation has been investigated. Non-linear vibration model of the bridge resonator in the pressure sensor is developed by Euler-Bernoulli beam modeling. Besides non-linear geometric effect, the model also takes into account the measured pressure, the heating effect of the electrothermal excitation, and the residual internal force in the bridge. An approximate solution for this model is found via Galerkin procedure and the multiple scales method. From the solution, a bending parameter is extracted to quantitatively evaluate the bending of the bridge amplitude-frequency response. Dependence of the bending parameter on the sensor operating conditions is examined analytically and experimentally. Analytical calculations and experimental observations show good agreement qualitatively. An improved prediction for the critical vibration amplitude of the bridge is proposed, beyond which frequency hysteresis occurs. The critical vibration amplitude presented in this work is compared with those in other literatures. Moreover frequency output error of the sensor due to non-linear vibration is also examined. And finally several suggestions on resonant sensor design and operation with respect to non-linear effects are proposed.

Keywords: Pressure sensor, Micro resonator, Non-linear vibration, Non-linear modeling

1 Introduction

The resonant pressure sensor relates the measured pressure to the natural frequency shift of the resonator. Resonant pressure sensors have found wide applications in avionics, industrial automation, and instrumentation for their unique merits: frequency output (without A/D converter) and easier interface to computers, long term stability, and high repeat-

---

① Sensors and Actuators A: Physical, 2012, 188(SI): 19-28.

ability. In 1984, Greenwood developed a silicon micro resonant pressure sensor as an alternative to its conventional macro counterpart[1]. Silicon micro manufacturing brings on the advantages such as compact size and feasibility of integration with IC technologies. However, when vibrating with large amplitude, the resonator in the pressure sensor usually takes on non-linearities due to its small size, which is known as "hard spring effect". This hard spring effect is highly undesirable in most of the resonant sensors[2].

In 1987, Andres et al. reported experimentally observed nonlinear vibration and hysteresis in the silicon micro resonant pressure sensor developed by Greenwood aforementioned[3]. They pointed out that the non-linearity arose from the geometry of the resonator rather than the material. It is believed to be the earliest literature on the non-linear vibration of micro resonators. Subsequent to the publication of that work, a number of research papers reported non-linear behavior of micro resonators. Effort of Pratt et al. in 1991 is of particular note, as they found that the non-linear response of a micro resonator could be well modeled by Duffing's equation[4]. However, almost all of the published efforts on non-linearity of micro resonators were based on lumped parameter models, until Abdel-Rahman and Nayfeh approached the micro resonator from the classical non-linear vibration perspective[5]. Their modeling emphasized the development of the non-linear distributed parameter model, the subsequent reduced order model and the perturbation solution. From 2003, the combined analytical and experimental investigations surged. The 2008 effort of Mahmoodi et al. noted soft spring effect in a piezoelectrically actuated microcantilever sensor and attributed it to the quadratic material non-linearity in the piezoelectric layer[6]. The 2009 effort of Kacem et al. dealt with both non-linear geometric effect and electrostatic non-linearities of an electrostatically actuated microbeam to improve the sensor performance[7]. The 2009 effort of Pinto examined the non-linear limitations on the design of PLL for a resonant accelerometer[8]. In 2010, Nayfeh et al. reported the non-linear dynamics of a resonant gas sensor[9]. A more comprehensive review up to 2010 about the work on the non-linear dynamics of micro resonators is presented in literature[10].

Non-linear dynamics of electrostatically and piezoelectrically actuated micro resonators have been well examined. However, quite few of the literatures about that of the electrothermally excited resonators could be found. Electrothermal excitation is very attractive for its compatibility with the standard IC process, whereas an inherent problem is the heating effect induced natural frequency shift[11]. In this work, it is shown that this heating effect also affects the non-linearity. Besides that heating effect, the influence of the measured pressure and the residual internal force on the non-linear vibration of the pressure sensor is also investigated. An improved prediction for the critical vibration amplitude of the resonant bridge is proposed, beyond which frequency hysteresis occurs. The critical amplitude presented in this work is compared with those in other literatures. Finally, several suggestions on the sensor design and operation with respect to non-linear effects are proposed. Due to the similar mechanisms of the electrothermal and photothermal excitations, the results in this work are readily generalized to the photothermally excited resonant sensors.

2 Considered Pressure Sensor

Fig. 1 gives the considered resonant pressure sensor. The sensing structure is mainly composed of a square diaphragm and a clamped-clamped bridge. The square diaphragm is the preliminary sensing element, and the bridge attached to the upper surface of the diaphragm is the final sensing component. The measured pressure directly applied on the diaphragm deforms the diaphragm shape and alters the stress in the bridge. So the effective stiffness and the natural frequency of the bridge change with the measured pressure. By measuring the natural frequency shift of the bridge, the measured pressure is obtained. A diffused drive resistor is located at the mid-section of the bridge. With the exciting signal applied on the resistor, a temperature gradient in the thickness direction will develop. This in turn creates a bending moment driving the bridge into vibration which is detected by a piezoresistor at the bridge end.

The sensing structure shown in Fig. 1 was realized starting from a ⟨100⟩ p-type Si substrate with resistivity of 7～10 Ω · cm. Firstly, a 1μm thick layer of thermal oxide was grown onto the wafer surface, and followed by a 0. 1 μm thick Si_3N_4 layer; see Fig. 2(a). Then, a cavity was etched using anisotropically etching from the wafer back-side to form the diaphragm; see Fig. 2(b). The predefined diaphragm thickness was achieve by the timed etch stop. The third step was to etch a shallow cavity under the resonant bridge for its vibration by deep RIE (reactive ion etching) process from the wafer front-side; see Fig. 2(c). Next, to fusion bond the etched substrate wafer to a ⟨100⟩ n-type structural wafer, from which the resonant bridge was formed. The bonded wafers were immediately annealed at 1 150℃ for 3 h, and followed by thinning, grinding and polishing process to obtain a 15μm thickness film; see Fig. 2(d). The fifth step was to make the drive resistor and the sense resistor by boron diffusion with nominal resistance of 200 Ω for each resistor, and the contact holes, the connecting wires and electrode pads were made by Al sputtering and wet etching; see Fig. 2(e). Finally, the resonant bridge was released by deep RIE process; see Fig. 2(f).

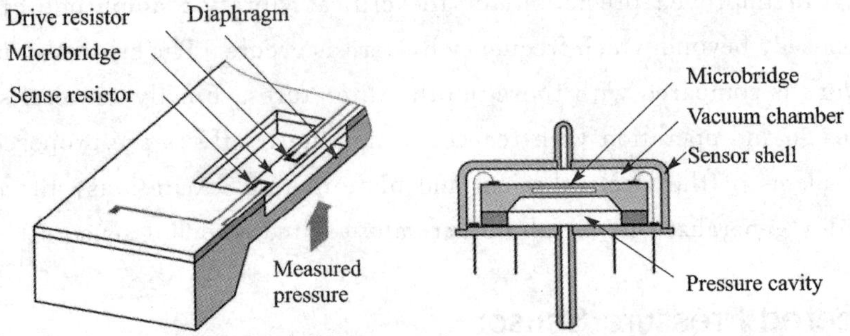

(a) Structure diagram

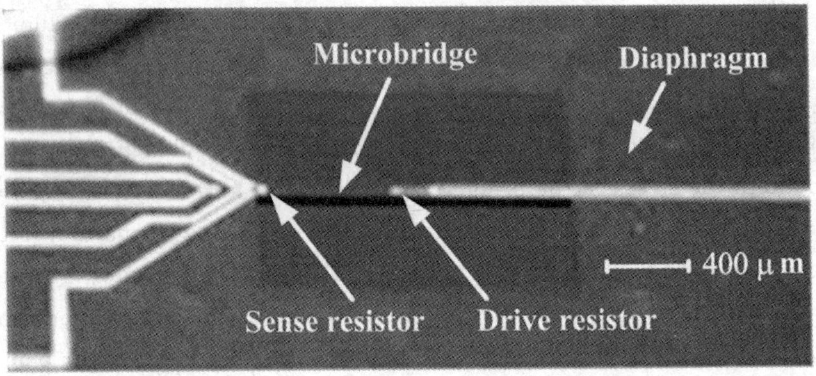

(b) Sectional view of the sensing structure assembly drawing of the sensor

Fig. 1 Structure diagram and photograph of the resonant silicon bridge pressure sensor

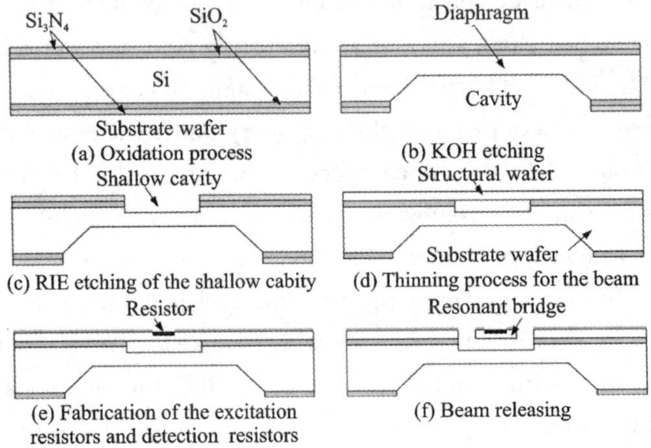

Fig. 2 Fabrication processes for the pressure sensor

The heat generated by the drive resistor is proportional to the square of the applied voltage. The frequency of the exciting signal can be same as the natural frequency of the bridge, if the exciting ac signal is superimposed on a dc bias. Assuming a temperature-independent resistance R of the drive resistor, the generated thermal power is given by

$$P_{th}(\hat{t}) = \frac{(U_{dc} + U_{ac} \cos \hat{\omega} \hat{t})^2}{R} \tag{1}$$

where

U_{dc} is the DC bias.

U_{ac} and $\hat{\omega}$ are the amplitude and frequency of the ac exciting signal respectively.

The thermal power $P_{th}(\hat{t})$ is composed of a static component P_s and two harmonic components $p_1(\hat{t})$ and $p_2(\hat{t})$ with frequencies of $\hat{\omega}$ and $2\hat{\omega}$. respectively:

$$P_s = \frac{U_{dc}^2 + 0.5U_{ac}^2}{R} \tag{2a}$$

$$p_1(\hat{t}) = \frac{2U_{dc}U_{ac}\cos\hat{\omega}\hat{t}}{R} \tag{2b}$$

$$p_2(\hat{t}) = \frac{0.5U_{ac}^2 \cos 2\hat{\omega}\hat{t}}{R} \tag{2c}$$

The static thermal power P_s develops a constant axial temperature rise, and the consequent thermal stress causes natural frequency shift of the bridge, which is known as heating effect of thermal excitation. Since the frequency of the harmonic component $p_1(\hat{t})$ is set approximately to the natural frequency of the bridge, the thermal bending moment induced by $p_1(\hat{t})$ drives the bridge into resonance. The thermal power $p_2(\hat{t})$ is a disturbance component which can be restrained if $U_{dc} \geqslant U_{ac}$ is satisfied.

The frequency of the exciting signal should track the fundamental natural frequency of the bridge in the pressure measurement. To this end, a PLL is utilized which is shown in Fig. 3.

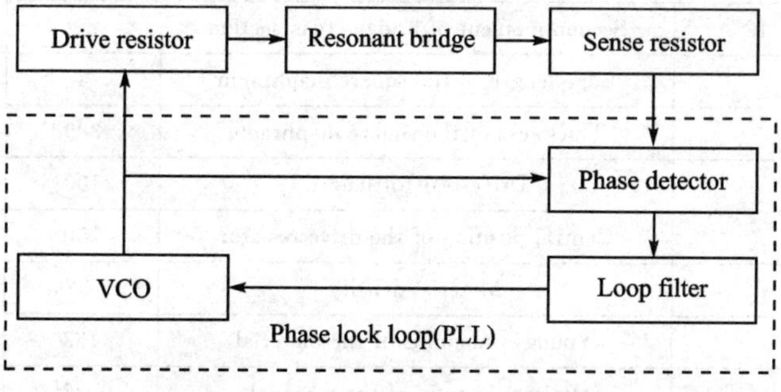

Fig. 3 Schematic of the PLL for natural frequency tracking

3 Modeling and Analysis

3.1 Non-linear modeling for the clamped-clamped bridge in the pressure sensor

The resonator in the pressure sensor is regarded as a prismatic wide clamped-clamped bridge with a rectangular cross section shown in Fig. 4. A summary of geometric, elastic and thermal parameters of the bridge and diaphragm is listed in Tab. 1. Since the bridge thickness h is much smaller than the length l, it is reasonable to model the bridge along the line of Euler-Bernoulli beam modeling with neglecting the rotational inertia and lateral shear deformation.

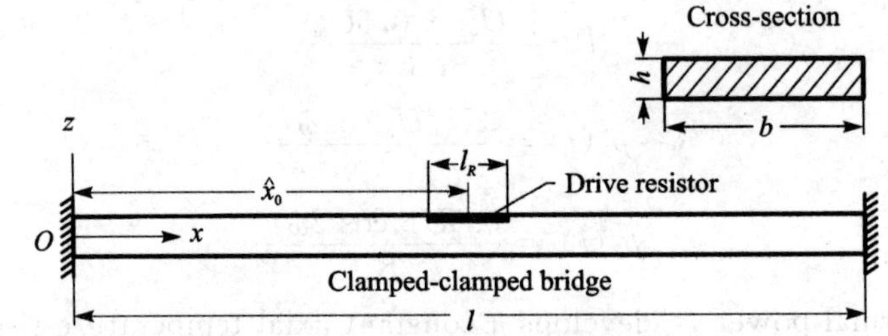

Fig. 4 Schematic of the clamped-clamped bridge resonator

Tab. 1 List of geometric, elastic and thermal parameters of the bridge and the diaphragm used in the modeling

Symbol	Description	Value	Unit
l	Bridge length	1 500	μm
h	Bridge thickness	15	μm
b	Bridge width	90	μm
$J=bh^3/12$	Second moment of bridge cross section		
L	Edge length of the square diaphragm	4	μm
H	Thickness of the square diaphragm	300	μm
l_R	Drive resistor length	100	μm
$\hat{x}_0$	Central position of the drive resistor	750	μm
ρ	Material density	2 329	kg/m^3
$\widetilde{E}$	Young's modulus of the material	133	GPa
υ	Poisson's ratio of the material	0.278	

continued

Symbol	Description	Value	Unit
α_L	Thermal expansion coefficient of the material	2.4×10^{-6}	K^{-1}
C	Specific heat of the material	733	J/(kg · K)
λ	Heat conductivity of the material	150	W/(m · K)

The equation of motion that governs the lateral deflection $\hat{w}(\hat{x},\ \hat{t})$ of the bridge is given by

$$EJ\frac{\partial^4\hat{w}}{\partial\hat{x}^4}+\rho bh\frac{\partial^4\hat{w}}{\partial\hat{t}^2}+c\frac{\partial\hat{w}}{\partial\hat{t}}=\left[\frac{Ebh}{2l}\int_0^l\left(\frac{\partial\hat{w}}{\partial\hat{x}}\right)^2\mathrm{d}\hat{x}+N_t+N_p+N_r\right]\times$$
$$\frac{\partial^2\hat{w}}{\partial\hat{x}^2}-\frac{\partial^2 M_{\mathrm{th1}}(\hat{x},\hat{t})}{\partial\hat{x}^2}-\frac{\partial^2 M_{\mathrm{th2}}(\hat{x},\hat{t})}{\partial\hat{x}^2}\tag{3a}$$

And the boundary conditions are

$$\hat{w}(0,\hat{t})=\hat{w}(l,\hat{t})=\frac{\partial\hat{w}(0,\ \hat{t})}{\partial\hat{x}}=\frac{\partial\hat{w}(l,\ \hat{t})}{\partial\hat{x}}=0\tag{3b}$$

where

$\hat{x}$ is the distance along the bridge from a clamped end.

$\hat{t}$ is time. The hat denotes dimensional quantities.

c is the damping coefficient. The damping coefficient is dependent on the vibration frequency when the resonator vibrating in liquid or air[12,13]. The bridge in the pressure sensor is encapsulated in a vacuum chamber with low pressure of 5～10 Pa. Therefore the damping coefficient is simplified to be frequency-independent in the neighborhood of resonance. And the value of c can be determined by the measured quality factor as shown below in Eq. (15b). An effective modulus of elasticity $E=\widetilde{E}/(1-v^2)$ is used. N_t is the axial compressive force caused by the heating effect of thermal excitation

$$N_t=-Ebh\alpha_L\Delta T\tag{4}$$

where ΔT is the average temperature elevation integrated over the bridge given by[14].

$$\Delta T=\frac{lPs}{2bh\lambda}\left[\frac{\hat{x}_0}{l}-\left(\frac{\hat{x}_0}{l}\right)^2\right]\tag{5}$$

The internal force N_p in the bridge is induced by the measured pressure p[15].

$$N_p=\frac{49bh(1-v^2)}{96}\left(\frac{L}{H}\right)^2\left[1-\left(\frac{1}{L}\right)^2\right]p\tag{6}$$

N_r in Eq. (3a) is the residual stress induced axial force.

$M_{\mathrm{th1}}(\hat{x},\ \hat{t})$ and $M_{\mathrm{th2}}(\hat{x},\ \hat{t})$ in Eq. (3a) are distributed time-dependent driving moments developed respectively by the dynamic thermal components of $p_1(\hat{t})$ and $p_2(\hat{t})$:

$$M_{th1}(\hat{x},\hat{t}) = H(\hat{x})M_{d1}(\hat{t}) \tag{7a}$$

$$M_{th2}(\hat{x},\hat{t}) = H(\hat{x})M_{d2}(\hat{t}) \tag{7b}$$

where

$H(\hat{x})=U[\hat{x}-0.5(l-l_R)]-U[\hat{x}-0.5(l+l_R)]$ is constructed by the unit step function $U(\hat{x})$.

$M_{d1}(\hat{t})$ and $M_{d2}(\hat{t})$ are the mechanical moments developed by the temperature gradient in the thickness direction of the bridge, which are given by[16]

$$M_{d1}(\hat{t}) = \frac{24U_{dc}U_{ac}EJ\alpha_L M(\hat{\omega})}{\lambda Rbl_R}\cos[\hat{\omega}\hat{t} - \varphi_d(\hat{\omega})] \tag{8a}$$

$$M_{d2}(\hat{t}) = \frac{6U_{ac}^2 EJ\alpha_L M(2\hat{\omega})}{\lambda Rbl_R}\cos[2\hat{\omega}\hat{t} - \varphi_d(2\hat{\omega})] \tag{8b}$$

where

$$M(\hat{\omega}) = \left|\frac{i}{2\gamma^2}\left\{\frac{1}{2}+\frac{1}{\gamma(1+i)}-\frac{2}{\gamma(1+i)[1+\exp(-\gamma(1+i))]}\right\}\right| \tag{8c}$$

$$\varphi_d(\hat{\omega}) = -\arctan\frac{1-\gamma-(1+\gamma)e^{-2\gamma}-2e^{-\gamma}[\gamma\cos\gamma-\sin\gamma]}{1-e^{-2\gamma}-2e^{-\gamma}\sin\gamma} \tag{8d}$$

In Eqs. (8c) and (8d), $\gamma=h/\delta$, where $\delta=\sqrt{2\lambda/\rho Cw}$ is the penetration depth of the temperature distribution in the bridge thickness direction.

In this work, the non-linear governing equations (Eqs. (3a) and (3b)) is solved analytically. Galérkin procedure is used to discretize that partial differential equation into a reduced order model. Then the method of multiple scales is utilized to find a uniform approximate solution. It is convenient for the multiple scales method to introduce dimensionless quantities, since such formulation will facilitate the identification of the order of magnitude of some variables. Hence let

$$w = \frac{\hat{w}}{r},\quad x = \frac{\hat{x}}{l},\quad t = \hat{\omega}_1\hat{t}_1,\quad \omega = \frac{\hat{\omega}}{\hat{\omega}_1},\quad r = \sqrt{\frac{J}{bh}} \tag{9}$$

where

$\hat{\omega}_1$ is the fundamental natural frequency of the bridge underlinear vibration[11].

$$\hat{\omega}_1 = \frac{4.73^2}{l^2}\left(\frac{EJ}{\rho bh}\right)^{0.5}\left[1+\frac{0.295(N_t+N_p+N_r)l^2}{12EJ}\right]^{0.5} \tag{10}$$

In non-dimensional form, Eqs. (3a) and (3b) become

$$\frac{EJr}{l^4}\frac{\partial^4 w}{\partial x^4}+\rho bhr\hat{\omega}_1^2\frac{\partial^2 w}{\partial t^2}+cr\hat{\omega}_1\frac{\partial w}{\partial t} =$$

$$\frac{r}{l^2}\left[\frac{Ebhr^2}{2l^2}\int_0^1\left(\frac{\partial w}{\partial x}\right)^2 \mathrm{d}x + N_t + N_p + N_r\right]\frac{\partial^2 w}{\partial x^2} -$$

$$\left[\delta_{-1}\left(\frac{2xl - l + l_R}{2}\right) - \delta_{-1}\left(\frac{2xl - l - l_R}{2}\right)\right][M_{\mathrm{d1}}(t) + M_{\mathrm{d2}}(t)] \tag{11a}$$

$$w(0,t) = w(1,t) = \frac{\partial w(0,\ t)}{\partial x} = \frac{\partial w(1,t)}{\partial x} = 0 \tag{11b}$$

where

$$M_{\mathrm{d1}}(t) = \frac{24U_{\mathrm{dc}}U_{\mathrm{ac}}EJ\alpha_L M(\hat{\omega})}{\lambda Rbl_R}\cos[\omega t - \varphi_{\mathrm{d}}(\hat{\omega})] \tag{12a}$$

$$M_{\mathrm{d2}}(t) = \frac{6U_{\mathrm{ac}}^2 EJ\alpha_L M(2\hat{\omega})}{\lambda Rbl_R}\cos[2\omega t - \varphi_{\mathrm{d}}(2\hat{\omega})] \tag{12b}$$

and $\delta_{-1}(x)$ is the first derivative of Dirac function. The solution to Eqs. (11a) and (11b) can be expressed as

$$w(x,\ t) = \sum_{i=1}^{n} u_i(t)\phi_i(x) \tag{13}$$

where

$\phi_i(x)$ is the ith linear undamped mode shape of the bridge governed by

$$EJ\phi_i^{iv}(x) = \rho bhl^4\hat{\omega}_i^2\phi_i(x) + (N_t + N_p + N_r)l^2\phi_i''(x) \tag{14a}$$

$$\phi_i = 0 \text{ and } \phi_i' = 0 \quad \text{at } x = 0 \text{ and } x = 1 \tag{14b}$$

where

$\hat{\omega}_i$ is the ith natural frequency of the bridge.

Substituting Eq. (13) into Eq. (11a), using Eq. (14a) to eliminate ϕ_i^{iv}, multiplying the result by $\phi_i(x)$, and integrating the outcome from $x=0$ to 1, a set of coupled ordinary differential equations in time are obtained.

Since the frequency $\hat{\omega}$ of the driving moment $M_{\mathrm{th1}}(\hat{x},\ \hat{t})$ is set approximately to the fundamental natural frequency $\hat{\omega}_1$ of the bridge by the PLL circuit, it is reasonable to assume that the fundamental mode should be dominant in the vibration (This assumption will be numerically validated later.). Hence let $n=1$ in Eq. (13) and then a reduced order model of the bridge is obtained:

$$\ddot{u}_1(t) + \frac{1}{Q_1}\dot{u}_1(t) + u_1(t) + \varepsilon u_1^3(t) =$$

$$F_1(\hat{\omega})\cos[\omega t - \varphi_{\mathrm{d}}(\hat{\omega})] + F_2(2\hat{\omega})\cos[2\omega t - \varphi_{\mathrm{d}}(2\hat{\omega})] \tag{15a}$$

where

$$Q_1 = \frac{\rho bh\hat{\omega}_1}{c} \tag{15b}$$

$$\varepsilon = \frac{\zeta E r^2}{2\rho \hat{\omega}_1^2 l^4} \tag{15c}$$

$$F_1(\hat{\omega}) = -\frac{24 U_{ac} U_{dc} E J \alpha_L \phi_1''(x_0)}{\lambda P b^2 l^3 \rho h r \hat{\omega}_1^2 \int_0^1 \phi_1(x)^2 \mathrm{d}x} M(\hat{\omega}) \tag{15d}$$

$$F_2(2\hat{\omega}) = -\frac{6 U_{ac}^2 E J \alpha_L \phi_1''(x_0)}{\lambda P b^2 l^3 \rho h r \hat{\omega}_1^2 \int_0^1 \phi_1(x)^2 \mathrm{d}x} M(2\hat{\omega}) \tag{15e}$$

The coefficient ζ in Eq. (15c) is a constant given by

$$\zeta = -\frac{\int_0^1 \phi_1''(x)\phi_1(x)\mathrm{d}x \int_0^1 \phi_1'^2(x)\mathrm{d}x}{\int_0^1 \phi_1^2(x)\mathrm{d}x} \approx 60.008\ 7 \tag{16}$$

Typically, ε is small ($\varepsilon = 0.109\ 8$, when $\hat{\omega}_1 = 2\pi \times 40\ 000$ rad/s). Thus, ε can be used as the small perturbation parameter in the subsequent perturbation analysis. Q_1 is the quality factor of the bridge for the fundamental mode. Note that since normal operation is under vacuum, $1/Q_1 \ll 1$ can be expressed in terms of the small perturbation parameter ε

$$\frac{1}{Q_1} = 2\varepsilon\mu \tag{17}$$

where

$$\mu = \frac{c\hat{\omega}_1 l^4}{\zeta E b h r^2} \tag{18}$$

Since near-resonant behavior is the operating regime of the pressure sensor, a detuning parameter σ is introduced, as is given by

$$\omega = 1 + \varepsilon\sigma \tag{19}$$

In addition, set $F_1(\hat{\omega}) = \varepsilon K_1(\hat{\omega})$ and $F_2(2\hat{\omega}) = \varepsilon K_2(2\hat{\omega})$, where

$$K_1(\hat{\omega}) = \frac{48 U_{ac} U_{dc} J \alpha_L l \phi_1''(x_0)}{\lambda R b^2 h r^3 \int_0^1 \phi_1''(x)\phi_1(x)\mathrm{d}x \int_0^1 \phi_1'^2(x)\mathrm{d}x} M(\hat{\omega}) \tag{20a}$$

$$K_2(2\hat{\omega}) = \frac{12 U_{ac}^2 J \alpha_L l \phi_1''(x_0)}{\lambda R b^2 h r^3 \int_0^1 \phi_1''(x)\phi_1(x)\mathrm{d}x \int_0^1 \phi_1'^2(x)\mathrm{d}x} M(2\hat{\omega}) \tag{20b}$$

so that the generalized load, the damping and the non-linear term in Eq. (15a) have the same order with respect to the perturbation parameter ε. Expanding $F_1(\hat{\omega})$ and $F_2(2\hat{\omega})$ in Taylor series around $\hat{\omega}_1$ and $2\hat{\omega}_1$ respectively yields

$$F_1(\hat{\omega}) = \varepsilon[K_1(\hat{\omega}_1) + \varepsilon K_1'(\hat{\omega}_1)\hat{\omega}_1\sigma + O(\varepsilon^2)] \tag{21a}$$

$$F_2(2\hat{\omega}) = \varepsilon[K_2(2\hat{\omega}_1) + 2\varepsilon K_2'(2\hat{\omega}_1)\hat{\omega}_1\sigma + O(\varepsilon^2)] \tag{21b}$$

Substituting Eqs. (17), (19), (21a) and (21b) into (15a) yields

$$\begin{aligned}&\ddot{u}_1(t)+2\varepsilon\mu\dot{u}_1(t)+u_1(t)+\varepsilon u_1^3(t)=\\&[\varepsilon K_1(\hat{\omega}_1)+\varepsilon^2K_1'(\hat{\omega}_1)\hat{\omega}_1\sigma+O(\varepsilon^3)]\cos[(1+\varepsilon\sigma)t-\varphi_d(\hat{\omega})]+\\&[\varepsilon K_2(2\hat{\omega}_1)+2\varepsilon^2K_2'(2\hat{\omega}_1)\hat{\omega}_1\sigma+O(\varepsilon^3)]\cos[2(1+\varepsilon\sigma)t-\varphi_d(2\hat{\omega})]\end{aligned}\tag{22}$$

The multiple scales method[17] is used to find the approximate solution to Eq. (22). To this end, the solution is expressed in terms of different time scales as

$$u_1(t;\varepsilon)=u_{10}(T_0,T_1)+\varepsilon u_{11}(T_0,T_1)+\cdots\tag{23}$$

where $T_0=t$ and $T_1=\varepsilon t$. Substituting Eq. (23) into Eq. (22) and equating the coefficients of ε^0 and ε on both sides gives

$$D_0^2u_{10}+u_{10}=0\tag{24a}$$

$$\begin{aligned}D_0^2u_{11}+u_{11}&=-2D_0D_1u_{10}-2D_0u_{10}-u_{10}^3+K_1(\hat{\omega}_1)\cos(T_0+\sigma T_1-\varphi_d\hat{\omega})+\\&K_2(2\hat{\omega}_1)\cos[2T_0+2\sigma T_1-\varphi_d(2\hat{\omega})]\end{aligned}\tag{24b}$$

The general solution of Eq. (24a) can be written as

$$u_{10}=A(T_1)\exp(\mathrm{i}T_0)+\widetilde{A}(T_1)\exp(-\mathrm{i}T_0)\tag{25}$$

where $A(T_1)$ will be determined later by eliminating the secular terms from u_{11}. Substituting Eq. (25) into Eq. (24b) yields

$$\begin{aligned}D_0^2u_{11}+u_{11}=&-[2i(A'+\mu A)+3A^2\widetilde{A}]\exp(iT_0)-A^3\exp(3iT_0)+\\&\frac{1}{2}K_1(\hat{\omega}_1)\exp\{\mathrm{i}[T_0+\sigma T_1-\varphi_d(\hat{\omega})]\}+\frac{1}{2}K_2(2\hat{\omega}_1)\cdot\\&\exp\{\mathrm{i}[2T_0+2\sigma T_1-\varphi_d(2\hat{\omega})]\}+cc\end{aligned}\tag{26}$$

where cc stands for the complex conjugate of the preceding terms. For a bounded solution the secular terms must be cancelled in Eq. (26). This gives

$$2\mathrm{i}(A'+\mu A)+3A^2\widetilde{A}-\frac{1}{2}K_1(\hat{\omega}_1)\exp\{\mathrm{i}[\sigma T_1-\varphi_d(\hat{\omega})]\}=0\tag{27}$$

where A can be expressed as $A=1/2(a\exp(\mathrm{i}\beta))$, in which a and β are real. Substituting $A=1/2(a\exp(\mathrm{i}\beta))$ into Eq. (27) and then separating the result into its real and imaginary parts yields

$$a'=-\mu a+\frac{1}{2}K_1(\hat{\omega}_1)\sin[\sigma T_1+\varphi_d(\hat{\omega})-\beta]\tag{28a}$$

$$a\beta'=\frac{3}{8}a^3-\frac{1}{2}K_1(\hat{\omega}_1)\cos[\sigma T_1+\varphi_d(\hat{\omega})-\beta]\tag{28b}$$

Substituting $A=1/2(a\exp(\mathrm{i}\beta))$ into Eq. (25) and substituting that result into Eq. (23) obtains the first approximation

$$u_1=a\cos(t+\beta)+O(\varepsilon)\tag{29}$$

where a and β are given by Eqs. (28a) and (28b).

Let $\varphi=\sigma T_1-\beta$ and substitute φ into Eqs. (28a) and (28b)

$$a' = -\mu a + \frac{1}{2}K_1(\hat{\omega}_1)\sin[\varphi+\varphi_d(\hat{\omega})] \tag{30a}$$

$$a\varphi' = \sigma a - \frac{3}{8}a^3 + \frac{1}{2}K_1(\hat{\omega}_1)\cos[\varphi+\varphi_d(\hat{\omega})] \tag{30b}$$

Steady-state motions occur when $a'=\varphi'=0$, which correspond to the solutions of

$$\mu a = \frac{1}{2}K_1(\hat{\omega}_1)\sin[\varphi+\varphi_d(\hat{\omega})] \tag{31a}$$

$$\sigma_a - \frac{3}{8}a^3 = -\frac{1}{2}K_1(\hat{\omega}_1)\cos[\varphi+\varphi_d(\hat{\omega})] \tag{31b}$$

Squaring and adding Eqs. (31a) and (31b), solving the consequent equation for σ as a function of a, and then transforming the result into a dimensional form gives

$$\hat{\omega} = \hat{\omega}_1 + \frac{3\hat{a}^2}{8r^2}\hat{\omega}_1\varepsilon \pm \left[\frac{K_1^2(\hat{\omega}_1)r^2}{4\hat{a}^2} - \mu^2\right]^{0.5}\hat{\omega}_1\varepsilon \tag{32}$$

Eq. (32) is an implicit equation for the amplitude response $\hat{a}$ of the bridge as a function of the exciting frequency $\hat{\omega}$.

Eliminating a in Eq. (31b) by Eq. (31a), and transforming that result into a dimensional form gives

$$\hat{\omega} = \hat{\omega}_1 + \frac{3\varepsilon\hat{\omega}_1 K_1^2(\hat{\omega}_1)}{32\mu^2}\sin^2[\varphi-\varphi_d(\hat{\omega})] - \mu\varepsilon\hat{\omega}_1\cot[\varphi-\varphi_d(\hat{\omega})] \tag{33}$$

Eq. (33) is an implicit equation for the phase response φ of the bridge as a function of the exciting frequency $\hat{\omega}$.

Substituting $\varphi=\sigma T_1-\beta$ and Eq. (19) into Eq. (29) and transforming the result into a dimensional form gives the first approximation to the steady-state solution:

$$\hat{u}_1(\hat{t}) = \hat{a}\cos(\hat{\omega}\hat{t}-\varphi) + O(\varepsilon) \tag{34}$$

Note that the fundamental mode shape $\phi_1(x)$ is normalized such that $\phi_1(0.5)=1$, therefore, the Eq. (34) represents the displacement at the bridge midpoint.

3.2 Numerical simulation

Numerical simulations were performed to justify utilizing only the foundational mode in the Galérkin procedure and to validate the multiple scales method in respect of accuracy. The first five modes of the bridge have been taken into account. Since the driving moments of $M_{th1}(\hat{x}, \hat{t})$ and $M_{th2}(\hat{x}, \hat{t})$ are distributed symmetrically with respect to the centre of the bridge, it is theoretically impossible to excite the anti-symmetric mode 2 and mode 4. Therefore the equation of motion was reduced to three coupled ordinary differential equations and solved numerically using medium order Runge-Kutta solver ode 45 integrated into the Matlab package. Then, the maximum steady-state displacement at certain frequency was extracted from the numerical solution to compare with that calculated by Eq. (32). Fig. 5 shows a good agreement between numerical integration and the analytical prediction, which numerically validates the analytical results.

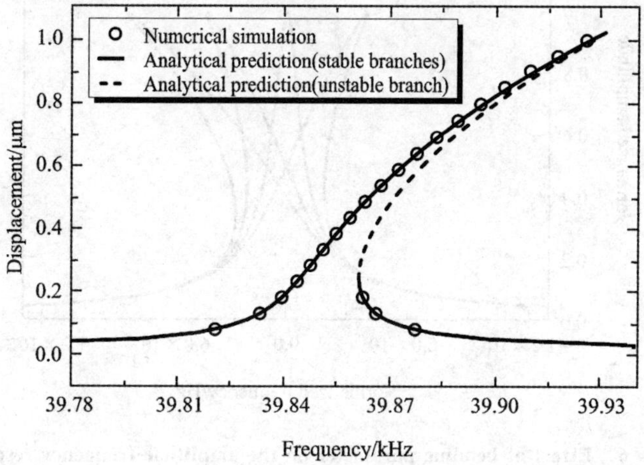

Fig. 5 Numerical validation of the approximate solution to the frequency response. DC bias $U_{dc}=1$ V, AC amplitude $U_{ac}=0.5$ V, resonant frequency $\hat{f}_1=39.8$ kHz, quality factor $Q_1=10^4$

3.3 Bending parameter of the frequency response

Non-linear vibration of resonators results in asymmetric frequency response, which interferes with the operation of resonant sensors. In order to quantitatively evaluate the bending of the asymmetric frequency response under various sensor operating conditions, a bending parameter is introduced. From Eq. (32), the backbone curve (see for instance the dashed line in Fig. 8) of the asymmetric amplitude-frequency response is obtained:

$$\hat{\omega} = \hat{\omega}_1 + \tilde{\alpha}\,\hat{a}^2 \tag{35}$$

where,

$$\tilde{\alpha} = \frac{3\hat{\omega}_1 \varepsilon}{8r^2} \tag{36}$$

The backbone curve is in a parabolic form. The coefficient $\tilde{\alpha}$ is referred to as bending parameter in this paper. The sign of the bending parameter implies bending directions. The amplitude-frequency response with positive bending parameter bends towards higher frequency and the one with negative bending parameter bends towards lower frequency. The absolute value of the bending parameter is in direct proportion to the bending degree of the amplitude-frequency response. Particularly, $\tilde{\alpha}=0$ implies linear vibration. This behavior is illustrated in Fig. 6.

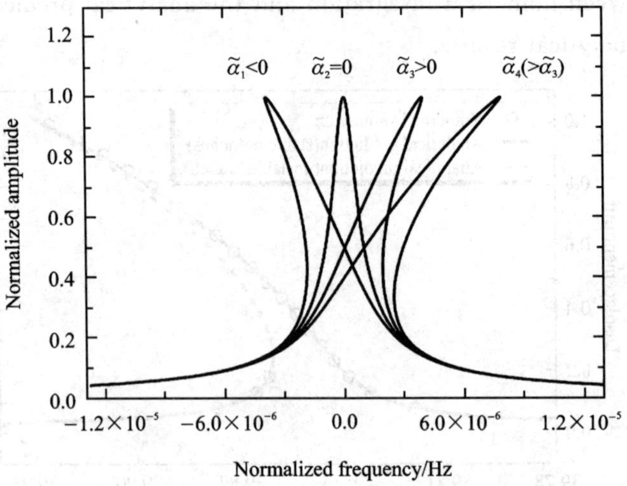

Fig. 6 Effect of bending parameter on the amplitude-frequency response

For the convenience of comparing analytical predictions and experimental observations, rewrite the backbone curve and the bending parameter as

$$\hat{f} = \hat{f}_1 + \alpha I^2 \tag{37}$$

$$\alpha(P_s, p, N_r) = \frac{\tilde{\alpha}}{2\pi S_e^2} = \frac{0.004\zeta E}{\pi^2 l 2 S_e^2}\left[\frac{12A}{12EJ\rho + 0.295(N_t + N_p + N_r)l^2\rho}\right]^{0.5} \tag{38}$$

where

$\hat{f}=\hat{\omega}/2\pi$ and $\hat{f}_1=\hat{\omega}_1/2\pi$.

I is the electric current amplitude in the sense resistor.

S_e is the transfer constant from $\hat{a}$ to I.

From Eq. (38), the bending parameter α is dependent on the sensor operating conditions: It increases with increasing static thermal power P_s, with decreasing measured pressure p and with decreasing axial tensile residual force N_r. The dependence of bending parameter on the static thermal power P_s and the measured pressure p is experimentally validated.

For the resonant sensor with electrothermal excitation and piezoresistive detection, experimental errors could arise from the excitation-to-detection cross-talks. Such cross-talks can be electrical as well as thermal, which has been investigated by Ferrari et al. [18], Aikele et al. [19]. The thermal components of the crosstalks are neglectable or can be easily eliminated by ac coupling. And the electrical components of the cross-talks are found to be dominant especially for a resonator with low quality factor. Refs. [18,19] adopted compensating circuits to eliminate the electrical cross-talks. In the following experiments, a properly designed excitation scheme was utilized to decrease the coupling capacitance crosstalk between the drive and the sense resistors. To this end, the ac exciting signal $U_{ac}\cos\hat{\omega}\hat{t}$ is split into two components of $0.5U_{ac}\cos\hat{\omega}\hat{t}$ and $-0.5U_{ac}\cos\hat{\omega}\hat{t}$ which were respectively applied to the two ends of the drive resistor. This excitation scheme was proven to be able to reduce the coupling capacitance crosstalk by one order of magnitude.

In Fig. 7(a), the dc bias and the ac exciting signal applied on the drive resistor are adjusted in a way in which the amplitude of the harmonic thermal power $p_1(\hat{t})$ is maintained at

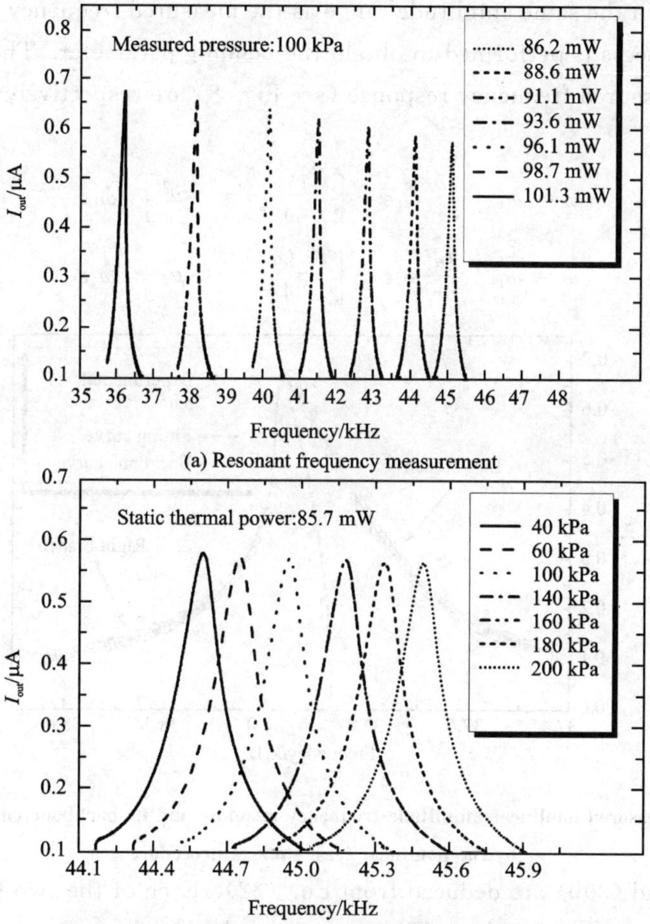

(a) Resonant frequency measurement

(b) Experimental result when pressure ranges from 40 to 200 kPa applied

Fig. 7 Measured amplitude-frequency responses of the bridge under different static thermal power and measured pressure

constant 91.6 mWand the static thermal power P_s is increased from 86.2 mW to 101.3 mW. In Fig. 7(b), the excitation is fixed and the measured pressure ranges from 40 kPa to 200 kPa. The bending parameters of these measured frequency responses could be obtained by the non-linear least squares procedure which is illustrated in Fig. 8. The natural frequency $\hat{\omega}_1$, the small parameter ε, the parameter μ as a function of the quality factor, the generalized load $K_1(\hat{\omega}_1)$ in Eq. (32), and the transfer constant S_e in Eq. (38) are selected as the fitted parameters. The initial magnitudes of $\hat{\omega}_1$, ε and $K_1(\hat{\omega}_1)$ used in the fitting process are respectively evaluated by Eqs. (10), (15c) and (20a) and the parameters listed in Tab. 1. The initial magnitude of μ is found by Eq. (17) and the measured quality factor. The initial magnitude of S_e is determined so that the theoretical resonant peak in terms of electric current is located within the same amplitude range as the measured frequency response. Then, a two-step fitting process is performed to obtain the bending parameter. The left and the right branches of the measured frequency response (see Fig. 8) are respectively fitted to their analytical expressions:

$$\hat{\omega} = \hat{\omega}_1 + \frac{3\hat{a}^2}{8r^2}\hat{\omega}_1\varepsilon - \left[\frac{K_1^2(\hat{\omega}_1)r^2}{4\hat{a}^2} - \mu^2\right]^{0.5}\hat{\omega}_1\varepsilon \tag{39a}$$

$$\hat{\omega} = \hat{\omega}_1 + \frac{3\hat{a}^2}{8r^2}\hat{\omega}_1\varepsilon + \left[\frac{K_1^2(\hat{\omega}_1)r^2}{4\hat{a}^2} - \mu^2\right]^{0.5}\hat{\omega}_1\varepsilon \tag{39b}$$

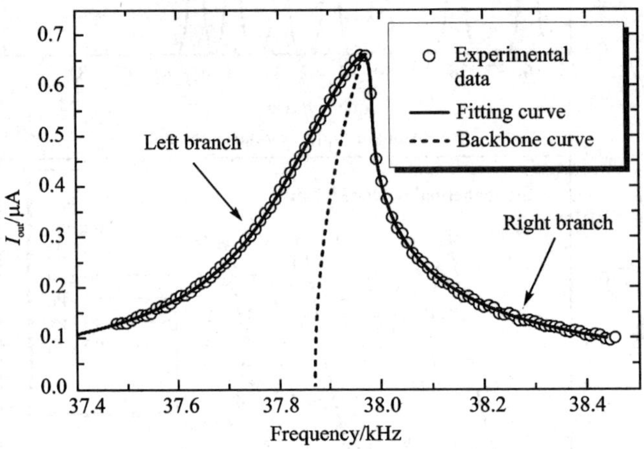

Fig. 8 Measured nonlinear amplitude-frequency response and its backbone curve obtained by the nonlinear least squares procedure

Eqs. (39a) and (39b) are deduced from Eq. (32). Each of the two fitting steps gives a group of the fitted parameter values. The average of the fitted parameter values given by the two fitting steps is utilized to plot the fitting curve and to determine if the desired level of fitting precision has been obtained. If the fitted parameter values are not determined to sufficient precision, the fitting process is performed again by using the current parameter values for the initial values. If the desired fitting precision is achieved, the value of the bending parameter is obtained by substituting these fitted parameters into Eq. (38).

Figs. 9 and 10 indicate that the fitted bending parameter increases with increasing static thermal power and with decreasing measured pressure, showing good agreement qualitatively with the analytical results. This behavior draws some interesting points on the sensor design and operation, which will be discussed later in Section 4.

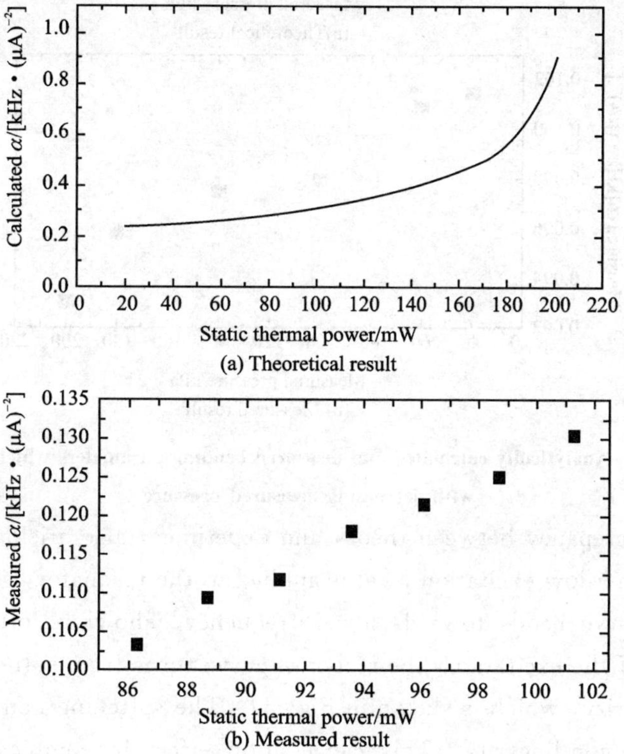

Fig. 9 Analytically calculated and measured bending parameters which increase with increasing static thermal power

Quantitative discrepancy in Figs. 9 and 10 may come from the non-ideal clamped boundary conditions of the bridge and the softening non-linearity. On the one hand, the axial degree of freedom at the clamped ends cannot be absolutely constrained in practice. Small axial displacements at the bridge ends are allowed in the vibration, which relieves the stretch of the bridge midplane, hence less hardening non-linearity than the model prediction[20]. On the

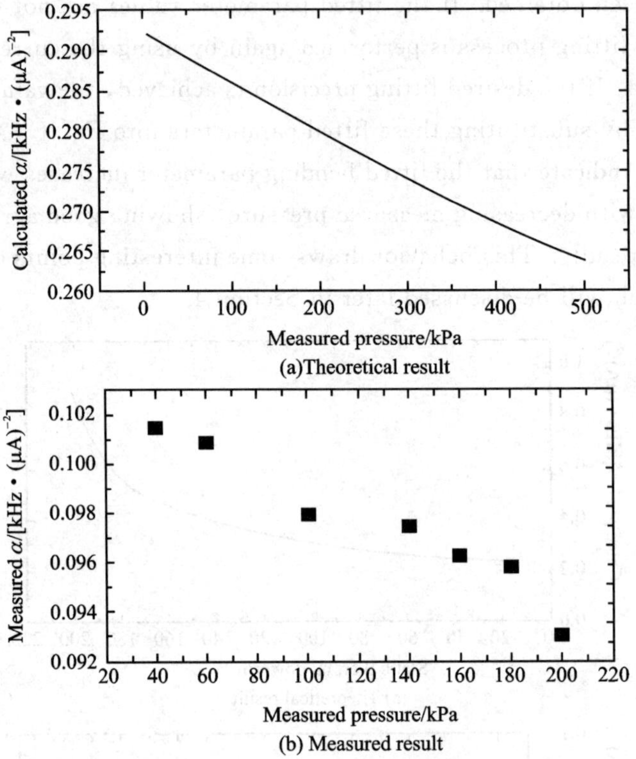

Fig. 10 Analytically calculated and measured bending parameters which increase with decreasing measured pressure

other hand, the discrepancy between theory and experiment also arising from the softening non-linearity. When a low excitation level is applied to the resonator, somewhat surprisingly, the response curve bends towards lower frequency, showing softening non-linearity. With the increase of the excitation, the response curve turns from softening non-linearity to hardening non-linearity, which is shown in Fig. 11. The softening non-linearity partly cancels the hardening non-linearity. This cancellation effect in combination with the axial boundary displacement at the bridge clamped ends makes the bending parameter measured from the experiments lower than that predicted by the model.

The softening non-linearity in the electrostatically excited resonators is attributed to the displacement dependence of the electrostatic force[7]. However the driving moment of thermal excitation is usually considered as displacement-independent[16]. Therefore, the observed softening non-linearity in Fig. 11 is unlikely caused by the transduction from the thermal power to the driving moment. Mahmoodi et al.[6] investigated the non-linearity of a piezo-

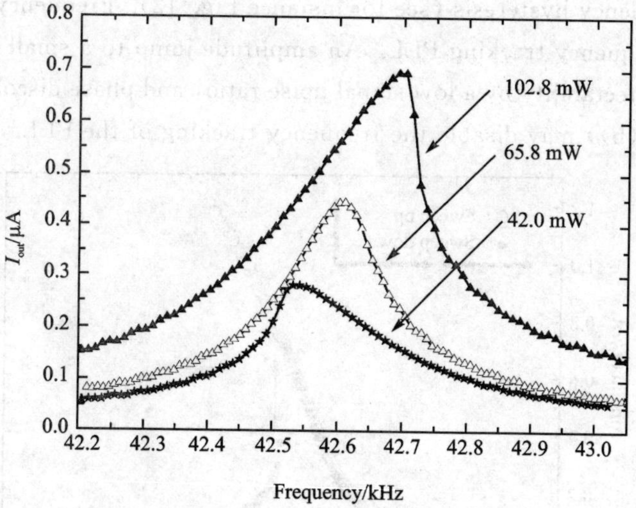

Fig. 11　Change from soft-spring non-linearity to hard-spring non-linearity with the increase of with excitation level. The dc bias and the ac voltage are adjusted in a way in which the static thermal power is fixed at 90 mW, and the dynamic thermal power with frequency $\hat{\omega}$ is increased from 42.0 mW to 102.8 mW. The measured pressure is 100 kPa

electrically actuated microcantiliever with softening behaviors, which was attributed to the quadratic material non-linearity of the piezoelectric layer. Whereas the material non-linearity in the silicon is usually regard as neglectable[3,6]. So there is little possibility of the measured softening response arising from the material non-linearity of the bridge. Comi[21] found that slightly changing the resonator geometry may cause completely different behavior. For an "L-shaped" beam resonator, the axial internal force is initially compressive. Then, under large vibration amplitude, the tensile axial internal force becomes dominant (see Fig. 10 in Ref. [21]). That is to say, the "L-shaped" beam resonator takes on softening non-linearity under small vibration amplitude and becomes hardening non-linearity when the amplitude is large. This transition from softening to hardening response with the increase of the amplitude is very similar to that observed in Fig. 11. So the measured softening non-linearity most likely comes from the geometrical defect of the bridge due to the manufacturing process. Though there exists quantitative discrepancy between the theory and the experiment due to the reasons aforementioned, the trend of the bending parameter changing with the thermal effect and the measured pressure was validated.

3.4　Critical vibration amplitude under electrothermal excitation

Beyond a critical vibration amplitude the response curves of the bridge become triple valued, known as frequency hysteresis (see for instance Fig. 12). Frequency hysteresis has significance for the frequency tracking PLL. An amplitude jump to a small value (Fig. 12(a)) reduces the PLL detectability by a low signal noise ratio, and phase discontinuity close to the resonance (Fig. 12(b)) may disable the frequency tracking of the PLL.

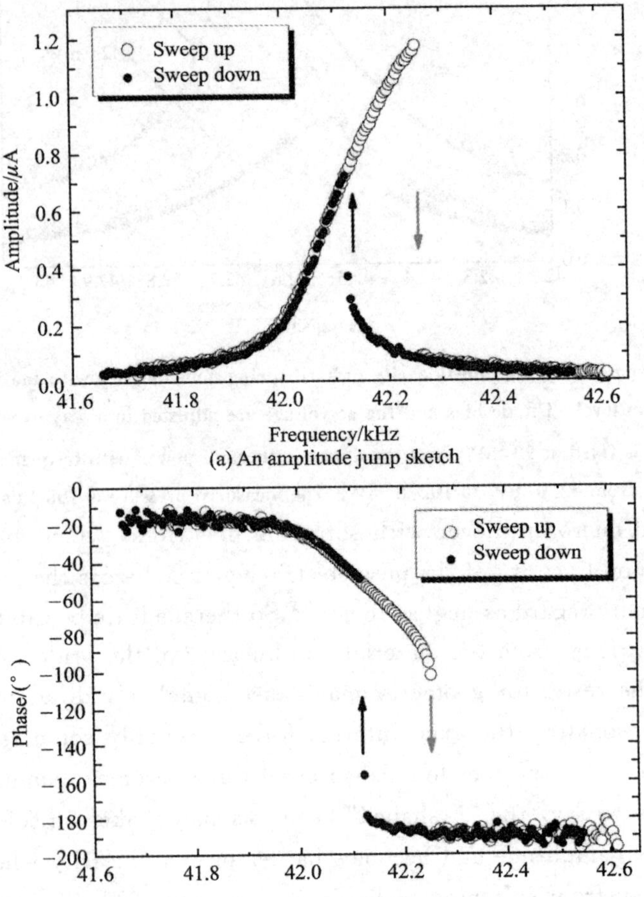

Fig. 12 Measured amplitude and phase jump phenomena of nonlinear frequency response of the bridge

Investigations on the critical amplitude of electrostatically excited beam resonators have been reported. Gui et al. [22] deduced a maximum allowable excitation level, beyond which frequency hysteresis occurs. Kaajakari et al. [23] examined the critical vibration amplitude of a beam resonator with the crystalline directions taken into account. Kacem et al. [7] calculated the critical amplitude with the complete contribution of electrostatic non-linearities. These efforts neglected the internal force in the beam. This is reasonable to some extent, for the internal force in an electrostatically excited resonator is usually tensile in natural. Tensile internal force leads to a smaller bending parameter according to Eq. (38), hence a higher actual critical amplitude than their predictions. Therefore their efforts define a safe operating region for the electrostatically excited resonators. However heating effect of the electrothermal excitation develops compressive axial force, which should be consideredin the critical amplitude prediction.

From Eq. (32), the maximum vibration amplitude can be expressed as

$$\hat{a}_{\max} = \frac{r}{2\mu} K(\hat{\omega}_1) \tag{40}$$

A critical value $K_C(\hat{\omega}_1)$ exists. For all $K(\hat{\omega}_1) > K_C(\hat{\omega}_1)$, frequency hysteresis occurs. The critical vibration amplitude $\hat{a}_C$ can be determined by substituting $K_C(\hat{\omega}_1)$ into Eq. (40). When $K(\hat{\omega}_1) = K_C(\hat{\omega}_1)$, the resonance curve exhibits one and only point of infinite slope, which means that $K_C(\hat{\omega}_1)$ can be determined by differentiation of Eq. (32) with respect to $\hat{a}^2$ and setting $\mathrm{d}\hat{\omega}/\mathrm{d}\hat{a}^2 = 0$. Calculating along this line gives

$$\hat{a}_C = 22.666 \frac{h}{\sqrt{Q_1}} \sqrt{\frac{1}{3\zeta} + \frac{0.295(N_t + N_p + N_r)l^2}{36\zeta EJ}} \tag{41}$$

Note that since the heating effect caused compressive force N_t is negative, Eq. (41) implies that the heating effect of electrothermal excitation decreases the critical vibration amplitude, making the resonator more susceptible to hysteresis. Moreover, high measured pressure brings on large critical amplitude by increasing the tensile force N_p, which expands the hysteresis-free operating region.

In a given sensor operating condition, for example $P_s = 85.7\,\mathrm{mW}$, $p = 100$ kPa, $N_r = 0.0237$ N and $Q_1 = 10^4$, the critical amplitude is about 0.233 μm. Neglecting the internal force in the bridge, i.e. $N_t = N_p = N_r = 0$ in Eq. (41), gives $\hat{a}_C = 1.689(h/\sqrt{Q_1})$ which is very close to $1.685(h/\sqrt{Q_1})$ presented in Ref. [7].

3.5 Frequency output error of the sensor caused by the non-linearity

In linear limits, the PLL tracks the natural frequency of the bridge by maintaining a phase difference of π/2 between excitation and bridge vibration. However, in the non-linear region the phase difference of π/2 between excitation and bridge vibration cannot guarantee the vibration at natural frequency accurately due to the distorted phase response. Moreover, for the electrothermal excitation, the phase lag in the transition from exciting voltage to mechanical moment also induces deviation from the natural frequency.

A comparison between the thermal power $p_1(\hat{t})$ (Eq. (2b)) and the bridge vibration $\hat{u}_1(\hat{t})$ (Eq. (34)) shows that φ denotes the phase shift from the excitation to the vibration. Assuming a neglectable phase shift of the sense resistor, the PLL maintains $\varphi=\pi/2$. Substituting $\varphi=\pi/2$ into the bridge phase response (Eq. (33)) yields

$$\hat{\omega} = \hat{\omega}_1 + \frac{3\varepsilon\hat{\omega}_1 K^2(\hat{\omega}_1)}{32\mu^2}\cos^2[\varphi_d(\hat{\omega})] - \frac{\tan[\varphi_d(\hat{\omega})]}{2Q_1}\omega_1 \tag{42}$$

The frequency output error can be written as

$$\frac{\hat{\omega}-\hat{\omega}_1}{\hat{\omega}_1} = E_{r1} + E_{r2} \tag{43a}$$

where

$$E_{r1} = \frac{3}{8}\varepsilon^3 Q_1^2 K^2(\hat{\omega}_1)\cos^2[\varphi_d(\hat{\omega})] \approx \frac{3}{8}\varepsilon^3 Q_1^2 K^2(\hat{\omega}_1) \tag{43b}$$

$$E_{r2} = -\frac{\tan[\varphi_d(\hat{\omega})]}{2Q_1} \tag{43c}$$

E_{r1} and E_{r2} are the error components caused by the nonlinearity and phase shift of the thermal excitation respectively. For a resonant pressure sensor with high quality factor Q_1, the error component E_{r2} is small and E_{r1} is dominant in the sensor frequency output error. Fig. 13 shows E_{r1} which increases with increasing static thermal power P_s, with decreasing measured pressure p and with decreasing tensile residual force N_r in the bridge.

4 Suggestions and Conclusions

Several suggestions on the resonant pressure sensor design and operation respect to non-linear effect are proposed:

- Since high measured pressure results in small bending parameter, moderately large vibration amplitude of the bridge is allowed in high-pressure measurement to improve the detectability.
- Under the same vibration amplitude, i. e. the same detectability, increase in Q-factor of the bridge means decrease in static thermal power, hence small bending parameter. In this sense, high Q-factor leads to low non-linearity.
- Though large tensile residual force in the bridge brings on small frequency output error, increase in tensile residual force results in stiffening of the bridge, hence decrease in vibration amplitude. To maintain the detectability, thermal excitation power should be increased, which in return increases the non-linear effects. Thus, high

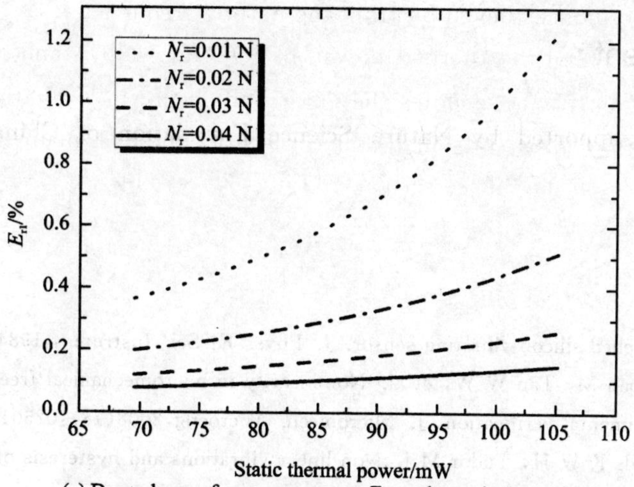

(a) Dependence of error component E_{r1} on the static thermal power

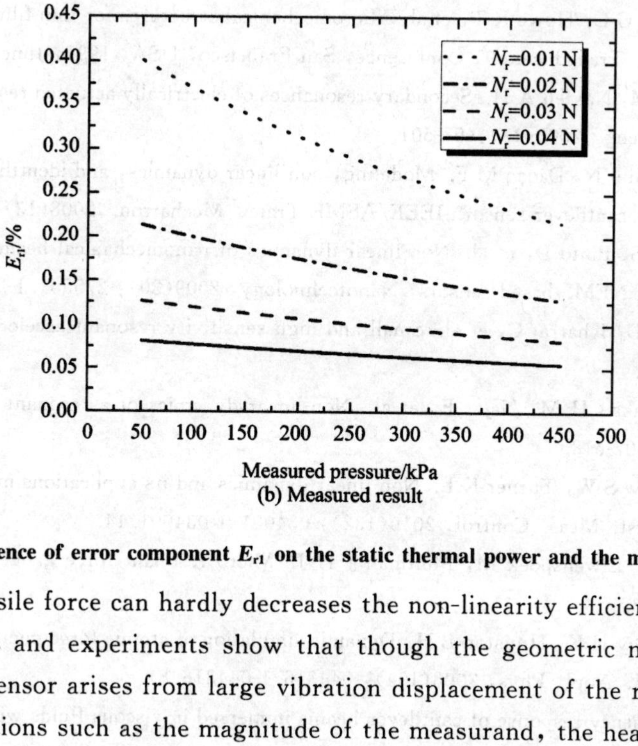

(b) Measured result

Fig. 13 Dependence of error component E_{r1} on the static thermal power and the measured pressure

residual tensile force can hardly decreases the non-linearity efficiently.

Model analyses and experiments show that though the geometric non-linearity of the resonant pressure sensor arises from large vibration displacement of the resonator, the sensor operating conditions such as the magnitude of the measurand, the heating effect of thermal excitation, and the residual internal force in the resonator also significantly influence the non-linear behavior of the resonant sensor. In order to reduce the non-linear effects, comprehensive considerations should be made in the sensor design and operation.

Acknowledgement

This study is supported by Nature Science Foundation of China (grant number: 60927005).

References

[1] Greenwood J C. Etched silicon vibrating sensor. J. Phys. E: Sci. Instrum., 1984(17): 650-652.

[2] Shao L C, Palaniapan M, Tan W W, et al. Nonlinearity in micromechanical free-free beam resonators: modeling and experimental verification, J. Micromech. Microeng,2008(18):025017. 1-025017. 11.

[3] Andres M V, Foulds K W H, Tudor M J. Non-linear vibrations and hysteresis of micromachined silicon resonators designed as frequency-out sensors, Electron. Lett. 1987(23): 952-954.

[4] Pratt R I, Johnson G C, Howe R T, et al. Micromechanical structures for thin film characterization// Digest Tech. Papers. Transducers 91 Conference, San Francisco, USA, 1991, June 24-27.

[5] Abdel-Rahman E M, Nayfeh A H. Secondary resonances of electrically actuated resonant microsensors. J. Micromech. Microeng. ,2003(13):497-501.

[6] Mahmoodi S N, Jalili N, Daqaq M F. Modeling, non-linear dynamics, and identification of a piezoelectrically actuated microcantilever sensor. IEEE/ASME Trans. Mechatron. ,2008(13): 58-65.

[7] Kacem N, Hentz S, Pinto D, et al. Non-linear dynamics of nanomechanical beam resonators: improving the performance of NEMS-based sensors, Nanotechnology, 2009(20): 275501. 1-275501. 11.

[8] Pinto D, Mercier D, Kharrat C, et al. Small and high sensitivity resonant accelerometer. Proc. Chem. , 2009(1): 536-539.

[9] Nayfeh A H, Ouakad H M, Najar F, et al. Non-linear dynamics of a resonant gas sensor. Non-linear Dyn, 2010(59): 607-618.

[10] Rhoads J F, Shaw S W, Turner K L. Non-linear dynamics and its applications in micro-and nanoresonators. J. Dyn. Syst. Meas. Control, 2010(132): 034001. 1-034001. 14.

[11] Tilmans H A C, Elwenspoek M, Fluitman J H J. Micro resonant force gauges. Sens. Actuators A, 1992 (30): 35-53.

[12] Martin M J, Fathy H K, Houston B H. Dynamic simulation of atomic force microscope cantilevers oscillating in liquid. J. Appl. Phys, 2008(104): 044316. 1-044316. 8.

[13] Sader J E. Frequency response of cantilever beams immersed in viscous fluids with applications to the atomic force microscope. J. Appl. Phys, 1998 (84): 64-76.

[14] Lammerink T S J, Elwenspoek M, Van Ouwerkerk R H, et al, Fluitman J H J. Performance of thermally excited resonators. Sens. Actuators A, 1990(21-23): 352-356.

[15] Xing W W. Parameter recognition layer of silicon MEMS resonant sensors: theory & realization. Doctor-

ate dissertation of Beihang University, 2007.

[16] Lammerink T S J, Elwenspoek M, Fluitman J H J. Frequency dependence of thermal excitation of micromechanical resonators, Sens. Actuators A,1991(25-27): 685-689.

[17] Nayfeh A H, Mook D T. Nonlinear Oscillations. John Wiley and Sons, 1979: 56-59.

[18] Ferrari V, Ghisla A, Marioli D, et al. Silicon resonant accelerometer with electronic compensation of input-output cross-talk. Sens. Actuators A, 2005(123-124): 258-266.

[19] Aikele M, Bauer K, Ficker W, et al. Resonant accelerometer with self-test. Sens. Actuators A,2001(92): 161-167.

[20] Pakdemirli M, Boyaci H. Non-linear vibrations of a simple-simple beam with a non-ideal support in between. J. Sound Vib., 2003(268): 331-341.

[21] Comi C. On geometrical effects in micro-resonators. Latin Am J. Solids Struct, 2009(6): 73-87.

[22] Gui C Q, Legtenberg R, Tilmans H A C, et al. Nonlinearity and hysteresis of resonant strain gauges. J. Microelectromech. Syst, 1998(7): 122-127.

[23] Kaajakari V, Mattila T, Lipsanen A, et al. Non-linear mechanical effects in silicon longitudinal mode beam resonators, Sens. Actuators A,2004(120): 64-70.

Appendix J
Design and FEM Simulation Study of the Electro-thermal Excitation Resonant Beam with Slit-structure[①]

Shi Huichao　Fan Shangchun　Xing Weiwei　Li Cheng　Sun Jinhao　Jing Zhongwu

Abstract: This paper proposes a slit-resonant beam based on the double clamped resonant beam theory of silicon micro-machined resonant pressure sensors. The slit structure can enhance surface alternating stress in the root area of beam while vibrating. Finite element method (FEM) is applied to calculate the stress concentration magnification, and we carry out a computational study of the effects of size and location of slit structure in terms of mechanical properties of resonant beam, such as stress concentration on resonant beam and natural frequency of resonant beam, as well as sensing performance, such as resonant beam amplitude sensitivity. Our simulations show that the slit structure could strengthen the stress concentration and increase the amplitude detection sensitivity, and the variations of these parameters can substantially influence the performance of slit structure. The research of stress concentration, caused by silt structure, could provide reference for further optimization and design for the resonant beam.

Abbreviations

L	Length of the resonant beam
B	Width of the resonant beam
H	Thickness of the resonant beam
l	Length of slit
b	Width of slit
h	Thickness of slit
s	Space distance of two slits
d	The distance of slit location from the resonant beam clamped end
l/b	Aspect ratio (length/width) of the slit

① Instrumentation and Control Technology (ISICT), 2012 8th IEEE International Symposium on, 2012.

R_{sc}　Average stress magnification radio
R_{rsc}　Reference average stress magnification radio
F_{nfr}　Natural frequency ratio
A_R　Single side cross-sectional area of the U-shape detection resistor
π_L　Longitudinal piezoresistive coefficient
π_T　Transverse piezoresistive coefficient
σ_L　Longitudinal stress of the detection resistor
σ_T　Transverse stress of the detection resistor
S_d　Amplitude detection sensitivity
$\varepsilon(\omega)$　Rate of detection resistance variation

1 Introduction

In recent years, the resonant sensor based on measuring resonant frequency attracts many researchers (Langdon 1985; Bouwstra et al. 1990). Silicon micro-machined resonant pressure sensor measures pressure by the change of nature frequency of the resonant beam. It has good characteristics of frequency output (without A/D converter) and easier interface to computers, long term stability, and high repeatability (Qingfeng et al. 2012); However, the resonator of the device must be excited into vibration and the vibration must be detected by measuring its resonant frequency (Zhangyang et al. 2011).

Excitation unit and detection unit are the key components. Since firstly been used to drive resonant devices (Wdfinger et al. 1968), electro-thermal excitation is believed to have certain advantages in the aspect of simply fabrication. Because of its inherent heat elevation, there are mechanical stresses that influence the resonance frequency and vibration amplitude (Lammerink et al. 1990). The signal obtained by detection resistor is very weak because of small vibration amplitude and being submerged in large background noise, which makes it difficult for signal detection and processing. Although amplification circuit works well, the noise introduced by the amplifier is still very impressive. Therefore, it is necessary to raise the amplitude detection sensitivity, and researching the structure is very meaningful, which could enhance the strength of useful signals and improve the sensitivity of the resonant devices in structural design.

Bashir et al. (2000) introduced width to improve the sensitivity of the micro-cantilever, and the sensitivity is raised by about 2～5 times through the optimization of the slit depth and size. Mo et al. (2003) utilized elliptical holes on the micro-cantilever to improve the detection sensitivity of biochemical stress. Joshi et al. (2007) processed different shapes of pore structure on a micro-cantilever to improve the acceleration detection sensitivity, and this micro-cantilever sensitivity increased by 2 times. Ahmed et al. (2010) put forward surface trenches to improve the sensor performance in the structure of a piezoresistive pressure sensor.

The above effortsmainly focus on the structure to improve the detection sensitivity of cantilever beam. However, there are few literatures studying about double clamped resonant beam. Furthermore, the stress concentration calculation method for rectangular holes like slit structure is very complicated.

In this article, firstly, we theoretically analyzed the mechanism of electrothermal excitation and piezoresistive detection and proposed a new structure to achieve the stress amplification. Then finite element method (FEM) was applied to investigate the stress concentration of silt structure on proposed slit-resonant beam. Finally, the author presented computational result of the effect of size and location of slits structure on the mechanical properties of resonant beam as well as sensing performance, and several suggestions on the further optimization and designof the resonant beam.

2 Working Principle of Resonant Beam

Fig. 1 illustrates the resonant beam schematically. The resonant beam studied in this article is electrothermally excited and piezoresistively detected. Excitation resistor is located in the center of the beam, and detection resistor is in the root area of beam. Alternating thermal power generated by the excitation electrical signal in the excitation resistor drives the beam into vibrating, which produces corresponding alternating stress on the root surface area. The piezoresistive effect of detection resistor is used to sensitive to this alternating stress.

Excitation resistor converts the excitation signal U_e into thermal power P_e, which has been investigated by Lammerink et al. (1990), and then thermal power produces excitation force $F(t)$ in the resonant beam. Set excitation signal to a voltage signal:

$$U_e = U_{dc} + U_{ac}\cos(\omega t + \varphi_e) \tag{1}$$

where

U_{dc} is DC bias,

U_{ac} is the amplitude of the AC component.

ω and φ_e are the frequency and the phase respectively.

Thermal power $P_e(t)$ generated in excitation resistor R_e is composed of a static component p_d and two harmonic components $p_a(\omega)$ and $p_a(2\omega)$ with frequencies of ω and 2ω respec-

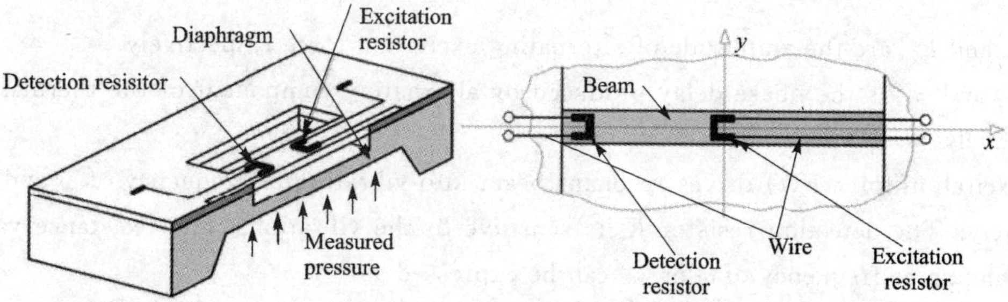

(a) Theoretical sketch

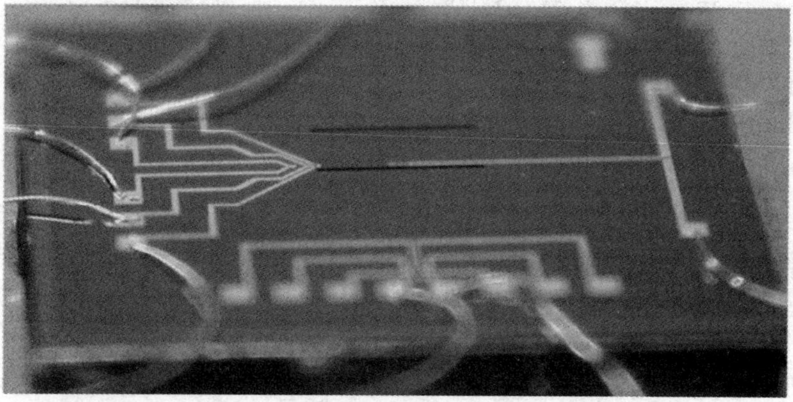

(b) Microscopy sketch of the fabricated beam

Fig. 1 Structure diagram and photograph of the resonant beam in silicon micromachined resonant pressure sensor

tively:

$$p_d = (U_d^2 + U_a^2/2)/R_e \tag{2a}$$

$$p_a(\omega) = 2U_dU_a\cos(\omega t + \varphi_e)/R_e \tag{2b}$$

$$p_a(2\omega) = U_a^2\cos2(\omega t + \varphi_e)/2R_e \tag{2c}$$

The constant component p_d products a constant thermal stress in the resonant beam, which is known as the heating effect of thermal excitation. This stress can also change the natural frequency of the resonant beam, but keeping this stress constant can reduce the measurement error caused by this factor.

The harmonic components $p_a(\omega)$ and $p_a(2\omega)$ generate two alternating excitation forces $F_\omega(t)$ and $F_{2\omega}(t)$. The excitation force $F(t)$ in the resonant beam is

$$F(t) = \begin{cases} F_{2\omega}\cos[2\omega t + 2\varphi_e + \varphi_{T2}(2\omega)] & (U_{dc} = 0) \\ F_\omega\cos[\omega t + \varphi_e + \varphi_{T1}(\omega)] & (U_{dc} \gg U_{ac}) \end{cases} \tag{3}$$

where

F_ω and $F_{2\omega}$ are the amplitude of alternating excitation force respectively.

φ_{T1} and φ_{T2} is the phase delay produced by alternating component of the thermal power respectively.

Excitation force $F(t)$ drives resonant beam into vibrating as frequency of ω and 2ω respectively. The detection resistor R_s is sensitive to the vibration. The resistance variation $r(t)$ changing as frequency of ω or 2ω can be expressed as

$$r(t) = \begin{cases} R_d \cos[2\omega t + 2\varphi_e + \varphi_{T2}(2\omega) + \varphi_s(2\omega)] & (U_{dc} = 0) \\ R_d \cos[\omega t + \varphi_e + \varphi_{T1}(\omega) + \varphi_s(\omega)] & (U_{dc} \gg U_{ac}) \end{cases} \tag{4}$$

where

R_d is the amplitude of the resistance variation $r(t)$,

φ_s is phase shift.

In order to make signal processing convenience, we select signal with the frequency of ω as output signal which is the same frequency to excitation signal. In actual, to avoid the heating effect, thermal power can not be too much, as usually taken $u_d = u_a$. In this case, the signal with the frequency of 2ω is regarded as a disturbance component. But we find that the magnitude of the exchange thermal power of frequency ω is 4 times of that of the frequency 2ω. The driving force of frequency 2ω is relatively small, and the disturbance signal with the frequency of 2ω could be ignored.

2.1 Vibration detection

The detection resistor is the key device to obtain the information of vibration frequency and amplitude. Ignore the phase, and we only consider the harmonic components $p_a(\omega)$ with frequency of ω. The resistance of detection resistor is expressed as

$$R(t) = R_s[1 + \varepsilon(\omega)\cos \omega t] \tag{5}$$

where $\varepsilon(\omega)$ is the rate of detection resistance variation, reflecting the vibration amplitude of the resonant beam, and is expressed as

$$\varepsilon(\omega) = \frac{R_d}{R_D} \tag{6}$$

The amplitude detection sensitivity S_d is

$$S_d = \frac{\varepsilon(\omega)}{W_m} \tag{7}$$

where W_m is the maximum amplitude of resonant beam (mid point position). Resonant sensor takes frequency as the output and always have a wide range of amplitude (to make sure the signal can be detected without obvious non-linear). The accurate value of amplitude is difficult to be calculated theoretically, so we measure the silicon micro pressure sensor sample when it works well and can provide reasonable output signal to get an approximation. Utilized white-light interferometry and measured open-loop characteristics curve, we estimate that the value of working amplitude is about 1 nm.

The following factors affect $\varepsilon(\omega)$ and S_d. From these factors, we can find ways to strengthen detection signal under the same working amplitude of resonant beam.

① The surface stress distribution on beam (modes) and location of detection resistor. In order to improve the sensitivity, conventional design always placed the resistor at root surface of beam where stress distributed is relatively great.

② The piezoresistive coefficient of detection resistor. The sensitivity can be properly improved by adjusting the design of the doping concentration and Crystallographic orientation. Axial ⟨100⟩ crystallographic orientation, n-type doping can achieve higher piezoresistive coefficient.

③ Increasing the excitation power. But we must avoid to producing a significantly nonlinear.

2.2 The introduction of the slit structure

According to the piezoresistive effect described by Smith (1954), that applying mechanical stress on silicon crystals results in resistance change, as a function of crystal orientation, dopant type, and doping concentration (Kovacs 1998). So effectively improving the resonant beam alternating stress in the root zone can increase $\varepsilon(\omega)$, and amplify detection signal. Finally, the amplitude detection sensitivity is improved.

Processing rectangular slit in the resonant beam root zone, the stress concentration emerge, which can achieve surface alternating stress amplification in resonant beam root. The designed slit-resonant beam is shown in Fig. 2. Rectangular slit make stress concentration in the surrounding area of slits.

According to the analysis of stress concentration caused by slit structure, U-shape resistor is placed around the slit layout. As the normalized resistance change presented in the effort of Bashir et al. (2000), and the piezoresistive sensitivity mentioned by Ahmed et al. (2010). The amplitude detection sensitivity S_d is given by

$$S_d = \frac{2}{A_R}\int_0^{A_R}(\pi_L\sigma_L + \pi_T\sigma_T)\partial A \tag{8}$$

where

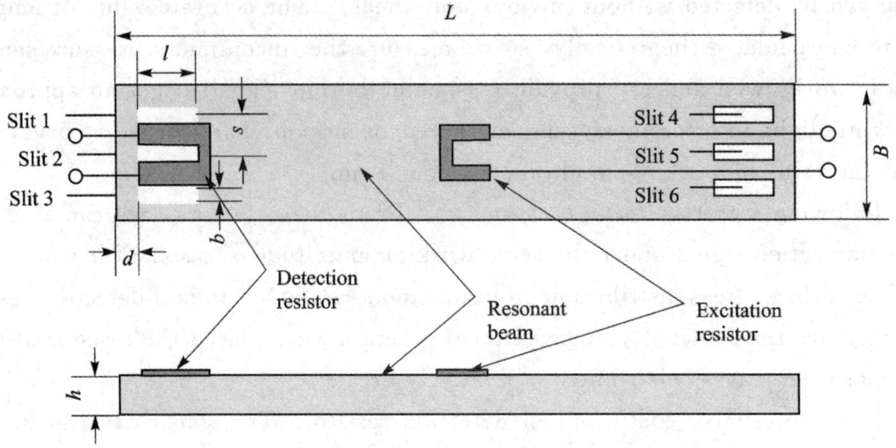

Fig. 2 A schematic drawing of slit-resonant beam

A_R is the single side cross-sectional area of the U-shaped detection resistor;

π_L is longitudinal piezoresistive coefficient;

π_T is the transverse piezoresistive coefficient;

σ_L is the longitudinal stress;

σ_T is the transverse stress of the detection resistor.

The axial piezoresistive coefficient is much larger than the transverse piezoresistive coefficient. And the axial stress is much larger than the horizontal stress in stress concentration region. So we ignore the transverse effect, and Eq. (8) can be expressed using the average stress values of σ_L in the stress concentration zone, which yields

$$S_d = 2\pi_L\sigma_L \tag{9}$$

Size of detection resistor affects the signal conversion, but that is not within the scope of this study. Subjected to the finite beam width, slit width was designated as the slit spacing distance. The length of the resistor was changes with the slit length.

3 Stress Concentration Calculation

Because slits represent openings in resonant beam, these give rise to stress concentrations at stress concentration region, which can be described by stress concentration factors (SCF=local maximum equivalent stress/far-Weld stress) (Pilkey 2008). Some methods can be used to solve stress concentration problem in two-dimensional infinite elastic structural with non-circular holes: the analytical method (Heller 1958) and FEM (Zienkiewicz 2000).

Heller (1958) proposed the conformal transformation and complex function method, by which the right angle of the hole can be transformed to a round angle and the straight lines to curves. This method can only be used to analyze the stress concentration of the approximate rectangular hole with round corners. However, for the case of an exact rectangular hole, relevant results have not appeared in the literature. In addition, Stress redistribution always happens because the peak stress is often more than the yield limit. Therefore the actual peak stress is often lower than the peak stress calculated by elasticity theory. Actual stress distribution is difficult to obtain theoretically.

With the emergence and popularity of the computer, the finite element method has been developed rapidly, and showed the advantages in mechanical calculations. FEM allows a wide variety of slit shapes to be modeled, and can be used to calculate the stress concentration magnification with the arbitrary size and position of the slit and under random loads.

The present work utilizes this flexibility of the Finite Element program ANSYS for studying the stress concentration caused by slits. All models with isotropic single crystal silicon material were meshed with Four-node quadrilateral plane stress elements. Convergence studies were performed and as a result the meshes were strongly refined in the vicinity of the slits. Loading applied in the form of displacement was equal to the working amplitude. With its powerful computing capabilities, we can get the stress magnification along the length direction of the slits, and the natural frequency of new structure. In order to reflect the stress concentration effect caused by slit structure, we use the average ratio of enhanced stress in the new structure to stress value in the same region of non-slit structure to illustrate the stress concentration effect. The average ratio is on behalf of the SCF, and is treated as the average stress magnification radio R_{sc}. Meantime, the ratio of the stress in the stress concentration region of slit structure whit certain size to stress on the same size area at beam root of non-slit structure (1 μm from the fixed end) is used to explain the actual stress values, and this radio is reference average stress magnification radio R_{rsc}. F_{nfr}, the natural frequency ratio of slit resonant beam to the non-slit beam indicates the natural frequency changing caused by slits.

4 Simulation Results and Discussion

4.1 Mechanical properties of the slit-resonant beam

Around slits the stress concentration occurs. Slit structure improves the stress distributed in the stress concentration area. Meanwhile, the size parameters and location of slits affect the mechanical properties of resonant beam such as stress concentration on resonant beam and natural frequency of resonant beam, as well as sensing performance such as resonant beam amplitude detection sensitivity. The structural size parameters include the slit length l and width b. d is the distance of slit from the resonant beam clamped end.

(1) Stress concentration distribution

Stress magnification along the length of the slit is shown in Fig. 3. Fig. 3(a) gives the upper edge stress magnification of slit 2, the lower edge stress magnification of slit 1 and the midline stress magnification between two slits. Fig. 3(b) shows stress magnification of slit 2 and slit 3. Symmetrically, stress magnification distribution around the other three slits on the other end of the beam is of the same. For calculating S_d, the average value of stress is calculated. Fig. 3(c) shows the average stress values distribution between two slits in the

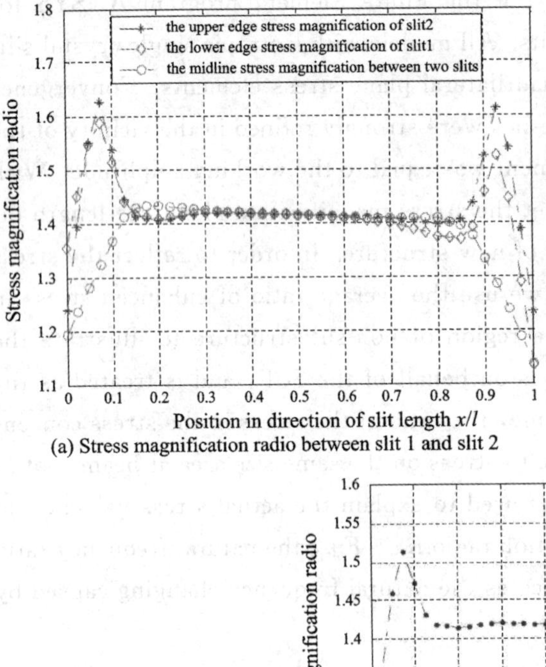

(a) Stress magnification radio between slit 1 and slit 2

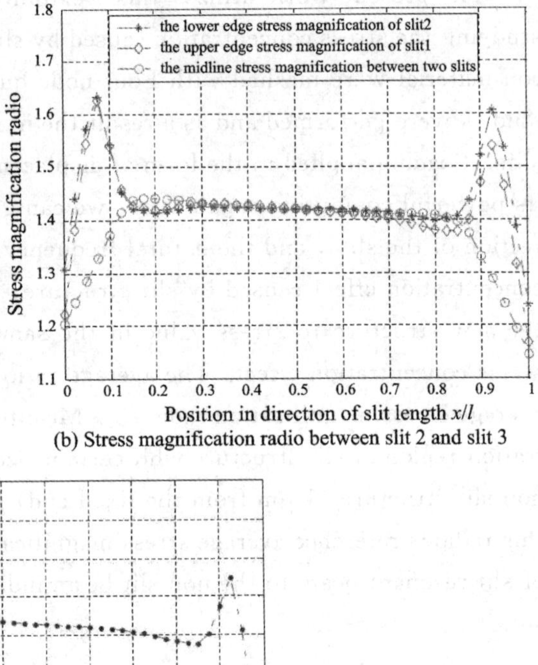

(b) Stress magnification radio between slit 2 and slit 3

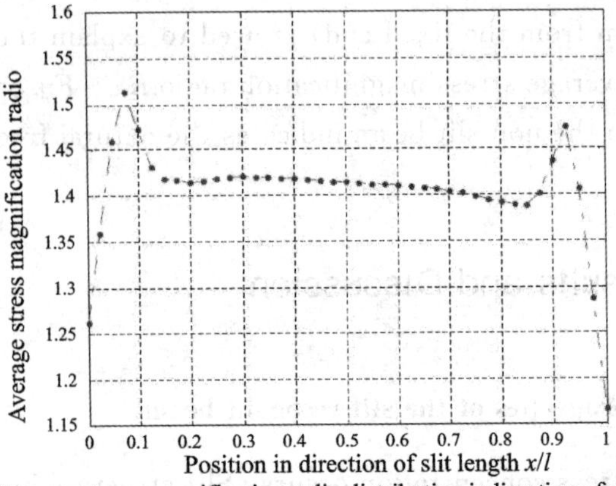

(c) Average stress magnification radio distribution indirection of slit length

Fig. 3 Stress magnification along the length of the slit

stress concentration region. And due to the symmetry of the structure, four stress concentration areas show the same average value of the stress concentration magnification.

In Fig. 3, two peaks of stress magnification near the right angles of slit 1, slit 2 and slit 3 show that the stress concentration magnification is relatively large in that areas. This result is consistent with literature calculations (Pilkey 2008). Except the right angles area, stress concentration magnification is a constant along the length, but stress gradually decreases duo to stress distribution on resonant beam.

Fig. 4 shows the average stress magnification radio of a given size slit structure under different amplitude (in arange around working amplitude). Seen by the curve, the average stress magnification radio is fixed. This shows that amplitude detection sensitivity of resonant beam is raised by a constant multiple and the multiple is determined by the size and location of the slit structure.

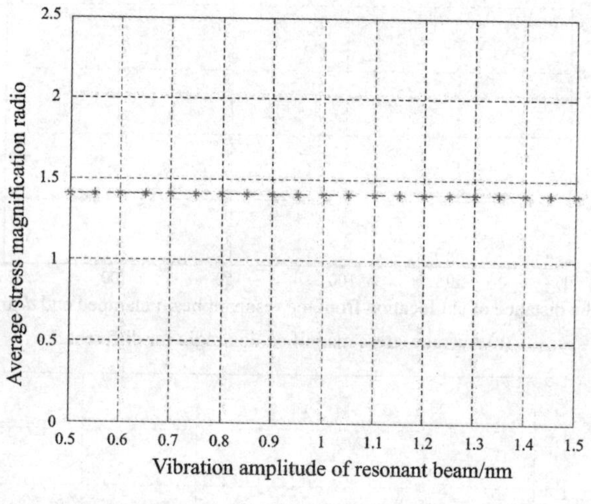

Fig. 4 Average stress magnification radio to different vibration amplitude of resonant beam

(2) Stress concentration

According to axial stress distribution of the non-slit resonant beam, we known that the stress values is small in the position where d is large, and detection resistor may even collect opposite sign stress there. Stress concentration is ineffective. d should not be too large, but not too small. If d is too small, the transverse stress is large enough to affect detection resistor sensing axial stress.

Fig. 5(a) shows that the average stress magnification radio of slit structure with fixed size rise with d increasing. Curves go up more quickly with large slit width. At the same time, for the certain length, larger width slit can get higher magnification radio. But for a certain width, smaller length slit do the same job. Therefore, we select the large width within the range allowed by the beam width and a relatively small length of the slits to get high stress magnification.

Fig. 5(b) shows that the reference average stress magnification radio decline with the increasing of the distance d, which implies the corresponding stress decreases in contrast to stress concentration magnification. Therefore, in order to obtain large stress, it is better to choose a small distance d.

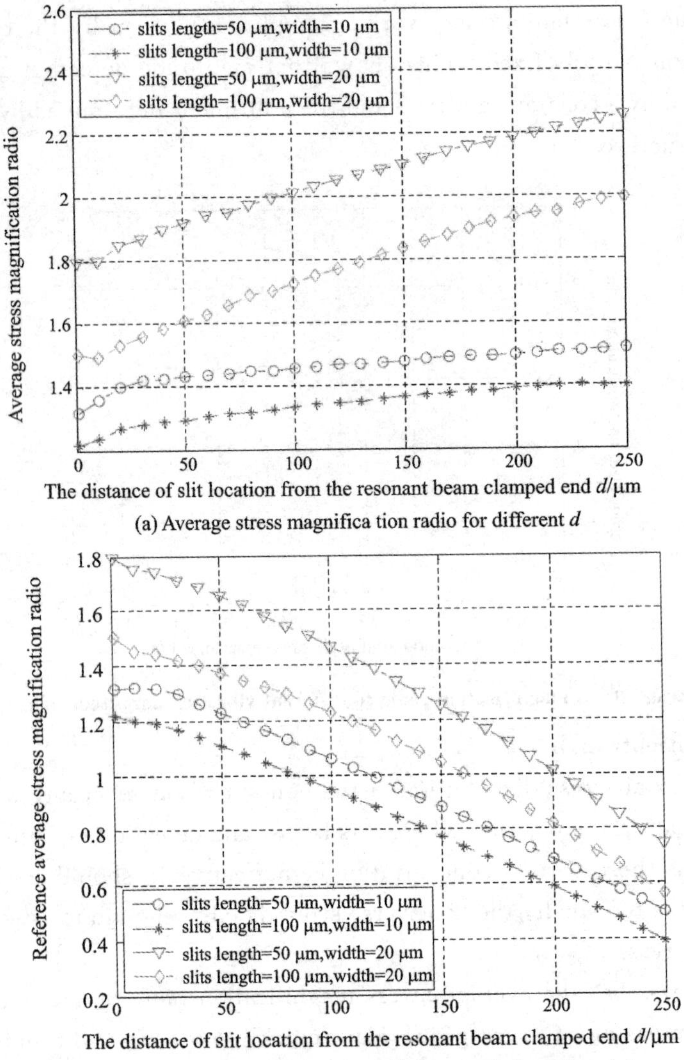

(a) Average stress magnifica tion radio for different d

(b) Reference average stress maghification radio for different d

Fig. 5 Sketch of tress magnification radio

Fig. 6 shows the impact of length l and width b on the stress concentration magnification. In Fig. 6(a), for fixed width b, the average stress magnification radio R_{sc} decreases with the growth of the length l, and as the length l increasing the decline speed becomes slow. The large width results in great R_{sc}. When length l is increases to a certain value, the

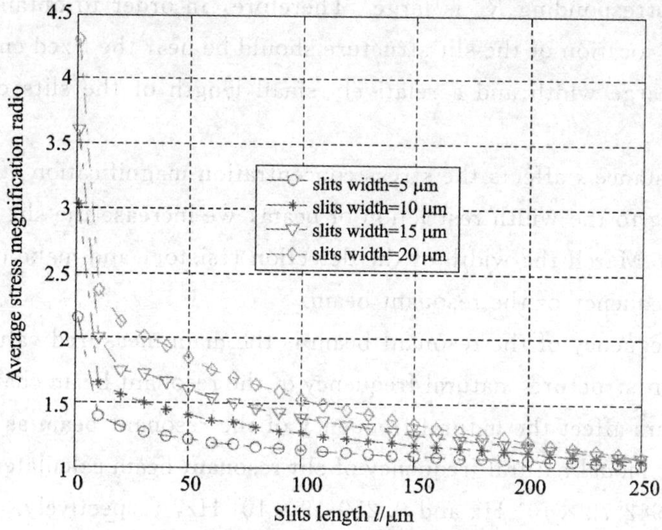

(a) Condition for the fixed width b

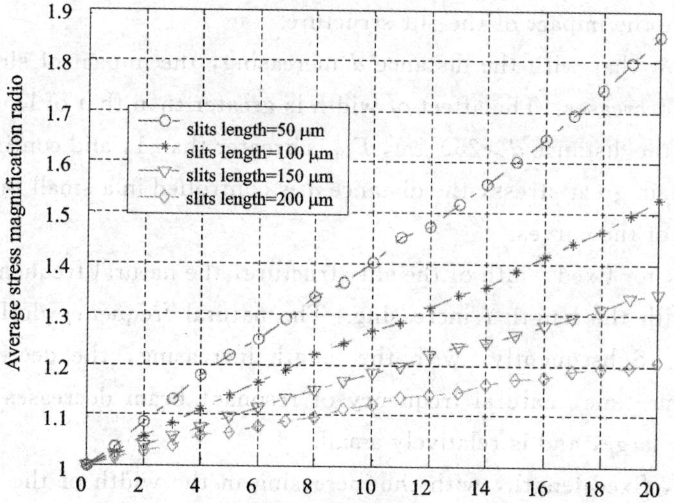

(b) Condition for the fixed length l

Fig. 6 Average stress magnification radio for different slit length l and average stress magnification radio for different slit width b

R_{sc} of different width slits tend to 1. In Fig. 6b, for fixed length l, the average stress magnification radio R_{sc} increases with slit widening. Although the R_{sc} of different length slits start at 1, the increasing is more quickly of the small length slit than of the large length slit. And R_{sc} is relatively large. From the above analysis we found that, when, l/b, the aspect ratio of the slit is small, corresponding R_{sc} is large. Therefore, in order to obtain large R_{sc} on the resonant beam, the location of the slit structure should be near the fixed end, which mines d should be small. Large width and a relatively small length of the slits could obtain large R_{sc} too.

Slit spacing distance s affects the stress concentration magnification at a certain degree. However, according to the width restriction of beam, we increase the slit spacing distance s in a permit range to March the width of the detection resistor, and make it fixed.

(3) Natural frequency of the resonant beam

The natural frequency of the resonant beam is the main measured value. Due to the introduction of the slit structure, natural frequency of the resonant beam changes. Slit size and positional parameters affect the natural frequency of the resonant beam as shown in Fig. 7. The first two order modal natural frequency of slit resonant beam calculated by the finite element method is $0.942\ 72 \times 10^8$ Hz and $0.259\ 19 \times 10^9$ Hz, respectively. The value is high consistency with that calculated by common theory formula. We use the first natural frequency to illustrate the impact of the slit structure.

Fig. 7(a) show that with the distance d increasing, the impact of slit structure on the natural frequency decreases. The affect of width is greater than that of length on the natural frequency. When the distance $d > 250\ \mu$m, F_{nfr} is greater than 1, and continues to grow. For the purpose to obtain great stress, the distance d is controlled in a small range, so we do not consider this part of the curves.

In Fig. 7(b), for fixed width of the slit structure, the natural frequency of the resonant beam decreases with the length l increasing. The natural frequency decline fast when the length l is small. Subsequently, with the length increasing, the decrease speed slowed down. At the same time, natural frequency of resonant beam decreases rapidly when the width is relatively large, and is relatively small.

In Fig. 7(c), fixed length, with the increasing in the width of the slit structure, the natural frequency of resonant beam decreases correspondingly, and the rate of decline is essentially the same.

According to the mechanism of resonant sensor, slit structure numerically affecting on the natural frequency of the resonant beam does not have a real impact on its performance.

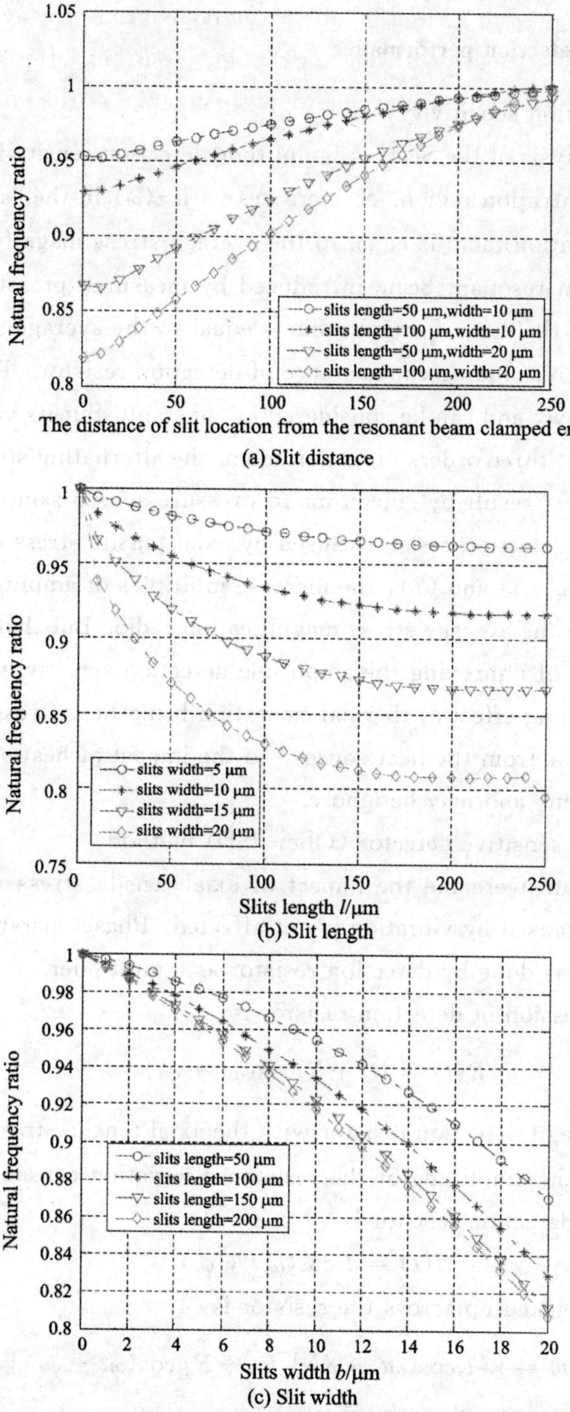

Fig. 7 Natural frequency changes of the resonant beam caused by slit distance *d*, slit length *l*, slit width *b*

4.2 Impact of the detection performance

(1) Amplitude detection sensitivity

According to the analysis of the Sect. 2, amplitude detection sensitivity is determined by the stress in stress concentration region. S_d increases with $\varepsilon(\omega)$ in the case of the same amplitude, and the magnification radio is equal to the average stress magnification radio R_{sc}.

Axial tensile stress in resonant beam introduced by measured pressure is also enlarged by the slit structure, and the magnification radio is equal to the average stress magnification radio R_{sc}. Axial tensile stress increases resistance of detection resistor. The changing rate of resistance is relatively slow, and can be considered as constant. But its value is so large that sometimes can reach about three orders of magnitude of the alternating stress in the root zone of the resonant beam. Test result by silicon micro pressure sensor sample demonstrate that resistance change rate of detection resistor caused by axial tensile stress is about 1%~2%.

According to the Eqs. (6) and (7), the increase multiples of amplitude detection sensitivity is slightly less than the average stress magnification radio. But that does not affect slit structure to play the role of improving the amplitude detection sensitivity.

Simultaneously, heating effect of thermal excitation bring heat stress to resonant beam. But the slit structure is far from the heat source, so the impact of heat stress is small in the stress concentration region, and may be ignore.

(2) The Ohm phase sensitive detector (Ohm-PSD) method

Ohm-PSD method can overcome the impact of axial tensile stress and thermal stress, and resistance variation caused by vibration is not affected. Phase operation of output signal with the reference signal is done by detection resistor as a multiplier.

The resistance expression of detection resistor is

$$R(t) = \overline{R}_s + R_d\cos(\omega t + \varphi_d) \tag{10}$$

where $\overline{R}_s$ is the DC value of detection resistor with the axial tensile stress and thermal stress being considered. R_d is the amplitude of the resistance variation caused by amplitude. The reference current in the detection resistor is

$$I(t) = I_r\cos(\omega_r t + \varphi_r) \tag{11}$$

The voltage signal picked up across the resistor is

$$V_d(t) = [(I_r\cos(\omega t + \varphi_r)][\overline{R}_s + R_d\cos(\omega t + \varphi_d)] \tag{12}$$

When beam is in the resonant state $\omega=\omega_r$,

$$V_d(t) = I_r \overline{R}_s \cos(\omega t + \varphi_r) + \frac{1}{2} I_r R_d \cos(\varphi_r - \varphi_d) + \frac{1}{2} I_r R_d \cos(2\omega t + \varphi_r + \varphi_d) \quad (13)$$

Through the low-pass filter, the voltage signal is

$$V_d(t) = \frac{1}{2} I_r R_d \cos(\varphi_r - \varphi_d) \quad (14)$$

We can find that the resistance variation of detection resistor caused by axial stress and thermal stress are filtered out. The resistance variation caused by the resonant beam vibration is used to reflect the amplitude of the resonant beam. Output signal intensity is improved by expected multiples. Signal to Noise Ratio (SNR) is improved accordingly.

(3) Finite element stress magnification calculation

When the structure size of resonant beam is L=1 600 μm, B=100 μm, H=30 μm, with rectangular slits placed in each ends of the resonant beam symmetrically. Slit size is l= 50 μm, b=10 μm, h=30 μm, d=20 μm and aspect ratio l/b=5. Three slits are placed in parallel at each end with spacing distance s=10 μm.

FEM calculation results show that the average stress magnification radio R_{sc} is 1.405 6. The rate of detection resistance variation $\varepsilon(\omega)$ increases by 1.4 times, and the amplitude detection sensitivity of slit-resonant beam raises by 1.4 times accordingly. By the help of Ohm-PSD method, SNR is improved by 1.96 times.

5 Conclusions

Slit structure leads to stress concentration in stress concentrated area, increases detection resistance variation, improves the amplitude sensitivity of the resonant beam and signal to noise ratio. Finite element method is used to numerically analysis the slit-resonant beam proposed in this paper. The results are as following.

① Slit location and structure size have a certain impact on the stress concentration as well as the natural frequency of the resonant beam. For the slit structure with a given size, with the increasing of l the average stress magnification radio increases, but the actual stress value decreases. For the slit structure at fixed position, average stress magnification radio decreases with the length increasing. In contrast, average stress magnification radio increases with width b. When the aspect of ratio l/b is small, the stress concentration magnification is relatively large. Therefore, in order to obtain larger stress values on the resonant beam, the location of the slit structure should be near the fixed end, d should be small, and large width b and a relatively small length l of the slit should be selected.

② Slit structure improve the detection performance of the amplitude detection sensitivity and output signal to noise ratio. Amplitude detection sensitivity multiples is equal to the stress concentration magnification. However, the axial stress and heat stress in resonant beam lead improving multiples of the amplitude detection sensitivity to decrease slightly, which is a little lower than the average stress magnification radio. Ohm-PSD method overcomes the impact of axial stress and heat stress in the resonant beam, and increases the signal to noise ratio of the useful output signal.

③ The simulation study in this paper is lack of experimental verification. The actual sample processing and performance verification will be the focus of future work.

Acknowledgments

This work is sponsored by the National Nature Science Fund of china with grant no. 60927005.

References

[1] Ahmed A S M, Walied A M, Edmond L. Optimization of geometric characteristics to improve sensing performance of MEMS piezoresistive strain sensor. J. Micromech Microeng 20:015015,2010. doi:10.1088/0960-1317/20/1/015015.

[2] Bashir R, Gupta A, Neudeck G W, et al. On the design of piezoresistive silicon cantilevers with stress concentration regions for scanning probe microscopy applications. J. Micromech Microeng,2000. 10:483-491. doi:10.1088/0960-1317/10/4/301.

[3] Bouwstra S, Legtenberg R, Tdmans H A C, et al. Resonating microbridge mass flow sensor. J. Sens. Actuators, 1990, A21-A23:332-335. doi:10.1016/0924-4247(90)85066-D.

[4] Heller S R, Brock J S, Bart R. The stresses around a rectangular opening with rounded corners in a uniformly loaded plate// Proceedings of the Third US National Congress of Applied Mechanics, ASME, 1958: 357-368.

[5] Joshi B P, Chaware A S, Gangal S A. Optimising performance of a cantilever-type micro accelerometer sensor. Def. Sci. J., 2007, 57(3): 261-269.

[6] Kovacs G. Micromachined transducers source book. New York: McGraw-Hill, 1998.

[7] Lammerink T S J, Elwensfqek M, Ouwerkerk R H V, et al. Performance of thermally excited resonators. J. Sens. Actuators,1990, 21:352-356. doi:10.1016/0924-4247(90)85070-K.

[8] Langdon R M. Resonator sensors a review. J. Phys. E Sci. Instrum,1985, 18:103-115. doi:10.1088/0022-3735/18/2/002.

[9] Mo Y, Xuan Z, Kambiz V, et al. High sensitivity piezoresistive cantilever design and optimization for an-

alytereceptor binding. J. Micromech Microeng,2003, 13:64-72. doi:10.1088/0960-1317/13/6/309.

[10] Pilkey W D, Pilkey D F. Peterson's Stress Concentration Factors. 3rd ed. New York: Wiley.

[11] L Qingfeng, F Shangchun, T Zhangyang, et al. Nonlinear dynamics of an electrothermally excited resonant pressure sensor. J. Sens Actuators A: Phys,2012. doi:10.1016/j. sna. 2012.01.006.

[12] Smith C S. Piezoresistance effect in germanium and silicon. J. Phys. Rev. ,1954, 94:42-49. doi:10.1103/PhysRev. 94.42.

[13] Wdfinger R J, Bardell P H, Chhabra D S. The resonistor: a frequency selective device utilizing the mechanical resonance of a silicon substrate. IBM J. Res. Develop. ,1968, 12:113-118. doi:10.1147/rd.121.0113.

[14] T Zhangyang, F Shangchun, X Weiwei, et al. An electrothermally excited dual beams silicon resonant pressure sensor with temperature compensation. J. Microsyst Technol. ,2011, 17:1481-1490. doi:10.1007/s00542-011-1319-6.

[15] Zienkiewicz O C, Taylor R L. The Finite element method. 7th ed. Elsevier Science and Technology Books,2000.

References

[1] Alper S E, Azgin K, Akin T. High-Performance SOI-MEMS Gyroscope with Decoupled Oscillation Modes[C]// Proceedings of the 19th IEEE International Conference on Micro Electro Mechanical Systems. 2006: 70-73.

[2] Beckwith T G, Marangoni R D. Mechanical Measurements[M]. 4th ed. Addison-Wesley Publishing Company, 1990.

[3] Bongsang Kim, Chandra Mohan Jha, Taylor White, et al. Temperature Dependence of Quality Factor in Mems Resonators. MEMS 2006, 2006. 1, Istanbul, Turkey.

[4] Budynas R G. Advanced Strength and Applied Stress Analysis [M]. 2nd ed. McGraw-Hill Book Company, 北京:清华大学出版社, 2001.

[5] Cai Chengguang, Fan Shangchun. Frequency Characteristics Testing of Micromachined Resonant Pressure Sensor [C]// Proceedings of the 3rd International Symposium on Instrumentation Science and Technology, Xi'an, China. 2004, 3:0512-0516,

[6] Lee Chiyuan, Lee Shuojen, Wu Guanwei. Fabrication of Micro Temperature Sensor on the Lexible Substrate. IEEE Review of Advancements in Micro and Nano Technologies, 2007: 1050-1053.

[7] Damir Ilic, Josip Butorac. Use of Precise Digital Voltmeters for Phase Measurements [J]. IEEE Transactions on Instrumentation and Measurements, 2001, 50(2):449-452.

[8] Doebelin E O. Measurement Systems Application and Design [M]. 3rd ed. McGraw-Hill Book Company, 1983.

[9] Fan Shangchun, Lee Man Hyung. The Frequency Characteristics of the Beam Resonator of the Thermal Excited Silicon Resonant Pressure Sensor[C]// The Proceedings of 2003 IEEE/ASME International Conference on Advanced Intelligent Mechatronics, Kobe, Japan. 2003, 7:460-464.

[10] Fan Shangchun, Li Yan, Guo Zhanshe, et al. Dynamic Characteristics of Resonant Gyroscopes Study Based on the Mathieu Equation Approximate Solution [J]. Chinese Physicas B, 2012, 21(5).

[11] Grandke T, KO W H. Sensors. Smart sensors, 1989, 1(12).

[12] Guenter Martin, Reinhard Kunze, Bert Wall. Temperature-Stable Double SAW Resonators[J]. IEEE Transactions on Ultrasonics, Ferroelectrics, and Frequency Con-

trol, 2008, 55,(1): 199-207.

[13] Istvan Kollar,Jerome J. Blair. Improved Determination of the Best Fitting Sine Wave in ADC Testing[J]. IEEE Transactions on Instrumentation and Measurement,2005,54(5):1978-1983.

[14] Luo R C. Sensor Technologies and Microsensor Issues for Mechatronics Systems [J]. IEEE/ASME Trans. on Mechatronics, 1996, 1(1):39-49.

[15] Manus Henry. Self-Validating Digital Coriolis Mass Flow Meter [J]. Computing & Control Engineering Journal, 2000:219-227.

[16] Nathan Siwak, Xiao Zhu Fan, Dan Hines, et al. Indium Phosphide MEMS Cantilever Resonator Sensors Utilizing a Pentacene Absorption Layer [J]. Journal of Microelectromechanical Systems, 2009, 18,(1): 103-110.

[17] McLACHLAN N W. Theory and Application of Mathieu Functions[M]. Dover Publicatons, Inc. , 1945.

[18] Enjeti Prasad N,Rahman Ashek,Ranjit Jakkli. Economic Single-Phase to Three-Phase Converter Topologies for Fixed and Variable Frequency Output [J]. Transactions on Power Electronics,1993,18(3).

[19] Li Qingfeng, Fan Shangchun, Tang Zhangyang. Non-Linear Dynamics of an Electrothermally Excited Resonant Pressure Sensor[J]. Sensors and actuators A,2012, 88:19-28.

[20] Li Qingfeng, Fan Shangchun, Tang Zhangyang, et al. Nonlinear Vibration in Resonant Silicon Bridge Pressure Sensor: Theory and Experiment[C]// 16th International Solid-State Sensors, Actuators and Microsystems Conference (TRANSDUCERS'2011). 2011: 1685-1688.

[21] Levy R, Janiaud D, Traon O L, et al. A New Analog Oscillator Electronics Applied to a Piezoelectric Vibrating Gyro[C]. IEEE International Proceedings of Frequency Control Symposium and Exposition, 2004:326-329.

[22] Said Emre Alper, Tayfun Akin. A Single-Crystal Silicon Symmetrical and Decoupled MEMS Gyroscope on an Insulating Substrate [J]. Journal of Microelectromechanical Systems, 2005, 14(4).

[23] Samer Guirguis, Fan ShangChun. Modeling of Coriolis Mass Flow Meter of a General Plane-shape Pipe[J]. Flow Measurement and Instrumentation, 2010, 21:40-47.

[24] Fan Shangchun, Liu Guangyu. Finite-Element Modelling and Simulation on Frequency Characteristics of the Silicon Beam Resonator Attached to an E-type Round Diaphragm for Measuring the Concentrated Force[J]. Sensors and Actuators A, 1997,63:169-176.

[25] Rajendran S, Liew K M. Design and Simulation of an Angular-Rate Vibrating Microgyroscope [J]. Sensors and Actuators A, 2004, 116: 241-256.

[26] Li Yan, Fan Shangchun, Guo Zhanshe. Frequency Measurement Study of Resonant Vibratory Gyroscopes [J]. Journal of Sound and Vibration, 2012, 331:4417-4424.

[27] Zhuang Haihan, Fan Shangchun. Dynamic Characteristics Analysis of Vibrating Cylinder Pressure Transducers (VCPT) [J]. Sensors and Actuators A, 2010,157:219-227.

[28] Tang Zhangyang, Fan Shangchun, Xing Weiwei. An Electrothermally Excited Dual Beams Silicon Resonant Pressure Sensor with Temperature Compensation [J]. Microsyst Technol, 2011, 17:1481-1490.

[29] Guo Zhanshe, Feng Zhou, Cao Le, et al. Theoretical and Experimental Study of Capacitance Considering Fabrication Process and Edge Effect for MEMS Comb Actuator [J]. Microsystem Technology, 2011, 11:71-76.

[30] Zheng Dezhi, Wang Shuai, Fan Shangchun. Nonlinear Vibration Characteristics of Coriolis Mass Flowmeter[J]. Chinese Journal of Aeronautics, 2009, 22(2):198-205.

[31] Zheng Dezhi,Fan Shangchun. A Novel Digital Coriolis Mass Flowmeter[C]// The 12th International Conference on Flow Measurement. 2004:455-460.

[32] (美)铁木辛柯 S,沃诺斯基 S. 板壳理论[M]. 北京:科学出版社,1997.

[33] (苏)马洛夫 B B. 压电谐振传感器[M]. 翁善臣,等译. 北京:国防工业出版社,1984.

[34] 程鹏. 自动控制原理[M]. 北京:高等教育出版社,2003.

[35] 蔡武昌,孙怀清,纪纲. 流量测量方法和仪表的选用[M]. 北京:化学工业出版社,2001.

[36] 蔡晨光. 硅微机械谐振式压力传感器闭环的研究与实现[D]. 北京:北京航空航天大学,2007.

[37] 樊大钧,刘广玉. 新型弹性敏感元件设计[M],北京:国防工业出版社,1995.

[38] 樊尚春,刘广玉. 热激励谐振式硅微结构压力传感器[J]. 航空学报,2000(9):474-476.

[39] 樊尚春. 科里奥利直接质量流量计[J]. 中国学术期刊文摘,1999(12):1551-1554.

[40] 樊尚春,刘广建,宋治生. 谐振式科里奥利质量流量计[J]. 北京航空航天大学学报, 2000(12):653-655.

[41] 樊尚春. 传感器技术及应用[M]. 2 版. 北京:北京航空航天大学出版社,2010.

[42] 樊尚春,周浩敏. 信号与测试技术[M]. 2 版. 北京:北京航空航天大学出版社,2010.

[43] 樊尚春,刘广玉. 新型传感器技术及应用[M]. 2 版. 北京:中国电力出版社,2011.

[44] 黄俊钦. 测试系统动力学[M]. 北京:国防工业出版社,1996.

[45] 黄俊钦,樊尚春,刘广玉. 微机械传感器最新发展[J]. 航空计测技术,2003(1):1-8.

[46] 刘广玉,樊尚春,周浩敏. 微机械电子系统及其应用[M]. 北京航空航天大学出版社,2003.

[47] 刘广玉. 几种新型传感器——设计与应用[M]. 北京:国防工业出版社,1988.
[48] 秦自楷. 压电石英传感器[M]. 北京:电子工业出版社,1980.
[49] 邢维巍. 硅微机械谐振式传感器参数辨识层的理论与实现[D]. 北京:北京航空航天大学,2007.
[50] 郑德智. 科氏质量流量计非线性影响因素的研究[D]. 北京:北京航空航天大学,2006.